CAPACITORS,

MAGNETIC CIRCUITS,

AND TRANSFORMERS

PRENTICE-HALL INTERNATIONAL, INC., *London*
PRENTICE-HALL OF AUSTRALIA, PTY., LTD., *Sydney*
PRENTICE-HALL OF CANADA, LTD., *Toronto*
PRENTICE-HALL OF INDIA (PRIVATE) LTD., *New Delhi*
PRENTICE-HALL OF JAPAN, *Tokyo*

PREFACE

Because of the tremendous rate at which scientific developments are expanding and the extremely short time lag before they are applied in engineering practice, the engineering curriculum has become increasingly science oriented. In addition, the student is introduced, at an early stage, to advanced and sophisticated analytical methods so that an increasing amount of subject matter is covered in the short space of four or five years of the undergraduate program. As a result, he often acquires a facility for manipulating mathematical expressions at the expense of understanding the underlying physical principles. This sometimes engenders a distaste for "hardware."

This text places major emphasis on the physical concepts, and uses relatively simple analytical approaches to the study of capacitors, magnetic circuits, and transformers. Principles of energy storage and conversion are applied in derivations where this approach seems most effective. The first chapter, for example, deals with energy relationships in rather simple systems.

Practical aspects of electric fields and the significance of electric field intensity as related to dielectric strength, corona, and the construction of capacitors and single-conductor cables are discussed in Chapter 2. The treatment of magnetic fields in Chapter 3 uses the concept of the unit magnetic pole to emphasize the aspect of forces in the magnetic field in a manner similar to that which makes use of the test charge in treating the electric field. The concept of inductance is reinforced by relating it to simple magnetic circuits; the inconsistencies among the various definitions of inductance are discussed for cases of circuits containing magnetic materials. Although the student has probably studied electric and magnetic fields in earlier courses from a more mathematical viewpoint, the treatment in this text is intended to strengthen his physical concepts. While permanent magnets, electromagnets, and transformers play a most vital part in many engineering applications, a chapter on saturable reactors and frequency multipliers is considered necessary and has been included. Much of the material in this text has been used successfully in mimeograph form for a three-hour course

CONTENTS

1 ENERGY — 1

1-1 Work and Energy, 2
1-2 Potential and Kinetic Energy, 3
1-3 The Law of Conservation of Energy, 3
1-4 The Law of Degradation of Energy, 4
1-5 Thermal Energy, 9
1-6 Electrical Energy, 9
1-7 Power, 10
1-8 Torque and Tangential Force, 10
1-9 Efficiency, 11
1-10 Energy and Power Relations in Mechanical and Electrical Systems, 12
1-11 Mass and Viscous Friction, 12
1-12 Resistance and Self-inductance, 15
1-13 Viscous Friction and Spring, 17
1-14 Resistance and Capacitance, 18
1-15 Spring, Mass, and Viscous Friction, 21
1-16 Series R-L-C Circuit, 24
1-17 Energy Stored in a Rotating Flywheel, 29
1-18 Power and Torque, 30
1-19 Mechanical and Electrical Analogies, 31
1-20 Chemically Stored Energy, 32
1-21 Storage Batteries, 34
1-22 Atomic Energy, 34

2 CAPACITANCE AND RELATED EFFECTS — 39

2-1 The Electrical Field, 39
2-2 Electric Flux, 40
2-3 Electric Field Intensity, 42
2-4 Voltage and Potential, 43
2-5 Gauss's Theorem, 45
2-6 Charge Within a Conductor That Has a Static Charge, 47
2-7 Uniformly Distributed Charge on an Isolated Sphere, 48
2-8 Capacitance, 49

CONTENTS

2-9 Capacitance of Concentric Spheres, 50
2-10 Parallel-plate Capacitor, 52
2-11 Relative Dielectric Constant, 53
2-12 Concentric Cylinders, 54
2-13 Electric Field Intensity between Concentric Cylinders, 56
2-14 Graded Insulation, 56
2-15 Energy Stored in a Capacitor, 58
2-16 Energy Stored in a Dielectric, 59
2-17 d-c Energy Storage Capacitors, 62
2-18 Kva Rating of Capacitors, 63
2-19 Dielectric Strength, 65
2-20 Types of Capacitors, 66
2-21 Polarization and Dielectric Constant, 68
2-22 Mechanism of Polarization, 71
2-23 a-c Characteristics of Dielectrics, 72
2-24 Complex Dielectric Constant, 74
2-25 Corona, 75
2-26 Resistance of Dielectric Configurations, 77
2-27 Mechanical Energy and Force in a Capacitor, 78
2-28 Electrostatic Synchronous Machine, 81

3 MAGNETIC CIRCUITS 89

3-1 Magnetism, 89
3-2 Magnetic Field about a Straight Wire Carrying Current, 90
3-3 Magnetic Flux and Magnetic Lines of Force, 93
3-4 The Unit Magnet Pole, 95
3-5 Magnetomotive Force, mmf, 97
3-6 The Toroid, 102
3-7 Comparison of the Magnetic Circuit with the Electric Circuit, 105
3-8 Other Common Systems of Magnetic Units, 106
3-9 Magnetic Materials, 107
3-10 Calculation of Magnetic Circuits without Air Gaps, 111
3-11 Magnetic Leakage, 114
3-12 Correction for Fringing at Short Air Gaps, 115
3-13 Iron and Air, 116
3-14 Graphical Solution for Simple Magnetic Circuit with Short Air Gap, 118
3-15 Flux Linkages, 121
3-16 Induced emf. Lenz's Law, 122
3-17 Energy Stored in Magnetic Circuits, 123
3-18 Magnetic Force in Terms of Flux Density, 126
3-19 Hysteresis Loop, 130
3-20 Permanent Magnets, 131
3-21 Demagnetization Curve, 133
3-22 Energy Product, 137
3-23 Operating Characteristics of Permanent Magnets, 140
3-24 Core Losses, 142
3-25 Hysteresis Loss, 142

CONTENTS　　　　　　　　　　　　　　　　　　　　xi

    3-26 Rotational Hysteresis Loss, 146
    3-27 Eddy-current Loss, 146
    3-28 Factors Influencing Core Loss, 151
    3-29 Magnetic Circuits in Series and in Parallel, 152

4 INDUCTANCE-ELECTROMAGNETIC ENERGY CONVERSION   163

    4-1 Inductive Circuits, 163
    4-2 Self-inductance, 163
    4-3 Variable Self-inductance, 166
    4-4 Force and Torque in a Circuit of Variable Self-inductance, 167
    4-5 Inductance in Terms of Magnetic Reluctance and Magnetic Permeance, 169
    4-6 Mutual Inductance, 172
    4-7 Torque and Force in Inductively Coupled Circuits, 178
    4-8 Forces in Nonlinear Magnetic Circuits, 180
    4-9 Inductive Reactance, 190
    4-10 Reactive Power, 191
    4-11 Effective Resistance and Q-Factor, 192

5 EXCITATION CHARACTERISTICS OF IRON-CORE REACTORS AND TRANSFORMERS   203

    5-1 Iron-core Reactors, 203
    5-2 Voltage Current and Flux Relations, 204
    5-3 Harmonics, 209
    5-4 Power, 212
    5-5 Effective Current, 214
    5-6 Core-loss Current and Magnetizing Current, 215
    5-7 Equivalent Circuits, 217
    5-8 Effect of Air Gaps, 219
    5-9 Time Constant and Rating of Reactors as Functions of Volume, 229

6 THE TRANSFORMER   237

    6-1 Induced emfs, 238
    6-2 The Two-winding Transformer, 239
    6-3 Voltage Ratio, Current Ratio, and Impedance Ratio in the Ideal Transformer, 241
    6-4 Equivalent Circuit of the Transformer, 246
    6-5 Open-circuit and Short-circuit Tests, Exciting Admittance, and Equivalent Impedance, 259
    6-6 Transformer Losses and Efficiency, 262
    6-7 Voltage Regulation, 266
    6-8 Autotransformers, 268
    6-9 Instrument Transformers, 270
    6-10 Variable-frequency Transformers, 270

6-11 3-Phase Transformer Connections, 276
6-12 Per Unit Quantities, 292
6-13 Multicircuit Transformers, 295

7 SATURABLE REACTORS 309

7-1 Magnetic Frequency Multiplies, 310
7-2 Frequency Tripler, 312
7-3 Relationship Between Applied Voltage and Conduction Angle for Noninductive Load, 313
7-4 Single-core Saturable Reactor with Premagnetization, 315
7-5 2-Core Saturable Reactor, 319
7-6 Gate Windings in Parallel, 319
7-7 Gate Windings in Series, 325
7-8 Operation with Free Even-harmonics, 326
7-9 Power Output, 328
7-10 Gains, 329
7-11 Steady-state Operation with Suppression of Even-harmonics, 330
7-12 Load Impedance Zero, 331
7-13 Finite Load Resistance, 335
7-14 Frequency Doubler, 338

INDEX 343

CAPACITORS,

MAGNETIC CIRCUITS,

AND TRANSFORMERS

ENERGY

1

The progress of man has been in almost direct proportion to his utilization of solar energy. Solar energy exists in chemical form in fossil fuels as well as in potential and kinetic form in waterfalls and in the flow of rivers. In the case of fossil fuels the sun's energy promotes the growth of plant life. This is followed by a long period of decomposition of the plants into fuels such as coal, oil, and gas. The process of storing energy in water that evaporates from the sea and other bodies, and then falls as rain on mountain slopes, maintaining waterfalls and flow of rivers, is a continuing one. Man has become highly skilled in the control and use of large amounts of energy to do his work for him. He has also acquired great skill and knowledge in developing means for detecting and amplifying minute amounts of power. The progress made in the understanding, development, and use of energy during the twentieth century exceeds all the previous progress made by man. This tremendous progress has been brought about largely by the electrical industry which had its beginnings about a century ago.

The birth of new industries and the accelerated growth of young industries has made and is continuing to make large demands on the electric power production facilities. Since the turn of the century giant industries such as radio, television, and commercial aviation have developed. The automobile industry has achieved tremendous growth since that time as well. The development of the electronic computer, along with the development of atomic and nuclear energy, gave a much greater impetus to the world's industrial, and consequently, economic development than did the industrial revolution of the eighteenth century. The modern trend is not only toward the utilization of more energy at greater efficiency but also toward the detection and use for control purposes of smaller and smaller quantities of energy.

1-1 WORK AND ENERGY

Work is performed by a force when the point of application undergoes a displacement in the direction of the force. Thus in Fig. 1-1, the work dW done by the force F in displacing the body an infinitesimal distance dx in the x direction is expressed by

$$dW = F \cos \theta \, dx \qquad (1\text{-}1)$$

or in vector notation by

$$dW = \mathbf{F} \cdot \mathbf{dx}$$

In the MKS (Meter, Kilogram, Second) system of units, the unit of work is the joule (watt-second). One joule equals the amount of work performed by a constant force of one newton when the point of application is displaced a distance of one meter in the direction of the applied force.

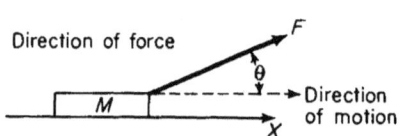

Figure 1-1. Force applied to moving body.

Work and energy are measured in the same units. Energy is a measure of the capability of doing work. Work and energy are scalar quantities, i.e., they have magnitude only and not direction. Force and distance are vector quantities or simply vectors as each has both magnitude and direction.

Generally, in the utilization of energy for the benefit of man, the stored energy present in nature is converted into mechanical, electrical, or thermal energy. The conversion usually results in all three of these forms. As an example, consider the case of a fuel such as coal, oil, or gas, when used for the generation of electric energy. First, the fuel undergoes combustion in a steam generator where its potential or chemical energy is converted into thermal energy. Part of the thermal energy is transferred to the water and steam in the steam generator. The steam enters the prime mover, usually a steam turbine, at a certain pressure and temperature and gives up part of this thermal energy. This part of the energy is converted into mechanical energy. The mechanical energy in turn is converted into electrical energy by the electric generator and into thermal energy by friction in the prime mover. In addition to the energy that is converted into other forms, there is the energy stored in the system as follows

1. The energy required to raise the temperature of the water and parts of the steam generator and prime mover.
2. The kinetic energy associated with the inertia and rotation of the generator and prime mover rotors.
3. The energy associated with the magnetic fluxes in the generator and with the electric fluxes in the insulation of the generator.

However, regardless of the changes in the form of energy, there is no gain or loss of energy. This is in accordance with the principle of conservation of energy.

If the energy were traced as it leaves the generator to the final point of utilization, it would be found that conversion and storage would continue in transformers, transmission lines, and finally in the motors or other types of devices that would convert the energy into the kind required by the user for mechanical work, heat, light, or sound.

1-2 POTENTIAL AND KINETIC ENERGY

Stored energy may occur in two general forms, namely, potential energy and kinetic energy.

Potential energy of a body is defined as the energy that results from the position of the body or its configuration. Potential energy is energy at rest. Examples are water at rest in a reservoir, a spring under tension or compression, a charged capacitor.

Kinetic energy is defined as the energy that results from the motion of a body. Examples of kinetic energy are water in a waterfall, a rotating flywheel, an electric current in an inductive circuit.

1-3 THE LAW OF CONSERVATION OF ENERGY

This law provides the basis for establishing relationships in all branches of engineering. It is also known as the First Law of Thermodynamics and postulates that energy can neither be created nor destroyed. Hence, for any system enclosed by boundaries as shown schematically in Fig. 1-2, the following relationship results from this law

$$\begin{bmatrix} \text{all the energy} \\ \text{that enters} \\ \text{the system} \end{bmatrix} = \begin{bmatrix} \text{all the energy} \\ \text{that leaves} \\ \text{the system} \end{bmatrix} + \begin{bmatrix} \text{the gain in} \\ \text{internal} \\ \text{energy} \end{bmatrix} \quad (1\text{-}2)$$

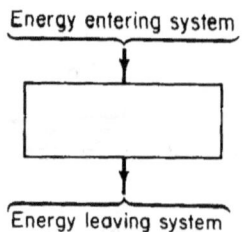

Figure 1-2. Energy balance in bounded system.

In Eq. 1-2 the gain in the internal energy may be positive or negative; part or all of the internal energy may be reversible, or part or all of it may be irreversible.

Reversible and irreversible processes

A reversible process is defined as one for which the system and surroundings can be completely restored to the respective initial states after a process has occurred.* Spontaneous changes that occur in nature are considered irreversible because they proceed in only one direction. To effect a reverse process identical in all respects with the spontaneous change, it is necessary to introduce an external source of some form. Water flows spontaneously through a pipe from a tank at one level to one at a lower level. However, to move the water through the pipe back into the tank at the higher level a pump is required and an external source of energy would have to be used to drive the pump to reverse the process. The following are examples of irreversible processes

1. Flow of ground water into wells.
2. Leakage in dams.
3. Flow of water in pipes.
4. Energy dissipation as heat in electrical circuits.
5. Combustion processes.
6. Diffusion processes.
7. Firing of a projectile.
8. Production of sound in a loud-speaker.
9. Braking of automobiles.
10. Corrosion.

1-4 THE LAW OF DEGRADATION OF ENERGY

This law, also known as the Second Law of Thermodynamics, governs the irreversible process. It states that differences in energy levels always tend to

---

* George A. Hawkins, *Thermodynamics for All Engineering Students.* Conference on Thermodynamics at Pennsylvania State University, June 27–29, 1955.

Sec. 1-4 THE LAW OF DEGRADATION OF ENERGY

disappear. Thus, water flows spontaneously from a higher level to a lower; the same is true for heat, which flows from the higher to the lower temperature. The same relationship is valid for the flow of electric charge, because it flows from a higher to lower potential. Equation 1-2 can be expanded into the following form

$$\begin{bmatrix} \text{all the energy} \\ \text{that enters} \\ \text{the system} \end{bmatrix} = \begin{bmatrix} \text{all the energy} \\ \text{that leaves} \\ \text{the system} \end{bmatrix} + \begin{bmatrix} \text{the gain in} \\ \text{reversible} \\ \text{energy} \end{bmatrix} + \begin{bmatrix} \text{the gain in} \\ \text{irreversible} \\ \text{energy} \end{bmatrix} \quad \text{(1-3)}$$

Consider a stationary electric motor as a system and assume this motor to be started while connected to its load. All of the energy that enters the motor is electrical as long as the ambient temperature does not exceed the temperature of the motor. The two major components of energy that leave the motor are

1. The energy supplied to the connected load.
2. The heat that escapes to surrounding space.

In addition, there is the energy of vibration and noise, or acoustic energy. The gain in reversible energy would be in the forms of

1. Kinetic energy of rotation.
2. Kinetic energy stored in the magnetic field.
3. Potential energy stored in the electrostatic field (in the insulation).

The gain in irreversible energy in the motor would be that which raises the temperature of the motor, i.e., heat resulting from electrical and magnetic losses as well as from friction and windage losses.

In some systems part or all of the stored reversible energy may change its form from potential to kinetic energy and vice versa. The simple pendulum affords a good example of the periodic change of stored energy. Suppose that a small body C of mass M is suspended by a thread of negligible mass and of length l from point O as shown in Fig. 1-3.

Consider the mass originally at rest with the position of the body at A. Assume further that this pendulum is enclosed in a completely evacuated boundary. At rest the body possesses a certain amount of potential energy with reference to some level in space at a distance S below A. If the body C is now displaced to the left through an angle θ_0, to position B, then the energy that entered the system is the increase in potential energy and is expressed by

$$W_{\text{in}} = gml(1 - \cos \theta_0) \quad \text{(1-4)}$$

where $g = 9.807$ mps², the acceleration of gravity.

If the body is now released and allowed to swing freely, and if there is no friction, then by the time C has traveled from its initial position at the angle θ_0 to the position corresponding to the angle θ it has given up some of its potential energy. This reduction in potential energy must be compensated by an equal amount of energy in some other form. In this case kinetic energy associated with the velocity of the pendulum provides the compensation. That part of the energy put into the system, which is in the form of potential energy when the angle has decreased from θ_0 to θ, is expressed by

$$W_p = gml(1 - \cos \theta) + gmS \quad (1\text{-}5)$$

and the kinetic energy is

$$W_k = \tfrac{1}{2}mv^2 \quad (1\text{-}6)$$

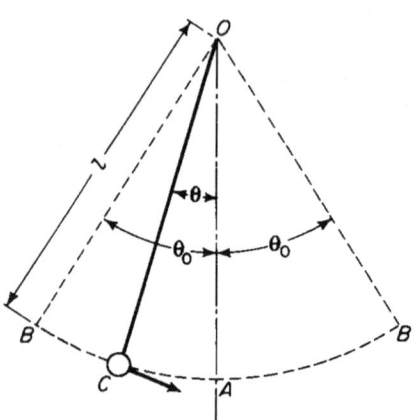

Figure 1-3. Simple pendulum.

It is evident that when angle θ is zero the body C is at its lowest level. This is also the original level before energy was added. At that instant the potential energy is gmS and the kinetic energy has its maximum value being now equal to W_{in}. However, as the mass C rises from positions to the right of A, it starts to regain its potential energy and its kinetic energy is reduced. By the time the angle θ has reached a value θ_0 to the right of center, the potential energy is again equal to W_{in} and the oscillation starts toward the left. In the absence of friction and viscosity the pendulum will again reach its original position. *This is a reversible process.* The period of the pendulum can be determined on the basis of the Law of Conservation of Energy as follows

$$W_{\text{in}} = W_p - gmS + W_k$$

$$gml(1 - \cos \theta_0) = gml(1 - \cos \theta) + \tfrac{1}{2}mv^2 \quad (1\text{-}7)$$

$$\tfrac{1}{2}v^2 = gl(\cos \theta - \cos \theta_0)$$

When Eq. 1-7 is differentiated with respect to time, the result is

$$v \frac{dv}{dt} = -gl \sin \theta \frac{d\theta}{dt} \quad (1\text{-}8)$$

but

$$v = l \frac{d\theta}{dt} \quad (1\text{-}9)$$

Sec. 1-4 THE LAW OF DEGRADATION OF ENERGY

Equation 1-8 divided by Eq. 1-9 yields

$$\frac{dv}{dt} = -g \sin \theta \tag{1-10}$$

The time derivative of Eq. 1-9 results in

$$\frac{dv}{dt} = l \frac{d^2\theta}{dt^2} \tag{1-11}$$

Equation 1-11 substituted in Eq. 1-10 yields

$$\frac{d^2\theta}{dt^2} + \frac{g}{l} \sin \theta = 0 \tag{1-12}$$

The solution of Eq. 1-12 is not elementary; it involves an elliptic integral. However, if θ is small, then

$$\sin \theta \approx \theta \tag{1-13}$$

and Eq. 1-12 can be approximated as follows

$$\frac{d^2\theta}{dt^2} + \frac{g}{l} \theta = 0 \tag{1-14}$$

It can be shown that the solution of Eq. 1-14 is

$$\theta = \theta_0 \cos \left(\sqrt{\frac{g}{l}} \right) t \tag{1-15}$$

The period T of a pendulum is the time in seconds for one complete oscillation as shown in Fig. 1-4(a). During the interval T the angle $(\sqrt{g/l})T$ in Eq. 1-15 has a value of 2π radians. Hence

$$\left(\sqrt{\frac{g}{l}}\right) T = 2\pi \quad \text{and} \quad T = 2\pi \sqrt{\frac{l}{g}} \tag{1-16}$$

Suppose that the boundary that encloses the pendulum is not evacuated, but the pendulum oscillates in air or in some other viscous medium. If the pendulum is again started from an initial angle θ_0 the amplitudes of successive oscillations become smaller as shown in Fig. 1-4(b) and eventually the pendulum comes to a stop. The energy put into the system when the pendulum is displaced from $\theta = 0$ to $\theta = \theta_0$ is converted into heat by the time the

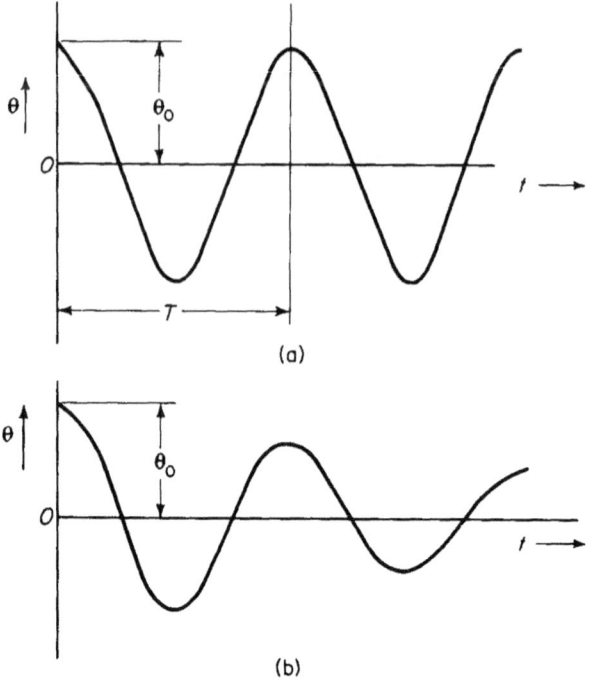

Figure 1-4. Angular displacement of a simple pendulum (a) without irreversible energy and (b) with irreversible energy.

pendulum comes to a complete stop. Were it possible to insulate the boundary in which the pendulum is enclosed so that no heat could escape, all the energy put into the system would remain in the system. Equation 1-3 then could be modified to represent the energy relations as follows

$$\begin{bmatrix} \text{all the energy} \\ \text{that enters} \\ \text{the system} \end{bmatrix} = \begin{bmatrix} \text{the gain in} \\ \text{potential} \\ \text{energy} \end{bmatrix} + \begin{bmatrix} \text{the gain in} \\ \text{kinetic} \\ \text{energy} \end{bmatrix} + \begin{bmatrix} \text{the gain in} \\ \text{thermal} \\ \text{energy} \end{bmatrix} \quad (1\text{-}17)$$

The first term in brackets in the right-hand side of Eq. 1-17 is associated with the position of the pendulum and the second term with the velocity of the pendulum. By the time the pendulum ceases to oscillate, the gain in potential and kinetic energy has been converted into the gain in thermal energy. Equation 1-3 as applied to this case would be reduced to the following simple relationship

$$\begin{bmatrix} \text{all the energy} \\ \text{that enters} \\ \text{the system} \end{bmatrix} = \begin{bmatrix} \text{the gain in} \\ \text{thermal} \\ \text{energy} \end{bmatrix} \quad (1\text{-}18)$$

The thermal energy is irreversible since it cannot impart an angular displacement to the pendulum without some agency external to the system.

In practical situations, heat would escape from the system and the thermal energy would leave the system if the surrounding space were at a lower temperature. Whether or not the irreversible energy, or thermal energy in this case, leaves the system, it is necessary to maintain an energy flow into the system. This flow of energy must equal the irreversible energy in order to sustain the oscillations of the pendulum. In some types of clocks a main spring, which needs to be wound from time to time, is used to furnish the energy; in others a system of weights is used.

1-5 THERMAL ENERGY

Potential mechanical and kinetic energy of random motion on a microscopic scale is known as thermal energy. In the case of solids this random motion is the result of atomic vibrations. However, in the case of liquids the molecules have sufficient energy of motion to overcome the forces that tend to restrict solids to a definite shape. In the case of a gas the velocities of the molecules are such that the distances between them become so great that the forces between molecules are negligible and the gas fills the space to which it is confined by an enclosing vessel.

1-6 ELECTRICAL ENERGY

Electrical energy results from forces between electric charges, from the motion of electric charges, and from both forces and motion associated with electric charges. Electrical energy may be stored in the electrostatic field. This occurs in capacitors where there is an accumulation of positive charges on one plate and an accumulation of an equal number of negative charges on the other plate. Such stored energy is potential energy because it results from the position of positive charges relative to negative charges giving rise to forces between the plates of the capacitor. Electrical energy may also be stored in the magnetic field, in which case electric charges in motion (current) produce a magnetic flux. This is a case of kinetic energy because it results from charges in motion.

Electric storage batteries and dry cells do not store electricity. Their operation depends upon energy conversion. During discharge, chemical energy is converted to electrical energy. In charging a storage battery the reverse process occurs, in that electrical energy is converted into chemical energy. Nuclear reactions that give rise to the transformation of mass to energy result in nuclear energy.

1-7 POWER

Power is the time rate of transferring or transforming energy. The unit of power in the MKS system is the watt or joule per second. If w = energy in joules then the power is expressed by

$$p = \frac{dw}{dt} \tag{1-19}$$

Equation 1-2 can be restated in terms of power as follows

$$\begin{bmatrix} \text{power input} \\ \text{to the system} \end{bmatrix} = \begin{bmatrix} \text{power out} \\ \text{of the system} \end{bmatrix} + \begin{bmatrix} \text{power absorbed} \\ \text{by the system} \end{bmatrix} \tag{1-20}$$

Part of the power out of the system may be useful power since the mechanical power output of an electric motor and part of the power out of the system may be in the form of heat produced by electrical and mechanical losses. Equation 1-1 gives the relationship between work dw, force F, and an infinitesimal displacement dx. If Eq. 1-1 is differentiated with respect to time, the following expression is obtained for the instantaneous power

$$p = \frac{dw}{dt} = \mathbf{F} \cdot \frac{\mathbf{dx}}{\mathbf{dt}} = \mathbf{F} \cdot \mathbf{dv} \tag{1-21}$$

The work dw associated with the motion of an infinitesimal charge dq, against the electrostatic force of a field, through a potential difference of v volts is expressed by

$$dw = v\, dq \tag{1-22}$$

and the power is

$$p = \frac{dw}{dt} = v\frac{dq}{dt} = vi \tag{1-23}$$

where $i = dq/dt$ is the electric current expressed in amperes.

1-8 TORQUE AND TANGENTIAL FORCE

Torque is a combination of force and distance that may be defined as twisting effort. This quantity is used in the analysis of rotating devices. In

Fig. 1-5 a tangential force of F newtons applied to the free end of the crank having an arm of r m produces a torque of

$$T = rF \quad \text{newton meters} \tag{1-24}$$

The amount of energy put into the crank in rotating it through an infinitesimal angle of $d\theta$ radians is, from Eq. 1-1

$$dw = \mathbf{F} \cdot \mathbf{dx} = Fr\, d\theta = T\, d\theta \tag{1-25}$$

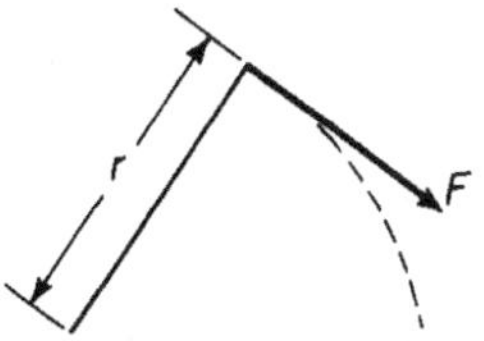

The power associated with torque is

$$p = \frac{dw}{dt} = T\frac{d\theta}{dt} = \omega T \tag{1-26}$$

Figure 1-5. Force applied to a crank.

Where ω is the angular velocity in radians per sec.
Hence, if n is the speed in rpm, the angular velocity is expressed by

$$\omega = \frac{2\pi n}{60} \quad \text{radians per sec} \tag{1-27}$$

1-9 EFFICIENCY

The efficiency of a device is defined as the ratio of the energy output to the energy input. It is also defined as the ratio of power output to power input. In these definitions the term "output" means *useful* or *desired* output. In that sense of the term, the output of a generator or transformer is the energy or the power transmitted to the connected load. The generator or transformer, as the case may be, in addition to supplying useful energy also gives up energy in the form of heat, vibration, and noise to surrounding space. According to the Law of Conservation of Energy there can be no loss of energy because energy can neither be created nor destroyed. However, in the case of the generator or transformer the energy given up in the form of heat, vibration, and sound is not put to use. Such energy, as far as utilization is concerned, is wasted and is therefore considered lost. Hence the terms, copper losses, core losses, friction and windage losses. The relationship between efficiency, input, and losses is expressed by

$$\text{Efficiency} = 1 - \frac{\text{losses}}{\text{input}} \tag{1-28}$$

The efficiencies of large power transformers are frequently greater than 99 percent at rated output. On the other hand the overall efficiency of commercial heat engines is below 40 percent. In many applications, such as in communication and control systems, efficiency is of minor importance in relation to quality and precision of performance. As an example, the efficiency of a cone loud-speaker is generally less than 5 percent.

1-10 ENERGY AND POWER RELATIONS IN MECHANICAL AND ELECTRICAL SYSTEMS

In the case of the mechanical system we are concerned with friction, mass (inertia), and elasticity. Motion in combination with friction gives rise to heat, i.e., mechanical energy is converted into thermal energy. Kinetic energy is stored from motion in combination with distortion (elasticity). Similarly, in the electrical system an electric current encountering resistance produces heat and electrical energy is converted into thermal energy. A current in an inductive circuit produces magnetic flux, thus storing kinetic energy. Potential energy is stored in the dielectric of a capacitive circuit when there is a flow of electric current.

1-11 MASS AND VISCOUS FRICTION

Consider a constant force F applied horizontally to a body of constant mass M moving along a stationary horizontal surface. Assume that the force required to overcome the frictional resistance alone is proportional to the velocity of the body. Such an assumption is valid to a good degree of approximation in the case of lubricated surfaces, once the moving body has attained an appreciable velocity. Then

$$F = R_F v + M \frac{dv}{dt} \qquad (1\text{-}29)$$

where $R_F v$ is the component of force required to overcome frictional resistance and $M\, dv/dt$ is the component of force required to overcome the inertia of the body. The mechanical power input to this system is

$$p = \mathbf{F} \cdot \mathbf{v} = Fv$$

as $\mathbf{v}$ and $\mathbf{F}$ are both in the same direction, the angle between these vectors thus being zero. Hence

$$p = R_F v^2 + Mv \frac{dv}{dt} \qquad (1\text{-}30)$$

In Eq. 1-30 the component of power represented by the term $R_F v^2$ is the one converted into heat. The term $Mv\,dv/dt$ represents the power expended in storing kinetic energy in the mass M. If v is constant, $dv/dt = 0$ and there is no energy storage because all of the power p is converted into heat. If dv/dt is positive the mass M undergoes acceleration and stores kinetic energy. On the other hand if dv/dt is negative the mass M decelerates and gives up its stored kinetic energy and the power $R_F v^2$ being converted into heat is greater than the applied power p. The energy input into this system is expressed by

$$\int p\,dt = \int R_F v^2\,dt + M \int v\,dv$$

Further, if the body starts from rest, i.e., $v = 0$, when $t = 0$ under the influence of the constant force F we have

$$\int_0^t p\,dt = \int_0^t R_F v^2\,dt + \frac{Mv^2}{2} \tag{1-31}$$

$\dfrac{Mv^2}{2}$ represents the stored energy and $\int_0^t R_F v^2\,dt$ represents the mechanical energy converted into thermal energy.

For this condition the solution of Eq. 1-29 expresses the velocity as a function of time as follows

$$M\frac{dv}{dt} + R_F v = F \tag{1-32}$$

Separation of the variables yields

$$\frac{dv}{v - \dfrac{F}{R_F}} = -\frac{R_F}{M}\,dt$$

from which

$$v - \frac{F}{R_F} = C\epsilon^{-(R_F/M)t} \tag{1-33}$$

when $t = 0$, and $v = 0$, hence

$$C = -\frac{F}{R_F}$$

and

$$v = \frac{F}{R_F}(1 - \epsilon^{-(R_F/M)t}) \tag{1-34}$$

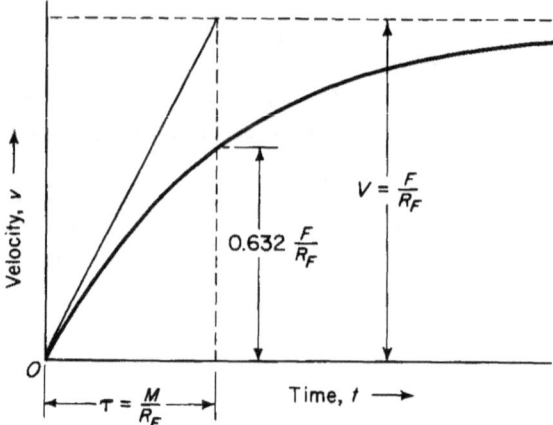

Figure 1-6. Velocity of a mass moving against viscous resistance under constant applied force.

Equation 1-34 is plotted graphically in Fig. 1-6 and shows the final value $v = F/R_F$ to be an asymptote that is approached as the time t approaches infinity.

Time constant

The reciprocal of the constant R_F/M in the exponent of Eq. 1-34 is called the time constant τ. Hence

$$\tau = \frac{M}{R_F} \quad \text{sec}$$

and Eq. 1-34 can be rewritten as

$$v = \frac{F}{R_F}(1 - \epsilon^{-t/\tau}) \tag{1-35}$$

The time constant also expresses the time required for the velocity v to attain the asymptotic value F/R_F if the acceleration were maintained constant at its initial value. This is indicated in Fig. 1-6 by the tangent to the curve at $t = 0$.

Since the friction absorbs some of the applied energy in irreversible form, i.e., in the form of heat, the acceleration is not constant but decreases with time. Hence the velocity when $t = \tau$ sec is determined from Eq. 1-34 or 1-35 as follows

$$v = \frac{F}{R_F}(1 - \epsilon^{-1}) = 0.632 \frac{F}{R_F} \tag{1-36}$$

where F/R_F = the velocity as the time t approaches infinity.

1-12 RESISTANCE AND SELF-INDUCTANCE

Figure 1-7 shows a schematic diagram of an *R-L* circuit. When there is current in such a circuit, the self-inductance has energy that is stored electromagnetically. The electrical energy in the resistance is converted into thermal energy. The mathematical relationships in this kind of circuit are very similar to those in the mechanical system involving mass and friction. In Fig. 1-7 the applied voltage V is constant and equals two components

$$V = e_R + e_L \qquad (1\text{-}37)$$

where $e_R = iR$, the resistance drop
$e_L = L\, di/dt$, the emf of self-induction
R = resistance of the circuit in ohms
L = self-inductance of the circuit in henries and is constant
i = current in amperes
V = applied voltage in volts
t = time in seconds

Equation 1-37 can be rewritten as

$$V = Ri + L\frac{di}{dt}$$

and the power is expressed by

$$Vi = Ri^2 + Li\frac{di}{dt}$$

The electric power converted into heat is Ri^2 and that stored electromagnetically is $Li\, di/dt$. The energy relationship is expressed by the following

$$\int Vi\, dt = \int Ri^2\, dt + L\int i\, di \qquad (1\text{-}38)$$

If the initial conditions are $i = 0$ when $t = 0$, then Eq. 1-38 becomes

$$\int_0^t Vi\, dt = \int_0^t Ri^2\, dt + \frac{Li^2}{2} \qquad (1\text{-}39)$$

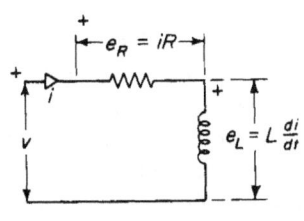

Figure 1-7. *R-L* circuit.

where $Li^2/2$ represents the energy stored in the self-inductance. If the initial conditions are such that $i = 0$ when $t = 0$, the current in the R-L circuit is expressed by

$$i = \frac{V}{R}(1 - \epsilon^{-Rt/L}) \qquad (1\text{-}40)$$

The reciprocal L/R of the constant term in the exponent of Eq. 1-40 is the time constant τ. It corresponds to the time constant M/R_F in the mechanical system.

A comparison of Eq. 1-34 with 1-40 shows that if voltage is taken as

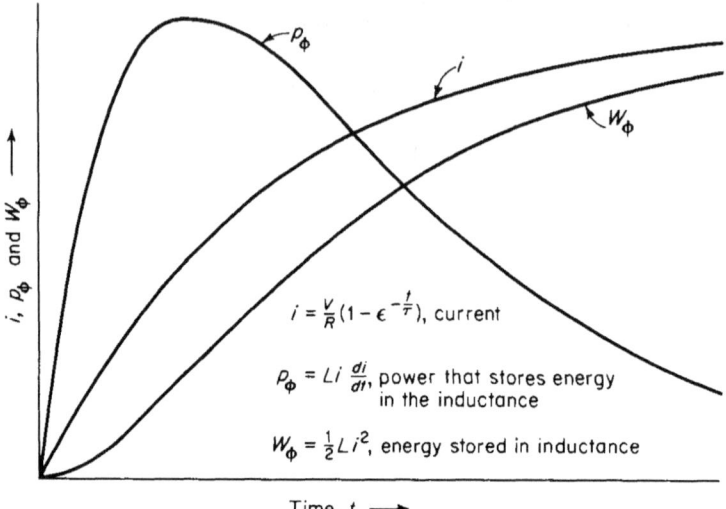

Figure 1-8. Characteristics of an R-L series circuit with constant applied voltage.

analogous to force and electrical resistance as analogous to frictional resistance in the mechanical system, current corresponds to velocity and self-inductance to inertia. Curves similar to those shown in Fig. 1-8 would portray the relationships for the mechanical system.

Thus the curve of current i vs time could be used to relate the velocity v to time in the mechanical system. Similarly, the curve of p_ϕ vs time could be used to represent the mechanical power $Mv\,dv/dt$, which stores energy in the moving mass of the mechanical system. The energy $\frac{1}{2}Mv^2$ stored in the mass could also be represented as a function of time by the curve of W_L vs time.

1-13 VISCOUS FRICTION AND SPRING

Consider a spring that has negligible mass and negligible losses. Suppose it is extended by a force applied through a rope sliding on a surface having a viscous friction such that the force required to overcome the friction is proportional to the velocity at which the rope slides across the surface. Assume the applied force F to be constant. In Fig. 1-9 let $S =$ coefficient of stiffness of the spring in newtons per m

$R_F =$ friction constant in newtons per meter per second

$x =$ extension of the spring in meters

then

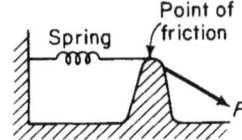

Figure 1-9. Mechanical system involving elasticity and friction.

$$F = R_F \frac{dx}{dt} + Sx \qquad (1\text{-}41)$$

where dx/dt is the velocity v with which the spring is extended, hence

$$F = R_F v + Sx \qquad (1\text{-}42)$$

The power is expressed by

$$p = Fv = R_F v^2 + Svx \qquad (1\text{-}43)$$

The power expressed by the term $R_F v^2$ is converted into heat and the power expressed by the term Svx is the power that stores energy in the spring.

The energy in this system is obtained by taking the time integral of Eq. 1-43 as follows

$$W = \int p \, dt = R_F \int v^2 \, dt + S \int vx \, dt \qquad (1\text{-}44)$$

If $x = 0$ when $t = 0$ there results

$$W = \int_0^t p \, dt = R_F \int_0^t v^2 \, dt + \frac{Sx^2}{2} \qquad (1\text{-}45)$$

NOTE: $dx = v \, dt$

The stored energy is $Sx^2/2$ and the converted energy is

$$R_F \int_0^t v^2 \, dt$$

To determine the relationship between the stored energy and that converted into heat, consider the total energy input for a final extension of the spring to a value X, then

$$W = F \int_0^X dx = FX$$

$$W = FX = R_F \int_0^X v\, dx + S \int_0^X x\, dx \qquad (1\text{-}46)$$

$$= R_F \int_0^X v\, dx + \frac{SX^2}{2}$$

However, the final value of x, namely X, is determined from the relationship

$$F = SX \qquad (1\text{-}47)$$

and

$$W = FX = SX^2 \qquad (1\text{-}48)$$

A comparison of Eq. 1-48 with 1-46 shows that the energy converted into heat is exactly equal to the energy stored in the spring, in that

$$R_F \int_0^X v\, dx = SX^2 - \frac{SX^2}{2} = \frac{SX^2}{2} \qquad (1\text{-}49)$$

This means that when a constant force is applied to such a system, it is capable of storing only one-half the applied energy. This is true regardless of the value of the friction constant R_F. If the parameter R_F is made low, the velocity v goes up correspondingly in such a manner that the frictional energy loss remains constant.

A similar situation exists for an electrical R-C circuit.

1-14 RESISTANCE AND CAPACITANCE

Figure 1-10 shows a circuit of capacitance in series with resistance. When a voltage is sustained across such a circuit the capacitance has energy that is stored electrostatically. The mathematical relationships in this kind of a circuit are quite similar to those in the mechanical system involving elasticity and friction. In Fig. 1-10 a constant voltage V is applied and

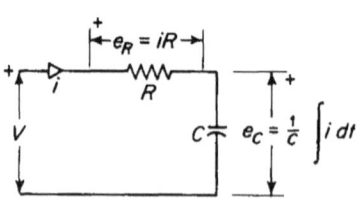

Figure 1-10. *R-C* circuit.

$$V = e_R + e_C$$

$$= iR + \frac{1}{C}\int i\, dt \qquad (1\text{-}50)$$

Sec. 1-14 RESISTANCE AND CAPACITANCE

and

$$p = Vi = i^2 R + \frac{i}{C} \int i \, dt \qquad (1\text{-}51)$$

is the expression for power.

The term $i^2 R$ expresses the power that is converted into heat and the term

$$\frac{i}{C} \int i \, dt$$

expresses the power that stores energy in the capacitance.

The solution of Eq. 1-51 for the condition that when $t = 0$ there is no energy stored in the capacitance, yields

$$i = \frac{V}{R} \epsilon^{-t/RC} \qquad (1\text{-}52)$$

In Eq. 1-51 the power that stores energy in the capacitance is expressed by the term

$$\frac{i}{C} \int_0^t i \, dt$$

If Eq. 1-52 is substituted in this term, the result is

$$p_c = \frac{V}{RC} \epsilon^{-t/RC} \int_0^t \frac{V}{R} \epsilon^{-t/RC} \, dt = \frac{V^2}{R} (\epsilon^{-t/RC} - \epsilon^{-2t/RC}) \qquad (1\text{-}53)$$

The energy stored in the capacitor at any time t after the constant voltage V is applied to the circuit is expressed by

$$W_c(t) = \int_0^t p_c \, dt = \frac{V^2}{R} \int_0^t (\epsilon^{-t/RC} - \epsilon^{-2t/RC}) \, dt$$
$$= \frac{V^2 C}{2} (1 - \epsilon^{-t/RC})^2 \qquad (1\text{-}54)$$

The energy in the capacitance approaches its final value as t approaches infinity. Hence, from Eq. 1-54

$$W_c = \frac{V^2 C}{2} \qquad (1\text{-}55)$$

The expression for the energy converted into heat is from Eq. 1-52

$$W_h = \int_0^\infty i^2 R \, dt = \int_0^\infty \frac{V^2}{R} \epsilon^{-2t/RC} = \frac{V^2 C}{2} \qquad (1\text{-}56)$$

Equations 1-55 and 1-56 show that the energy converted into heat in the R-C circuit with constant applied voltage is equal to the energy stored in the capacitor as time increases without limit. In the case of the spring under

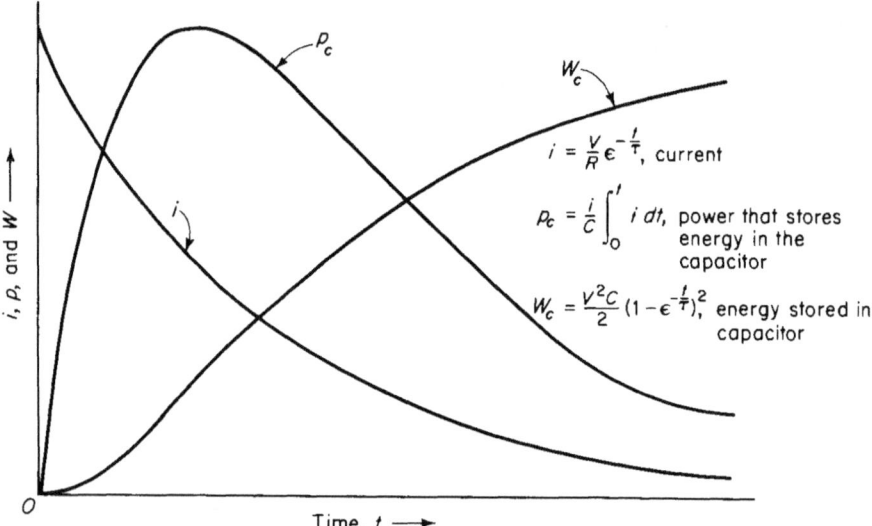

Figure 1-11. Characteristics of an R-C series circuit with constant applied d-c voltage.

constant applied force, the total energy converted into heat is also equal to the energy stored in the spring as time increases without limit.

The variation with respect to time of current, power and, energy in the capacitor is shown graphically in Fig. 1-11. The same graph can be used to express the relationships for the massless spring. In that case the curve that shows the relationship between current and time in Fig. 1-11 would show the variation of the velocity of the force F in Fig. 1-9. Similarly the curve p_o in Fig. 1-11 would indicate the power that stores energy in the spring as a function of time, whereas the curve w_o would show the amount of energy stored in the spring as a function of time.

The time constant for the R-C circuit is $\tau = RC$ sec. Similarly the time constant for the mechanical system with the massless spring is $\tau = R_F/S$ sec.

1-15 SPRING, MASS, AND VISCOUS FRICTION

Figure 1-12 shows a system in which a spring is extended by applying a force F to the mass M starting from rest. Let $v =$ the velocity at which the spring is extended, then

$$M \frac{dv}{dt} + R_F v + S \int v \, dt = F \qquad (1\text{-}57)$$

where $R_F =$ constant of viscous friction, newton seconds per meter
$S =$ stiffness coefficient of the spring, newtons per meter
$M =$ mass in kilograms

Differentiation of Eq. 1-57 with respect to time yields

$$M \frac{d^2 v}{dt^2} + R_F \frac{dv}{dt} + Sv = 0 \qquad (1\text{-}58)$$

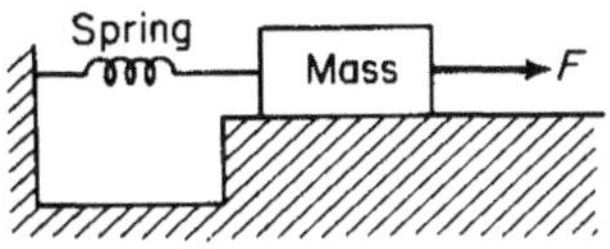

The relationship

$$v = A \epsilon^{mt} \qquad (1\text{-}59)$$

Figure 1-12. Elasticity, mass, and viscous friction.

is a solution that satisfies Eq. 1-58 and upon substitution of Eq. 1-59 in Eq. 1-58 there results

$$M m^2 \epsilon^{mt} + R_F m \epsilon^{mt} + S \epsilon^{mt} = 0 \qquad (1\text{-}60)$$

Equation 1-60 divided by ϵ^{mt} yields

$$m^2 + \frac{R_F}{M} m + \frac{S}{M} = 0 \qquad (1\text{-}61)$$

or

$$m_1 = -\frac{R_F}{2M} + j\sqrt{\frac{S}{M} - \left(\frac{R_F}{2M}\right)^2} = -a + jb$$

$$m_2 = -\frac{R_F}{2M} - j\sqrt{\frac{S}{M} - \left(\frac{R_F}{2M}\right)^2} = -a - jb$$

where $j = \sqrt{-1}$.

Since Eq. 1-57 is a second order differential equation, two constants of integration are involved in the general solution, and the velocity is expressed by

$$v = A_1 \epsilon^{m_1 t} + A_2 \epsilon^{m_2 t} \qquad (1\text{-}62)$$

To evaluate the constants of integration A_1 and A_2, let $t = 0$. Now at $t = 0$, $v = 0$ as the mass is started from rest. Furthermore, if

$$v = 0, \qquad R_F v \quad \text{and} \quad S\int v\, dt$$

must both be zero at $t = 0$. Therefore

$$M \frac{dv}{dt} = F \quad \text{at } t = 0$$

So

$$A_1 + A_2 = 0 \quad (v = 0, t = 0) \tag{1-63}$$

$$M \frac{dv}{dt} = M\, (m_1 A_1 \epsilon^{m_1 t} + m_2 A_2 \epsilon^{m_2 t}) \tag{1-64}$$

$$m_1 A_1 + m_2 A_2 = \frac{F}{M} \quad \text{at } t = 0 \tag{1-65}$$

From Eqs. 1-63 and 1-65 we have

$$A_1 = \frac{F}{M}\left(\frac{1}{m_1 - m_2}\right) = \frac{F}{j2bM}$$

$$A_2 = \frac{F}{M}\left(\frac{1}{m_2 - m_1}\right) = -\frac{F}{j2bM} \tag{1-66}$$

Substitution of Eq. 1-66 in Eq. 1-62 yields

$$v = \frac{F}{bM}\left(\frac{\epsilon^{m_1 t} - \epsilon^{m_2 t}}{j2}\right)$$

$$= \frac{F}{bM} \epsilon^{-at}\left(\frac{\epsilon^{jbt} - \epsilon^{-jbt}}{j2}\right) \tag{1-67}$$

$$= \frac{F}{bM} \epsilon^{-at} \sin bt$$

Frictional losses

The force required to overcome frictional resistance is

$$F_R = R_F v \tag{1-68}$$

Sec. 1-15   SPRING, MASS, AND VISCOUS FRICTION   23

and the power converted into heat, i.e., heat expended in overcoming friction, is

$$p_R = F_R v = R_F v^2 \qquad (1\text{-}69)$$

The energy converted into heat through friction is

$$W_R = \int_0^\infty R_F v^2 \, dt = \frac{F^2 R_F}{b^2 M^2} \int_0^\infty \epsilon^{-2at} \sin^2 bt \, dt$$

$$= \frac{F^2 R_F}{b^2 M^2} \frac{(b^2)}{[4a(a^2 + b^2)]} = \frac{F^2 R_F}{M^2} \frac{1}{4R_F/2M(S/M)} = \frac{F^2}{2S} \qquad (1\text{-}70)$$

The same relationship can be derived in a simpler manner on the basis of Eq. 1-3. The gain in reversible energy is the energy stored in the spring, whereas the irreversible energy is in the form of heat, some of which raises the temperature of part of the system, thus representing a gain in irreversible energy; the remainder of the heat is dissipated to media surrounding the system. When Eq. 1-3 is applied to this case the following relationship is valid

$$\begin{bmatrix} \text{the irreversible} \\ \text{energy} \end{bmatrix} = \begin{bmatrix} \text{all the energy} \\ \text{that leaves the} \\ \text{system} \end{bmatrix} + \begin{bmatrix} \text{the gain in} \\ \text{irreversible} \\ \text{energy} \end{bmatrix} \qquad (1\text{-}71)$$

Accordingly, for this situation Eq. 1-3 can be reduced to

$$\begin{bmatrix} \text{all the energy} \\ \text{that enters} \\ \text{the system} \end{bmatrix} = \begin{bmatrix} \text{the irreversible} \\ \text{energy} \end{bmatrix} + \begin{bmatrix} \text{the stored} \\ \text{energy} \end{bmatrix}$$

$$= \int_0^X F \, dx = FX = SX^2 \qquad (1\text{-}72)$$

where $X = F/S$ so that the total energy input

$$W_{\text{in}} = \frac{F^2}{S}$$

The gain in reversible energy is the energy stored in the spring

$$W_{\text{stored}} = \int_0^X Sx \, dx = \frac{F^2}{2S}$$

Then from Eq. 1-72 the irreversible energy is

$$W_{\text{irrev}} = W_{\text{in}} - W_{\text{stored}} = \frac{F^2}{2S} = W_R$$

The final energy stored in the spring is also $F^2/2S$. Hence the total energy input to the system is F^2/S. One-half of this is converted into heat.

1-16 SERIES R-L-C CIRCUIT

Figure 1-13 shows the schematic diagram of a series R-L-C circuit to which is applied a constant d-c voltage V. In this circuit the applied voltage satisfies the following equation

$$L\frac{di}{dt} + Ri + \frac{1}{C}\int i\, dt = V \tag{1-73}$$

where L = constant self-inductance in henries
R = resistance in ohms
C = capacitance in farads
i = current in amperes

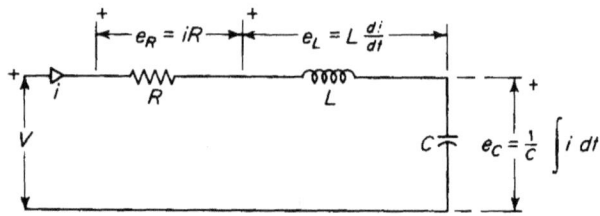

Figure 1-13. *R-L-C* circuit.

Equation 1-73 is of exactly the same form as Eq. 1-57, which applies to the mechanical system with mass, friction, and elasticity. The integral in the left-hand side of Eq. 1-71 is eliminated by differentiation with respect to time. Hence

$$\frac{d^2i}{dt^2} + \frac{R}{L}\frac{di}{dt} + \frac{i}{C} = 0 \tag{1-74}$$

If Eq. 1-58 is divided by M, the result is

$$\frac{d^2v}{dt^2} + \frac{R_F}{M}\frac{dv}{dt} + \frac{S}{M}v = 0 \tag{1-75}$$

A comparison of Eq. 1-75 with Eq. 1-74 shows these equations to be identical except for the values of the constant coefficients. This comparison then shows that solution of Eq. 1-72 must be of the same form as that of Eq. 1-75,

Sec. 1-16 SERIES R-L-C CIRCUIT

the only difference being in the constant coefficients. The solution of Eq. 1-75 is given by Eq. 1-62 as follows

$$v = A_1 \epsilon^{m_1 t} + A_2 \epsilon^{m_2 t}$$

On that basis the solution of Eq. 1-74 is

$$i = B_1 \epsilon^{n_1 t} + B_2 \epsilon^{n_2 t} \tag{1-76}$$

Because of the self-inductance the current must be zero at $t = 0$, the instant when the d-c voltage is applied; so from this boundary condition when $t = 0$, then $i = 0$, and

$$i = B_1 + B_2 = 0 \tag{1-77}$$
$$B_2 = -B_1$$

Equation 1-76 establishes the relationship between the unknown constants B_1 and B_2 but does not evaluate them in terms of known constants. Either of these constants, B_1 or B_2 can now be evaluated, however, in terms of the known constants R, L, and C and the applied voltage V for the same boundary condition as follows. The current is zero at $t = 0$. This means that the term Ri in Eq. 1-73 must be zero at $t = 0$ also. Furthermore, since at $t = 0$ no current has as yet been applied to the capacitor, the time integral $\int_0^t i\, dt$ of the current must also equal zero. Substitution of

$$t = 0 \quad Ri = 0 \quad i\, dt = 0$$

in Eq. 1-71 yields

$$L \frac{di}{dt} = V$$

and

$$L n_1 B_1 + L n_2 B_2 = V \tag{1-78}$$

From Eqs. 1-77 and 1-78 there results

$$B_1 = -B_2 = \frac{V}{(n_1 - n_2)} L \tag{1-79}$$

The process involved in the solution of Eqs. 1-58, 1-59, and 1-60 when applied to this case yields

$$n_1 = -\frac{R}{2L} + j\sqrt{\frac{1}{LC} - \left(\frac{R}{2L}\right)^2}$$

and

$$n_2 = -\frac{R}{2L} - j\sqrt{\frac{1}{LC} - \left(\frac{R}{2L}\right)^2} \qquad (1\text{-}80)$$

Equations 1-79 and 1-80, when substituted in Eq. 1-76 yield

$$i = \frac{V\epsilon^{-at}}{L\,b} \sin bt \qquad (1\text{-}81)$$

where $a = R/2L$ and

$$b = \sqrt{\frac{1}{LC} - \left(\frac{R}{2L}\right)^2}$$

From Eq. 1-81 it is evident that the current i may undergo two different kinds of variation with respect to time. The current may (1) oscillate or it may (2) rise to a maximum and then gradually die out without reversing its direction.

Current oscillatory

When the constant b in Eq. 1-81 is real, i.e., if

$$\frac{1}{LC} > \left(\frac{R}{2L}\right)^2$$

the current i oscillates, its successive amplitudes decreasing at a rate determined by the constant a. The current and energy relationships of a series R-L-C circuit are shown graphically as functions of time in Fig. 1-14. When the current in Fig. 1-14 is positive the circuit absorbs energy from the source of voltage and when the current is negative the circuit returns energy to the source.

If all of the energy in the circuit were reversible, i.e., if there were no heat losses, the negative current peaks would equal the positive current peaks and the circuit would return as much energy to the source when the current is negative as it received during the half cycle when the current was positive. To satisfy the condition of no heat losses the resistance R of the circuit must be zero. When $R = 0$, the constant a in Eq. 1-81 is also zero, and the equation then defines a steady alternating current.

Current nonoscillatory

When the constant b in Eq. 1-81 is imaginary, i.e., if

$$\frac{1}{LC} < \left(\frac{R}{2L}\right)^2$$

then the trigonometric sine in Eq. 1-81 becomes a hyperbolic sine and the current flows in one direction only, building up to a maximum value and then gradually dying out as the capacitor becomes charged to a potential difference that approaches the value of the applied voltage V. This is a case in which the reversible energy is so great in relation to the energy

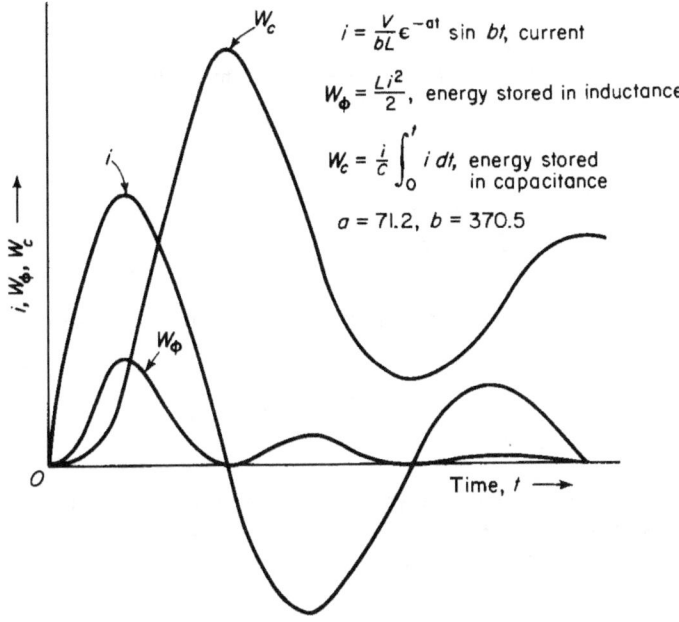

Figure 1-14. Characteristics of oscillating series R-L-C circuit.

stored in the self-inductance L and in the capacitance C that there is no energy returned to the source while the capacitor is charged.

The case that serves as a dividing line between an oscillating current and a nonoscillating current is a critically damped circuit where the constant $b = 0$, i.e.

$$\frac{1}{LC} = \left(\frac{R}{2L}\right)^2$$

If b is made zero in Eq. 1-81 the right-hand side of that equation becomes an indeterminate that can be evaluated on the basis that

$$\lim_{b \to 0} \sin bt = bt \qquad (1\text{-}82)$$

Substitution of Eq. 1-80 in Eq. 1-79 yields

$$i = \frac{V}{L} \epsilon^{-a_1 t} \qquad (1\text{-}83)$$

Whether the current oscillates or not, the irreversible energy, i.e., the energy converted into heat is expressed by

$$W_R = \int_0^\infty i^2 R \, dt = \frac{V^2 R}{L^2 b^2} \int_0^\infty \epsilon^{-2at} \sin^2 bt \, dt = \frac{V^2 C}{2} \qquad (1\text{-}84)$$

The total energy input to the circuit is

$$W_{in} = \int_0^\infty Vi \, dt$$

and since the applied voltage V is a constant it can be taken outside the integral, thus

$$W_{in} = V \int_0^\infty i \, dt = V^2 C$$

The final voltage across the capacitor determines the final energy stored in the capacitor and, in this case, this final value of voltage is the constant applied voltage V. The final energy stored in the capacitor is therefore given by Eq. 1-55 as

$$W_c = \frac{V^2 C}{2}$$

It therefore follows that, with a constant applied d-c voltage, the irreversible energy in the series R-L-C circuit equals the stored energy so that, in this case also, one-half of the energy input is irreversible.

Since Eq. 1-67 for the case of the mechanical system is similar to Eq. 1-81, it follows that conditions analogous to the cases of the oscillating and nonoscillating current must exist. Thus, if in Eq. 1-67 the constant b is real, the motion of the mass M reverses its direction periodically, and the mass gradually comes to rest. If on the other hand the constant b is imaginary, the mass M moves only in the direction of the applied force F, and gradually comes to rest without any reversal in the direction of motion.

It is significant that when a constant force is applied to a system in which the energy is finally stored in the deformation of a body such as a spring, one-half of the energy input to the system is converted into heat. This is true also in the case of an electric circuit in which the energy is finally stored in an ideal capacitor. Thus, under the condition of a constant applied force, a spring is capable of storing, at best, only 50 percent of the applied energy in reversible form. This is also true of a capacitor under the condition of constant applied voltage.

1-17 ENERGY STORED IN A ROTATING FLYWHEEL

There are many applications in which a flywheel is used alternately to store and give up kinetic energy. Examples are reciprocating engines, compressors in which there are power pulsations, and punch presses where the power is expended in a relatively short part of the duty cycle.

A body in pure translational motion has a kinetic energy of $mv^2/2$ because all parts of the body have the same velocity. However, in the case of a rotating body all parts do not have the same velocity. Parts near the axis of rotation have relatively low velocities and those further removed from the axis have greater velocities in direct proportion to their distances from the center of rotation. All parts, however, have the same angular velocity. Hence, the velocity of a particle of mass m_n located at a distance r_n from the axis of rotation would have a tangential velocity of

$$v_n = r_n \omega$$

where ω is the angular velocity in radians per sec.

The kinetic energy of rotation for such a particle is expressed by

$$\tfrac{1}{2} m_n v_n^2 = \tfrac{1}{2} m_n r_n^2 \omega^2$$

Hence, for the entire rotating body

$$W_K = \frac{1}{2} \left[\sum m_n r_n^2 \right] \omega^2$$

The quantity $\sum m_n r_n^2$ is known as the moment of inertia and is represented by the symbol I, hence

$$W_K = \tfrac{1}{2} I \omega^2 \tag{1-85}$$

If I is expressed in kilogram meters^2 and ω in radians per sec, then the kinetic energy is expressed in joules.

As an example, the moment of inertia of a solid cylinder rotating about its central axis is $I = \tfrac{1}{2} mR^2$ where R is the radius of the cylinder.

The moment of inertia, $I = \frac{1}{2}m(R_1^2 + R_2^2)$ is that of another common shape namely an annular cylinder with inside and outside radii of R_1 and R_2 respectively.

1-18 POWER AND TORQUE

The work done in accelerating a rotating body from rest to an angular velocity of ω radians per sec is equal to the kinetic energy of rotation, hence, from Eq. 1-85

$$W = \tfrac{1}{2}I\omega^2 \qquad (1\text{-}86)$$

However, power is the rate of doing work, and we have

$$p = \frac{dw}{dt} = I\omega \frac{d\omega}{dt} \qquad (1\text{-}87)$$

where $d\omega/dt$ is angular acceleration in radians per sec².

Suppose that, to develop the power expressed by Eq. 1-87, an external force F is applied to a point at a distance R from the center; then from $p = Fv$ it follows that

$$p = FR\omega \qquad (1\text{-}88)$$

because $v = R\omega$.

The product FR is called torque and in the rationalized MKS system is expressed in newton meters. Thus, if the torque is expressed by $T = FR$, Eq. 1-88 becomes $p = \omega T$, and we have

$$T = \frac{p}{\omega} = I \frac{d\omega}{dt} \qquad (1\text{-}89)$$

the torque required to accelerate a rotating mass. This also expresses the torque developed by a decelerating mass.

Suppose that a constant torque T is applied to a flywheel starting from rest and that the moment of resistance of the bearings and other moving parts can be represented by R_M. Then if the moment of inertia of the flywheel is I, we have

$$T = I \frac{d\omega}{dt} + R_M \omega \qquad (1\text{-}90)$$

the solution of which is

$$\omega = \frac{T}{R_M}\left[1 - \epsilon^{\frac{-R_M t}{I}}\right] \qquad (1\text{-}91)$$

Note that this expression for angular velocity has the same form as that for a current in an inductive circuit with constant applied d-c voltage, as in

Eq. 1-40. The moment of resistance R_M corresponds to the electrical resistance of the circuit, and the moment of inertia I corresponds to the self-inductance L. Furthermore, the energy stored in the flywheel is expressed by $\frac{1}{2}I\omega^2$, and the energy stored in the self-inductance is expressed by $\frac{1}{2}Li^2$.

1-19 MECHANICAL AND ELECTRICAL ANALOGIES

The behavior of the analogous mechanical and electrical systems is defined by the same types of differential equations as is evident from a comparison of Eq. 1-29 with Eq. 1-26 and of Eq. 1-57 with Eq. 1-72. Analogies based on these types of differential equations are shown in Table 1-1.* The force F

TABLE 1-1. DIRECT RELATION BETWEEN MECHANICAL AND ELECTRICAL QUANTITIES

Mechanical quantity	Electrical quantity	Mechanical energy stored or dissipated	Electrical energy stored or dissipated	Symbol employed for both systems
Force = F	Electromotive force = E			
Displacement = x	Charge = q			
Velocity = v	Current = i			
Mass = M	Inductance = L	$\frac{1}{2}Mv^2$	$\frac{1}{2}Li^2$	⏻⏻⏻
Mechanical resistance = R	Electrical resistance = R	$\int_0^t Rv^2\,dt$	$\int_0^t Ri^2\,dt$	⟋⟍⟋⟍
* Compliance = C	Capacitance = C	$\frac{1}{2}\frac{x^2}{C} = \frac{1}{2}CF^2$	$\frac{1}{2}\frac{Q^2}{C} = \frac{1}{2}CE^2$	⊣⊢

* Equations are sometimes expressed in terms of the stiffness factor S, which is the reciprocal of compliance, instead of in terms of compliance

$$\left[S = \frac{1}{C} \right]$$

and the voltage or electromotive force E, in Table 1-1 may be constants as well as functions of time. It should be noted that both V and E serve as symbols for voltage.

The term *compliance* is used in connection with the mechanical quantities listed in Table 1-1. Compliance denotes the inverse of stiffness and

* Tables 1-1, 1-2, and 1-3, are from W. P. Mason, *Electromechanical Transducers and Wave Filters*, 2nd Ed. Princeton, N.J.: D. Van Nostrand Co., Inc., 1948.

corresponds to capacitance in the electrical circuit when force in the mechanical system is considered analogous to voltage in the electrical system. The relation shown in Table 1-1 is sometimes called the direct relation. The inverse of these relations may also be used. These follow from comparing a parallel electrical circuit, as shown in Fig. 1-15, with the mechanical system. The differential equation for this parallel circuit is

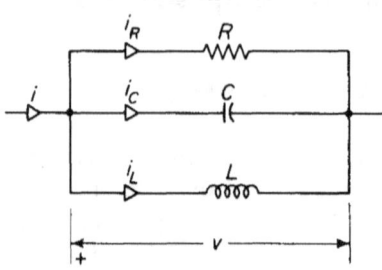

Figure 1-15. Parallel R-L-C circuit.

$$i = i_C + i_R + i_L$$
$$= C\frac{dv}{dt} + \frac{v}{R} + \frac{1}{L}\int v\,dt \quad (1\text{-}92)$$

The inverse relation between mechanical and electrical quantities are shown in Table 1-2.

TABLE 1-2. INVERSE RELATION BETWEEN MECHANICAL AND ELECTRICAL QUANTITIES

Mechanical quantity	Electrical quantity	Mechanical energy stored or dissipated	Electrical energy stored or dissipated	Symbol employed for both systems
Force = F	Current = i			
Impulse = $F\,dt$	Charge = q			
Velocity = v	Electromotive force = E			
Displacement = x	$\int E\,dt$			
Mass = M	Capacitance = C	$\frac{1}{2}mv^2$	$\frac{1}{2}CE^2$	⊣⊢
Mechanical resistance = R	Leakance = G	$\int_0^t Rv^2\,dt$	$\int_0^t GE^2\,dt$	⌇
Compliance = C	Inductance = L	$\frac{1}{2}CF^2$	$\frac{1}{2}Li^2$	⌇

A comparison of quantities for the torsional system is shown in Table 1-3.

1-20 CHEMICALLY STORED ENERGY

In processes where the combustion of fuels is used for obtaining mechanical energy, the chemical energy stored in the fuel is converted into thermal

TABLE 1-3. TORSIONAL SYSTEM

Mechanical quantity	Electrical quantity	Mechanical energy stored or dissipated	Electrical energy stored or dissipated	Symbol used
Moment or torque = M	Electromotive force = E			
Angle of twist = θ	Charge = q			
Angular velocity = ω	Current = i			
Moment of inertia = I	Inductance = L	$\tfrac{1}{2} I \omega^2$	$\tfrac{1}{2} L i^2$	⌒⌒⌒⌒
Moment of resistance = R_M	Electrical resistance = R	$\int_0^t R_M \omega^2 \, dt$	$\int_0^t R i^2 \, dt$	⋀⋀⋀
Moment of compliance = C_M	Capacitance = C	$\dfrac{1}{2}\dfrac{\theta^2}{C_M} = \dfrac{1}{2} C_M M^2 = \dfrac{1}{2}(\text{torque})^2 C_M$	$\dfrac{1}{2}\dfrac{Q^2}{C} = \dfrac{1}{2} C E^2$	⊣⊢

energy that in turn is converted into mechanical energy. The mechanical energy thus obtained may be converted into electrical energy by means of an electric generator. Also, when a storage battery or a dry cell is discharged, chemical energy is converted into electrical energy.

Typical values of energy for the three most common fuels, in terms of Btu per pound, are: coal = 14,000; gas = 20,000; and oil = 18,000.

These are large quantities of energy in relation to the amount of energy that can be stored in an inductance, capacitance, or flywheel. This is evident since 1 Btu = 1,055 j. Hence the energy in a pound of coal is approximately 1.5×10^7 j. It can be shown that a 5-kva, 60-cycle capacitor, the type that is used to correct power factor on distribution systems, stores a maximum of approximately 13.3 j under rated conditions. The kva requirements in terms of capacitance to store the amount of energy present in a pound of coal would then be

$$\text{kva} = \frac{5 \times 1.5 \times 10^7}{13.3} = 5.65 \times 10^6$$

The largest a-c generators today have ratings less than 1.5 million kva; this corresponds to about 1.71 million hp if the rated power factor of the generators is 0.85, which is quite common. In terms of capacitors, the one pound of coal contains stored energy equivalent to that of 1.13 million capacitors, each rated at 5 kva.

The energy in a pound of coal corresponds roughly to the kinetic energy in a flywheel 9 ft in diameter, weighing a ton, and rotating at a speed of 3,600 rpm.

1-21 STORAGE BATTERIES

Large amounts of energy can be stored in electric storage batteries. Consider a 6-v battery rated 100 amp-hr. If the drop in voltage is neglected as the battery gives up its charge, the energy given up by the battery is 600 whr. This amounts to $600 \times 3,600 = 2.18 \times 10^6$ j or the equivalent of about 14 percent of the energy stored in a pound of coal.

1-22 ATOMIC ENERGY

Official estimates indicate that the atomic bomb at Hiroshima was equivalent to 20,000 tons of TNT. On the basis of 3.8×10^9 j of energy being released by detonation of 1 ton of TNT, about 1 kg of U^{235} was consumed and the total mass decrease involved in the explosion was about 8.5×10^{-4} kg.

It should be remembered that the amount of energy stored chemically in a pound of fossil fuel is enormous when compared with the amount of energy that can be stored in the same amount of dielectric of a capacitor or in the same amount of magnetic material in an inductor. However, even this relatively large amount of energy stored chemically in fossil fuels or even storage batteries becomes almost insignificant when compared with the amounts of energy involved in atomic and nuclear reactions.

PROBLEMS

1-1 A body with a mass of 5 kg is elevated to a height of 100 m from a given solid surface and is then allowed to fall. Neglect air resistance and determine
   (a) The potential energy relative to the given surface when the body is 100 m above this surface.
   (b) The height at which the potential energy of this body equals the kinetic energy.
   (c) The velocity for the condition of part (b) above by equating the change in potential energy to the kinetic energy.
   (d) The irreversible energy for the condition of part (b) above if air resistance can be neglected.
   (e) The condition of the body and the surface such that the irreversible energy is zero after striking the given solid surface.

1-2 The body referred to in Problem 1-1 is raised on an inclined surface to a height of 100 m above a given solid surface. It is then allowed to slide down the inclined surface. At the instant the body has descended 50 m it has a velocity of 20 mps. When the body is at a height of 50 m, determine
   (a) The irreversible energy in the system, assuming that the amount of energy leaving the plane and the body to be negligible.
   (b) The stored energy that is reversible.

1-3 A 250-v, d-c motor delivers 10 hp at a speed of 600 rpm to a load. The voltage applied to the motor is 250 v and the current is 35.6 amp. Assume the motor to have been in operation for a sufficient period so that all its parts are at a steady temperature and determine
   (a) All the power into the motor.
   (b) All the power out of the motor.
   (c) The useful output.
   (d) The efficiency of the motor.
   (e) The output torque in newton meters.
   NOTE: 1 hp = 746 w.

1-4 A 3-phase, a-c generator delivers 100,000 kw at a speed of 3,600 rpm. The efficiency is 98.5 percent. Determine
   (a) The horsepower input.
   (b) The input torque in newton meters.
   (c) The power converted into heat.

1-5 The input to the voice coil of a loud-speaker is 200 mws. The acoustic output is 4 mw. Determine
   (a) The efficiency.
   (b) The power converted into heat.

1-6 When a constant d-c voltage V is applied to an inductive circuit of constant resistance R ohms and constant self-inductance of L h, the current is expressed by

$$i = \frac{V}{R}(1 - \epsilon^{-Rt/L}) \quad \text{amp}$$

Where t is the time following the application of the voltage and ϵ is the natural logarithmic base. Express as a function of time
   (a) The power input to this circuit.
   (b) The energy input to this system.
   (c) The irreversible energy.
   (d) The energy stored in the self-inductance.

1-7 The winding of an electromagnet has a resistance of 10 ohms and a self-inductance of 1.0 h. A constant d-c emf of 100 v is applied to this magnet. Determine
   (a) The initial current when the voltage is applied.
   (b) The final current.
   (c) The initial rate in amp per sec at which the current changes.
   (d) The time, after application of voltage, when the power that stores energy in the inductance is a maximum.
   (e) The maximum power taken by the inductance.
   (f) The time constant.

1-8 An inductive circuit has a resistance of 10 ohms and a self-inductance of 1 h. Determine the voltage applied to this circuit expressed as a function of time such that the current increases at a constant rate of 20 amp per sec. Assume

$$i = 0 \quad \text{at} \quad t = 0$$

1-9 When a constant d-c voltage is applied to a capacitive circuit having a resistance R ohms in series with an ideal capacitance C f, the current is expressed by

$$i = \frac{V}{R}\epsilon^{-t/RC} \quad \text{amp}$$

Where t is the time following the application of the voltage.
   (a) Express as functions of time
      (i) The power input to this circuit.
      (ii) The energy input to this circuit.
      (iii) The energy stored in the capacitance.
      (iv) The irreversible energy.

(b) What is the ratio of irreversible energy to the stored energy as time t approaches infinity?

(c) How much of the total supplied energy has been degraded by the time the capacitor is fully charged? Into what form is the energy degraded?

1-10 A constant d-c potential is applied to a series circuit having a resistance of 100 ohms and a capacitance of 1 μf. Determine

(a) The time expressed in seconds after application of voltage when the power absorbed by the capacitor is a maximum.

(b) Repeat part (a) above expressing the time in terms of the time constant τ.

(c) The amount of energy stored in the capacitor at the instant the power absorbed by the capacitor is a maximum if the voltage applied to the circuit is 600 v.

1-11 A flywheel in the form of a disc has a mass of 222 kg and a radius of 0.3 m.

(a) Determine the kinetic energy of rotation when this flywheel is rotating at a speed of 954 rpm.

(b) What is the average value of the power required to bring this flywheel from rest to a speed of 954 rpm in 1 min?

(c) The duty cycle is such that the flywheel gives up energy during 0.2 sec and recovers energy during 1.8 sec. If the speed drops from 954 rpm to 944 rpm in the 0.2-sec period, what is the average value of the power delivered by the flywheel?

(d) What is the average value of the power delivered to the flywheel during the recovery period of 1.8 sec, if the speed is returned to 954 rpm?

1-12 A projectile with a mass of 5 kg has a velocity of 300 mps. Determine the kinetic energy in joules.

1-13 (a) Determine the value of a current flowing in a magnetic circuit having a self-inductance of 9 h such that the stored energy is the same as the kinetic energy in Problem 1-12.

(b) Assume the magnetic circuit of (a) to be linear, i.e., the value of the self-inductance is constant for all values of current, and determine the inductive reactance for a frequency of 60 cps.

(c) If the value of the current determined in (a) is the maximum instantaneous value of a sinusoidal 60-cycle current, what is the rms value of the applied voltage if the resistance of the circuit can be neglected?

(d) What is the reactive power in kilovars consumed by the magnetic circuit under the above conditions (Kvar = $I^2 \omega L \times 10^{-3}$)?

1-14 A capacitor is rated 50 kvar, 4,160 v at 60 cycles. Determine

(a) The capacitance (Kvar = $E^2 \omega C \times 10^{-3}$).

(b) The maximum instantaneous energy stored in the dielectric of the capacitor when operating under rated conditions.

(c) The number of such capacitors that under rated conditions will store a maximum potential energy equal to the kinetic energy of the projectile of Problem 1-12.

NOTE: The approximate weight of the capacitor mentioned above is 60 lb and the price is approximately $160.

1-15 A self-inductance of L h and negligible resistance is connected in parallel with a noninductive resistance. At $t = 0$ a constant current of I is applied to the parallel combination. Express

(a) The current through the inductance as a function of time.
(b) The current through the resistance as a function of time.

1-16 In Problem 1-15 determine

(a) The final energy stored in the self-inductance.
(b) The total energy converted into heat.
(c) The value of time at the instant the power supplied to the resistance is equal to the power supplied to the self-inductance.
(d) The energy stored in the self-inductance when the power delivered to the resistance equals that delivered to the self-inductance.

1-17 Assume a 60-cell storage battery to have a constant terminal voltage of 120 v and to be rated 100 amp-hr. (Such a battery would be approximately equivalent to ten 12-v automobile storage batteries.) Determine

(a) The electrical energy stored chemically in this battery.
(b) The capacitance of a capacitor that would store the same amount of energy in its dielectric at a voltage of 120 v.
(c) The kvar rating of this capacitor for 60-cycle operation if the maximum instantaneous applied voltage is 120 v.
(d) The number of pounds of coal, having a Btu content of 14,000, that have stored the same amount of energy.

NOTE: 1 Btu = 1,055 j.

1-18 Determine the amount, expressed in feet, by which a spring, with a stiffness coefficient of 10 tons per inch, would be stretched to store the same amount of energy as the battery of Problem 1-17. How much force is required to stretch the spring to this extent?

1-19 (a) Express, as a function of time, the energy stored in the self-inductance L in R-L-C circuit, making use of Eq. 1-81.
(b) Plot the energy stored in the self-inductance during the interval from $bt = 0$ to $bt = 4\pi$ radians with $a = 2b$, where b is real.
(c) Plot the current as a function of time during the interval from $bt = 0$ to $bt = 4\pi$ radians with $a = 2b$, where b is real.

1-20 (a) Repeat Problem 1-19 when b is imaginary but a is real.
(b) Determine the time after voltage is applied when the energy stored in the self-inductance is a maximum.

1-21 (a) Repeat Problem 1-19 when b is zero.

HINT: $\sin bt$ approaches bt as bt approaches zero.

(b) Determine the time after voltage is applied when the energy stored in the self-inductance is a maximum.

CAPACITANCE AND RELATED EFFECTS

2

Aspects of energy storage and conversion in simple mechanical and electrical systems were discussed in Chapter 1. Mechanical systems generally have compliance, inertia, and friction. The counterparts of these in electrical systems are capacitance, inductance, and resistance, and are shown in Table 1-1. The energy associated with capacitance and inductance in electrical circuits where the capacitance and inductance are constant is reversible because the energy absorbed in a constant capacitance or in a constant inductance can be recaptured completely in electrical form. However, the energy absorbed by a resistance is irreversible because it is converted into heat and therefore cannot be recaptured in electrical form without the intervention of some external agency. It is with capacitance and related effects that this chapter is concerned. Since field theory forms the basis upon which the concept of capacitance rests, a brief review of some of the principles pertaining to electric fields may be helpful to the student.

2-1 THE ELECTRIC FIELD

The region surrounding any object that carries an electric charge is occupied by a field of force known as the electric field. This is evident from the force developed when an exploring particle carrying an electric charge is introduced into such a region. The exploratory charge is also known as a test charge. By convention the test charge is considered positive.

Analyses are made by evaluating effects by means of an imaginary test charge rather than by actually making use of a physical test charge. Thus, in Fig. 2-1 for point charges in free space

$q_1 = $ a charge that produces the field
$q_t = $ test charge that is used to explore the field
$r = $ distance between the charges q_1 and q_t

Then according to Coulomb's law the force between the charges is expressed by

$$\mathbf{F} = \frac{q_1 q_t}{4\pi\epsilon_0 r^2} \mathbf{i}_r \tag{2-1}$$

in any consistent rationalized system of units. If q_1 and q_t are expressed in coulombs, r in meters, and F in newtons, which are the units common to the rationalized MKS system, then

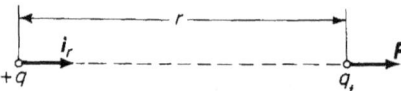

Figure 2-1. Direction of force vector—Coulomb's law.

$\epsilon_0 = 8.854 \times 10^{-12}$ farads per m

The factor ϵ_0 is the dielectric constant of free space. It also represents the capacitance of a meter cube of free space if the electric field intensity is uniform throughout the cube. The vector $\mathbf{i}_r$ is a unit vector in the direction along the line connecting the test charge with the source charge.

The charge q_1 produces an electric field of force. The test charge experiences a radial force with q_1 as a center. If the charge q_1 is positive, the force on the test charge is one of repulsion.

The forces resulting from the presence of electric charges are extremely powerful and play a vital role in nature. It is the nuclear charge within the atom which distinguishes one element from another and which resists the penetration and disintegration of the nucleus by impinging charged particles. The physical strength and other properties of matter are largely determined by the number and disposition of the electrons swarming about the nuclear charge. Electrostatic forces and principles are used in the experimental procedures of modern physics for deflecting charged particles in motion and for accelerating them to high energies. The characteristics of electrical devices such as vacuum tubes are largely dependent on electrostatic effects. Lightning is a large-scale phenomenon in nature that results from electrostatic forces.

2-2 ELECTRIC FLUX

Figure 2-2 shows the lines of force resulting from a positive charge q_1. It should be kept in mind that these lines radiate in all directions. The forces in a positive test charge are radially outward. A negative charge produces a field in which the lines of force are exactly the same as in Fig. 2-2, except that the forces are directed radially inward. Thus electrostatic lines of force emanate from positively charged bodies and terminate on negatively charged bodies as shown for the two point charges $+q$ and $-q$ in Fig. 2-3. The field of Fig. 2-3 results from superimposing the radial fields of equal and opposite charges $+q$ and $-q$.

The strength of the electric field due to an isolated

point charge varies inversely as the square of the distance from the charge. It is convenient to conceive of the space surrounding the charged body as being permeated by an electric flux emanating from the charged body and equal to the charge on the body. The concept of flux has found applications in other

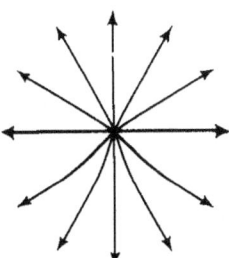

Figure 2-2. Electric field about a positive charge.

fields. For instance, in the field of illumination the quantity light flux is used. Here the inverse square law also holds because the intensity of illumination varies inversely as the square of the distance from a point source of light. Hence, the amount of flux emanating from a point source and passing through

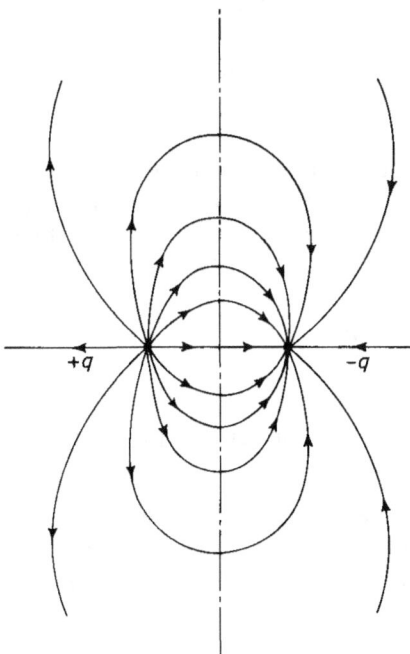

Figure 2-3. Electric field between equal and opposite charges.

a unit area varies inversely as the square of the distance. In fact, if the charge is considered at the center of an imaginary sphere, the flux passing through a unit area of the spherical surface is expressed by

$$D = \frac{\psi}{4\pi r^2} \quad \text{coulombs per sq m} \tag{2-2}$$

where ψ is the symbol for electric flux and

$$\psi = q \tag{2-3}$$

However, the flux density at any point on the surface of this imaginary sphere is expressed by

$$\mathbf{D} = \frac{q}{4\pi r^2} \mathbf{i}_r \quad \text{coulombs per sq m} \tag{2-4}$$

in a radial direction as indicated by the unit vector $\mathbf{i}_r$.

2-3 ELECTRIC FIELD INTENSITY

According to Coulomb's law, the force on a test charge in free space is given by Eq. 2-1 as

$$\mathbf{F} = \frac{q_1 q_t}{4\pi \epsilon_0 r^2} \mathbf{i}_r$$

This force is not only a function of the source charge q_1 but also of the test charge q_t. When both sides of Eq. 2-1 are divided by q_t the force per unit test charge is obtained. This quantity is a function of the source charge only and is known as the electric field intensity, expressed by

$$\mathbf{E} = \frac{\mathbf{F}}{q_t} = \frac{q_1}{4\pi \epsilon_0 r^2},$$

$$\mathbf{i}_r = \frac{\mathbf{D}}{\epsilon_0} \quad \text{v per m or newtons per coulomb} \tag{2-5}$$

If the source charge and test charge are in a medium other than free space as for example in a liquid dielectric that has dielectric polarization, the electric field intensity is lower than for free space in the ratio of k_r or relative dielectric constant. Thus

$$\mathbf{E} = \frac{\mathbf{D}}{k_r \epsilon_0} \tag{2-6}$$

Values of relative dielectric constants for several typical media are shown in Table 2-1.

2-4 VOLTAGE AND POTENTIAL

When the test charge q_t is moved in a direction *toward* the source charge, it is moved against the Coulomb force of repulsion and, energy is put into the system. On the other hand, if the test charge is moved in a direction

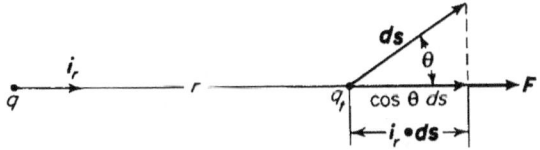

Figure 2-4. Projection of differential distance on force vector.

away from the source charge, the system gives up energy and the energy input to the system is negative. Hence, in moving the test charge q_t through a differential distance ds the differential energy input to the system is expressed by

$$dW = -\mathbf{F} \cdot \mathbf{ds} = -F\,\mathbf{i}_r \cdot \mathbf{ds} = -F \cos \theta \, ds \qquad (2\text{-}7)$$

where θ is the angle between the direction of the force and the direction of the differential distance ds as shown in Fig. 2-4.

The energy input to the system when the test charge is carried from a point P_1 at a distance r_1 from the source charge to a point P_2 at a distance r_2 from the source charge along any path whatsoever, as for example that illustrated in Fig. 2-5, is expressed by

$$W = \int_0^W dW = -\int_{r_1}^{r_2} \mathbf{F} \cdot \mathbf{ds} = -\frac{qq_t}{4\pi\epsilon_0} \int_{r_1}^{r_2} \mathbf{i}_r \cdot \mathbf{ds} \qquad (2\text{-}8)$$

From Fig. 2-4 it is evident that the scalar product $\mathbf{i}_r \cdot \mathbf{ds}$ is the projection of the differential distance vector ds upon r and is dr because in Eq. 2-7

$$dr = \cos \theta \, ds$$

Hence

$$W = -\frac{qq_t}{4\pi\epsilon_0} \int_{r_1}^{r_2} \frac{dr}{r^2} = \frac{qq_t}{4\pi\epsilon_0}\left(\frac{1}{r_2} - \frac{1}{r_1}\right) \qquad (2\text{-}9)$$

Equation 2-9 divided by q_t yields

$$V_{21} = \frac{W}{q_t} = \frac{q}{4\pi\epsilon_0}\left(\frac{1}{r_2} - \frac{1}{r_1}\right) \tag{2-10}$$

Equation 2-10 expresses the voltage between the points P_1 and P_2, the point P_2 being at a potential higher than that of P_1. The voltage V_{21} is positive. If

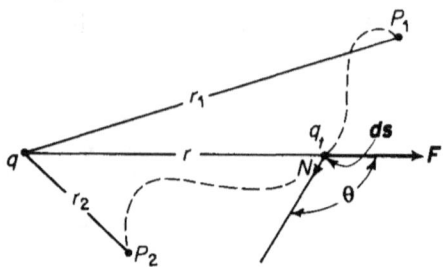

Figure 2-5. Difference of potential between two points.

the point P_1 in Fig. 2-5 were removed to an infinite distance then the voltage between the point P_2 and point P_1 would be expressed by

$$V_2 = \frac{q}{4\pi\epsilon_0}\frac{1}{r_2} \tag{2-11}$$

However, from Eq. 2-9 it is evident that the energy required to move the test charge from an infinite distance

$$W_2 = \frac{qq_t}{4\pi\epsilon_0}\frac{1}{r_2} \tag{2-12}$$

because r_1 approaches infinity. The energy associated with the test charge at an infinite distance is zero. Hence, if Eq. 2-12 is divided by q_t, Eq. 2-11 results. Equation 2-11 expresses the potential of the point P_2. Similarly the potential of point P_1 is expressed by

$$V_1 = \frac{q}{4\pi\epsilon_0}\frac{1}{r_1} \tag{2-13}$$

If Eq. 2-13 is subtracted from Eq. 2-11 the result is the difference of potential between point P_2 and point P_1 and Eq. 2-10 is obtained. From this it follows that

$$V_{21} = V_2 - V_1 \tag{2-14}$$

Thus the voltage between the points P_1 and P_2 is simply the difference of electrical potential, the point P_2 being at a higher potential than P_1. This means also that the voltage V_{21} is positive. Similarly the potential difference obtained by subtracting Eq. 2-11 from Eq. 2-13 results in

$$V_{12} = V_1 - V_2 = \frac{q}{4\pi\epsilon_0}\left(\frac{1}{r_1} - \frac{1}{r_2}\right) \qquad (2\text{-}15)$$

$$V_{12} = -V_{21}$$

and V_{12} is negative.

It is evident from Eq. 2-8 that the work done in carrying a test charge from one point to another in an electric field is independent of the path along which the charge is carried. By the same token if the test charge is carried from the second point back to the first along any path whatsoever, the same amount of energy is given up. Hence

$$W = -\int_{r_1}^{r_2} \mathbf{F} \cdot \mathbf{ds} - \int_{r_2}^{r_1} \mathbf{F} \cdot \mathbf{ds} = 0 \qquad (2\text{-}16)$$

This can also be expressed as a line integral as follows

$$\oint \mathbf{F} \cdot \mathbf{ds} = 0$$

from which

$$\oint \mathbf{E} \cdot \mathbf{ds} = \frac{1}{q_t}\oint \mathbf{F} \cdot \mathbf{ds} = 0 \qquad (2\text{-}17)$$

Equation 2-17 states that the summation of the electric field intensity around a closed path in an electrostatic field is zero.

2-5 GAUSS'S THEOREM

Gauss's theorem states that the net electric flux emanating from an imaginary closed surface is equal to the net electric charge enclosed within that surface.

Thus in Fig. 2-6 the electric flux that emanates from surface (1) is equal to $+q$ and that from surface (2) is equal to $-q$. The flux that emanates from surface (3) is zero because that surface does not enclose any charge. The flux that emanates from surface (4) is likewise zero because that surface encloses equal but opposite charges, the net charge being zero.

Consider a differential area da of a closed surface, as shown in Fig. 2-7, with an electric flux density of D coulombs per square meter. The flux emanating from the surface represented by this differential area is expressed by

$$d\psi = \mathbf{D} \cdot \mathbf{da} \qquad (2\text{-}18)$$

It must be kept in mind that the vector *da*, which represents the area of a surface, is normal to that surface. The product in Eq. 2-18 is a scalar product.

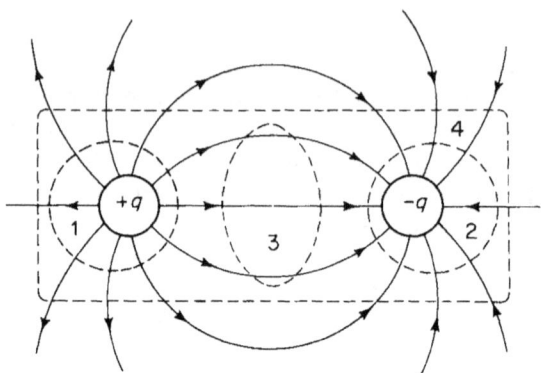

Figure 2-6. Gaussian surfaces.

The scalar product of the infinitesimal area and the normal component of flux density integrated over the entire enclosing surface yields the total flux emanating from this surface, i.e.

$$\psi = \oint \mathbf{D} \cdot \mathbf{da} = q \qquad (2\text{-}19)$$

This flux may also be expressed in terms of the surface integral of electric field intensity. If the enclosing surface lies completely within a medium having a fixed relative dielectric constant k_r, the relationship in Eq. 2-19 can also be expressed by

$$\psi = k_r \epsilon_0 \oint \mathbf{E} \cdot \mathbf{da} \qquad (2\text{-}20)$$

because

$$\mathbf{D} = k_r \epsilon_0 \mathbf{E}$$

On the other hand, if the enclosing surface lies in a medium in which the relative dielectric constant k_r is not fixed, then the flux is expressed in terms of electric field intensity by

$$\psi = \epsilon_0 \oint k_r \mathbf{E} \cdot \mathbf{da} \qquad (2\text{-}21)$$

2-6 CHARGE WITHIN A CONDUCTOR THAT HAS A STATIC CHARGE

Consider two wires insulated from each other. One is connected to the positive side of the d-c source, and the other to the negative side of the d-c source

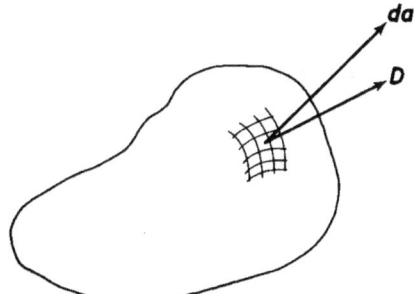

Figure 2-7. Flux density in differential area on a closed surface.

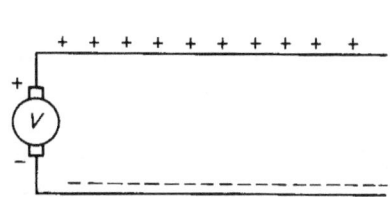

Figure 2-8. Linear conductors connected to a d-c source.

as shown in Fig. 2-8. The upper wire receives a positive charge, and the lower wire a negative charge. These charges are static, i.e., they are not in motion once they are established, and there is no difference of potential from one end of a wire to the other. In other words the electric field intensity E within the wires is zero. If the wires were #18 AWG copper, a size commonly used in lamp cord, a field intensity E along the wires of only one v per m would produce a current of about 50 amp. Therefore, when there is no current within a conductor the electric field intensity E everywhere within a conductor is zero.

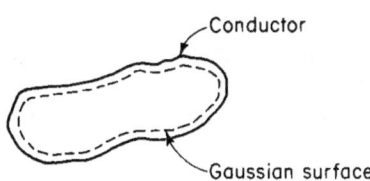

Figure 2-9. A Gaussian surface just inside the surface of a conductor with a static charge.

Accordingly, it can be shown by means of Gauss's theorem that the static charge is on the surface of a conductor. Figure 2-9 shows a conductor that carries a static charge. The broken line represents a Gaussian surface an infinitesimal distance inside the actual conductor surface. Since the charge on the conductor is static, there is no flow of current within the conductor. The electric field intensity E must therefore be zero within the Gaussian surface. Hence from Eq. 2-20

$$q = \psi = k_r \epsilon_0 \oint \mathbf{E} \cdot \mathbf{da} = 0 \tag{2-22}$$

This means that there is no charge within the Gaussian surface and the entire static charge must be on the surface of the conductor. This is true also for a hollow conductor. Therefore a charge outside of the conductor does not produce an electric field inside the hollow. However, a charge inside the hollow will produce an electric field that follows from Gauss's theorem outside the conductor.

2-7 UNIFORMLY DISTRIBUTED CHARGE ON AN ISOLATED SPHERE

The derivations of flux density and electric field intensity in this chapter so far have centered around an electrical charge concentrated at a point. And although the concept of the point charge is useful for analytical purposes, it is an unrealistic one because in actual physical arrangements charges are distributed within finite volumes over surfaces that are finite. Equation 2-13 shows that the potential of a finite electric charge concentrated to a point would be infinite because r would need to approach zero to satisfy the mathematical concept of a point.

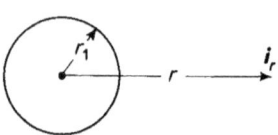

Figure 2-10. Conducting sphere with uniformly distributed charge.

The sphere is one of the simplest configurations on the surface of which an electric charge might be distributed. Consider a conducting sphere, isolated in free space, with a charge of q coulombs uniformly distributed over its surface. If the radius of the sphere is r_1 m as indicated in Fig. 2-10 then the density of charge on the sphere's surface is

$$\boldsymbol{\sigma} = \frac{q}{4\pi r_1^2} \mathbf{i}_{r_1} \quad \text{coulombs per sq m} \tag{2-23}$$

The right-hand side of Eq. 2-23 is the same as the right-hand side of Eq. 2-4. This shows that the electric field intensity at a point an infinitesimal distance external to the surface of the sphere is exactly the same as it would be if the sphere were removed and the charge formerly on the sphere's surface were concentrated at the point formerly occupied by the center of the sphere.

From this it follows that the field external to the sphere due to a uniformly distributed charge on the surface is the same as that for a charge concentrated at the center of the sphere. Thus in Fig. 2-10 the electric flux density at a distance r from the center is expressed by Eq. 2-4.

The potential of the sphere with respect to an infinitely remote point must be the same as that of the point at an infinitesimal distance from the surface as expressed by Eq. 2-13, i.e.

$$V = \frac{q}{4\pi\epsilon_0 r_1}$$

If the sphere were isolated in a medium of infinite extent and had a relative dielectric constant k_r, the voltage of the sphere would be

$$V = \frac{q}{4\pi k_r \epsilon_0 r_1} \tag{2-24}$$

2-8 CAPACITANCE

Equations 2-23 and 2-24 show that the charge and voltage of the sphere are proportional to each other. The ratio of charge to voltage is called the capacitance. Hence

$$q = Cv \tag{2-25}$$

where q = the charge in coulombs
  v = the voltage in volts
  C = the capacitance in farads

Thus the capacitance of the isolated sphere is expressed by

$$C = 4\pi k_r \epsilon_0 r_1$$

Energy is required to charge the sphere to the potential v and the energy is considered to reside in the electric field that occupies the medium, known as the dielectric, external to the sphere. The capacitance of a certain configuration depends in general upon the dimensions and the dielectric constant that is a property of the dielectric in which the energy is stored. The dielectric constant and the dielectric strength are the properties of the dielectric that determine the energy that can be stored electrically in a given volume of material.

2-9 CAPACITANCE OF CONCENTRIC SPHERES

Suppose that a charge of $+q$ coulombs is uniformly distributed over the surface of a sphere having a radius r_1. Also, this sphere is concentric with a sphere of larger radius r_2, which carries a charge of $-q$ coulombs, as shown in Fig. 2-11. Consider a point P in the space between spheres at a distance r from the common center. The electric field intensity at P due to the charge $+q$ on the surface of the inner sphere is according to Eq. 2-5

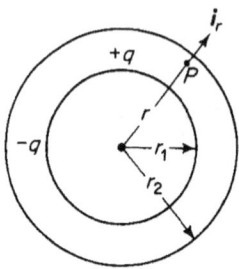

Figure 2-11. Concentric spheres.

$$\mathbf{E} = \frac{q}{4\pi\epsilon_0 r^2}\mathbf{i}_r \quad \text{v per m} \qquad (2\text{-}26)$$

The electric field intensity due to the charge $-q$ on the inner surface of the outer sphere is zero in accordance with Gauss's theorem.

The field within the two concentric spheres then is taken completely into account by considering only the positive or negative charge, as the case may be, on the inner sphere. Then the electric field intensity at a point between the spheres at a distance r from the center is

$$\mathbf{E} = \frac{q}{4\pi\epsilon_0 r^2}\mathbf{i}_r \quad \text{v per m}$$

and the difference of potential between the two concentric spheres is

$$V_{12} = -\int_{r_2}^{r_1} \mathbf{E} \cdot d\mathbf{r} = \frac{q}{4\pi\epsilon_0 r}\bigg]_{r_2}^{r_1}$$

$$= \frac{q}{4\pi\epsilon_0} \frac{r_2 - r_1}{r_1 r_2} \qquad (2\text{-}27)$$

The capacitance between spheres is

$$C_{12} = \frac{q}{V_{12}} = 4\pi\epsilon_0 \frac{r_1 r_2}{r_2 - r_1} \quad \text{farads} \qquad (2\text{-}28)$$

EXAMPLE 2-1: Figure 2-12(a) shows the sketch of a cell for making measurements of dissipation factor and dielectric constant of liquid dielectrics. The plates of the cell are two concentric spherical sections mounted and spaced from each other on a conical Pyrex glass plug. The conical plug has a generating angle of 30°. The liquid dielectric occupies the space

Sec. 2-9   CAPACITANCE OF CONCENTRIC SPHERES   51

between the spherical sections, i.e., between the radii r_2 and r_1. Within the glass plug is a gold foil guard electrode having the same radius of curvature as the inner sphere, namely r_1, but insulated by a very small gap from the inner sphere. The purpose of the guard electrode is to prevent fringing; it is so connected that the capacitive and leakage current

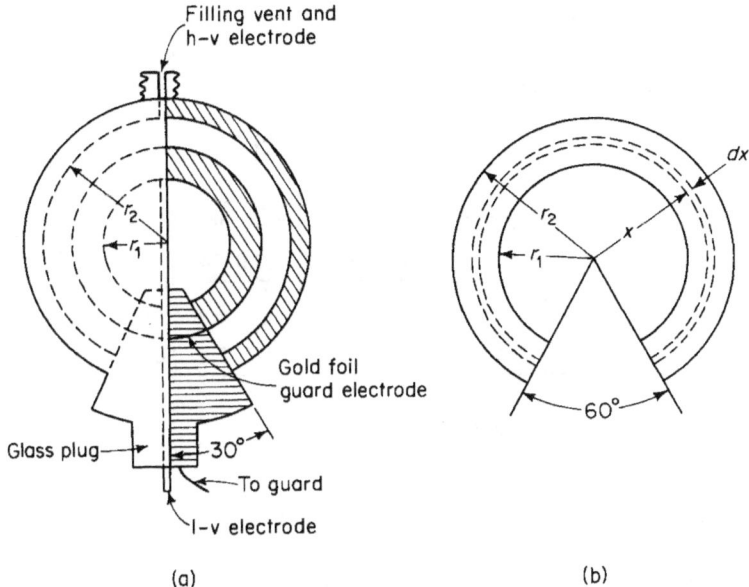

Figure 2-12. (a) Dielectric test cell; (b) elemental shell in test cell.

to it by-pass the measuring circuit. Such cells are used in a-c bridge circuits. When the bridge is balanced the guard circuit is at the same potential as the inner sphere.

The inside radius r_2 of the outer sphere is 1.027 in. and the outside radius of the inner sphere is 1.000 in. Determine the capacitance of the empty cell not including the gold foil guard electrode.

Solution: Figure 2-12(b) shows an elemental section of radius x and thickness dx within the space between the spherical plates of the cell.

Let q = charge uniformly distributed on the inner spherical surface. Then the electric flux density in the elemental shell is

$$\mathbf{D}_x = \frac{q\mathbf{i}_x}{4\pi x^2(\tfrac{1}{2} + \tfrac{1}{2}\cos 30°)}$$

where $4\pi x^2(\tfrac{1}{2} + \tfrac{1}{2}\cos 30°)$ is the area of the elemental shell.

The electric field intensity in the elemental shell with air as a dielectric is

$$E_x = \frac{D_x}{\epsilon_0} = \frac{qi_x}{4\pi\epsilon_0 x^2(\frac{1}{2} + \frac{1}{2}\cos 30°)}$$

and the voltage between the electrodes (spherical surfaces) of the cell is

$$V = \int_{r_1}^{r_2} E_x \cdot d_x = \frac{q}{2\pi\epsilon_0(1 + \cos 30°)} \frac{r_2 - r_1}{r_1 r_2}$$

The capacitance according to Eq. 2-25 is found to be

$$C = \frac{q}{V} = \frac{2\pi\epsilon_0(1 + \cos 30°)r_1 r_2}{r_2 - r_1}$$

$$\epsilon_0 = 8.854 \times 10^{-12} \text{ farads per m}$$

$$\frac{r_1 r_2}{r_2 - r_1} = \frac{1.000 \times 1.027 \times 0.0254}{1.027 - 1.000} \text{ m}$$

and

$$C = \frac{2\pi \times 8.854 \times 10^{-12} \times 1.866 \times 1.027 \times 0.0254}{0.027}$$

$$= 100.4 \times 10^{-12} \text{ farads}$$

2-10 PARALLEL-PLATE CAPACITOR

Suppose that the distance $d = r_2 - r_1$ between the inner and outer sphere in Fig. 2-11 is held constant while the radii r_1 and r_2 are increased without limit. Then the ratio r_1/r_2 approaches unity.

When both sides of Eq. 2-28 are multiplied by the ratio r_1/r_2 the product is

$$\frac{r_1}{r_2} C_{12} = 4\pi\epsilon_0 \frac{r_1^2}{r_2 - r_1} = 4\pi\epsilon_0 \frac{r_1^2}{d} \qquad (2\text{-}29)$$

As the ratio r_1/r_2 approaches unity the left-hand term of Eq. 2-29 approaches C_{12} and Eq. 2-29 becomes reduced to

$$C_{12} = 4\pi\epsilon_0 \frac{r_1^2}{d} \qquad (2\text{-}30)$$

The quantity $4\pi r_1^2$ in Eq. 2-30 represents the area of the dielectric, between the infinite spheres, normal to the electric field intensity. Hence, the capacitance per unit area of dielectric at right angles to the electric field, from Eq. 2-30, is

$$\frac{C_{12}}{4\pi r_1^2} = \frac{\epsilon_0}{d} \quad \text{farads per sq m} \qquad (2\text{-}31)$$

Sec. 2-11   RELATIVE DIELECTRIC CONSTANT   53

Since the voltage across the dielectric between spheres is V_{12}, the charge per unit area is

$$\sigma = \frac{V_{12} C_{12}}{4\pi r_1^2} = V_{12} \frac{\epsilon_0}{d} \quad \text{coulombs per sq m}$$

Then, for any area A between the infinite spheres and normal to the field, the charge is expressed by

$$q_A = \sigma A = V_{12} \frac{\epsilon_0 A}{d}$$

The capacitance for this area A and thickness d of dielectric is found from Eq. 2-25 to be

$$C_A = \frac{q_A}{V_{12}} = \frac{\epsilon_0 A}{d} \tag{2-32}$$

Since the surfaces of the spheres approach those of planes as the radii r_1 and r_2 approach infinity, Eq. 2-32 expresses the capacitance between plane parallel plates of area A and separation d as shown in Fig. 2-13. Equation 2-32 is valid if there is no fringing of electric flux at the edges of the parallel

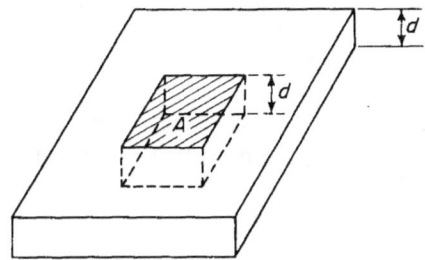

Figure 2-13. Parallel-plate capacitor.

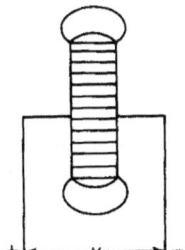

Figure 2-14. Flux fringing at the edges of parallel-plate capacitor.

plates. Fringing is appreciable unless the area A is large in relation to the separation d. Figure 2-14 shows a parallel-plate capacitor with flux fringing at the edges.

2-11   RELATIVE DIELECTRIC CONSTANT

The factor ϵ_0 is the dielectric constant of free space and has a value of 8.854×10^{-12} farads per m in the rationalized MKS system of units. This is also the capacitance of a meter cube of free space when the field is uniform.

The dielectric constant of air and other gaseous dielectrics is slightly higher than that of free space and in nearly all cases this slight difference is neglected. However, the dielectric constant of liquid and solid dielectrics is appreciably greater than that of free space. The ratio of the dielectric constant of a material to that of free space is known as the relative dielectric constant k_r. Therefore, Eq. 2-32 becomes modified to

$$C = k_r \epsilon_0 \frac{A}{d} \qquad (2\text{-}33)$$

Typical values of dielectric constants are shown in Table 2-1.

TABLE 2-1. RELATIVE DIELECTRIC CONSTANTS

Material	State	k_r	Material	State	k_r
Glass	Solid	3.9–5.6	Steatite	Solid	5.9
Isolantite	Solid	3.8	Pyranol	Liquid	5.3
Mica	Solid	5.45	Oil	Liquid	2.2
Polyethelene	Solid	2.25	Water	Liquid	78.0
Polystyrene	Solid	2.25	Air	Gas	1.00009
Quartz	Solid	3.9	CO_2	Gas	1.000985
Paper	Solid	3.5	H_2	Gas	1.000264

Liquid and solid dielectrics have higher dielectric constants than free space because of an orientation of electric dipoles, a phenomenon known as dielectric polarization, and explained in Section 2-22.

2-12 CONCENTRIC CYLINDERS

The conductor and sheath of coaxial cables, such as telegraph cable and single-conductor lead-sheath power cable, are concentric cylinders. The capacitance between the inner cylinder, the conductor, and outer cylinder, the sheath or shield, has an important effect on the characteristics of such cables. Consider the two concentric cylinders in Fig. 2-15(a).

Let r_1 = the radius of the inner cylinder in meters
   r_2 = the radius of the outer cylinder in meters
   l = length of the cylinders in meters
   q = charge on the inner cylinder in coulombs

Sec. 2-12 CONCENTRIC CYLINDERS

Then

$$\mathbf{D} = \frac{q}{2\pi l x} \mathbf{i}_x \quad \text{electric flux density coulombs per sq m} \quad (2\text{-}34)$$

In the elemental shell of radius x and thickness dx

$$\mathbf{E} = \frac{\mathbf{D}}{k_r \epsilon_0} \quad \text{electric field intensity at a distance } x \text{ from the center} \quad (2\text{-}35)$$

$dv = \mathbf{E} \cdot d\mathbf{x}$ voltage drop across the elemental shell

The charge q on the inner cylinder is stationary. Therefore, the electric field intensity must be normal to the surface of the conductor because any tangential component or axial component would cause a flow of current in a tangential or an axial direction.

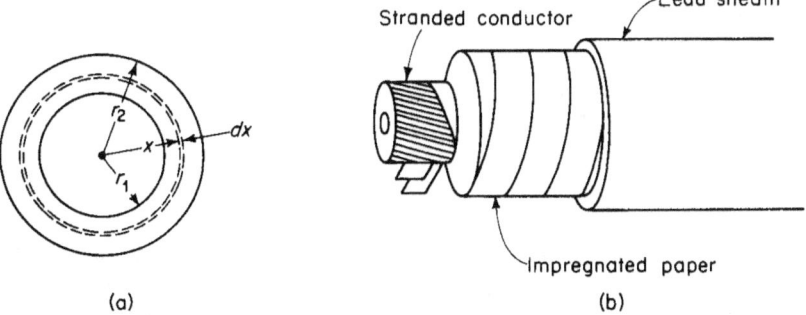

Figure 2-15. (a) Concentric cylinders; (b) single-conductor cable.

The angle between E and dx is zero, therefore

$$dv = E\, dx$$

$$= \frac{q}{2\pi l k_r \epsilon_0 x} dx$$

The total voltage between the inner and outer cylinders is

$$V = \int_{r_1}^{r_2} dv = \frac{q}{2\pi l k_r \epsilon_0} \bigg]_{r_1}^{r_2} \frac{dx}{x} = \frac{q \ln r_2/r_1}{2\pi l k_r \epsilon_0} \quad (2\text{-}36)$$

but $C = q/V$, hence

$$C = \frac{2\pi l k_r \epsilon_0}{\ln r_2/r_1} \quad (2\text{-}37)$$

2-13 ELECTRIC FIELD INTENSITY BETWEEN CONCENTRIC CYLINDERS

The electric field intensity, or its opposite the voltage gradient, is of great importance in the operation of high-voltage single-conductor cable. If the electric field intensity or voltage gradient is excessive the life of the cable is shortened. In fact, a high enough electric field intensity may cause the dielectric to puncture almost instantaneously. The expression for the electric field intensity at a distance x from the center is obtained from Eqs. 2-34, 2-35, and 2-36 and is as follows

$$E = \frac{V}{x \ln r_2/r_1} \tag{2-38}$$

If x is in meters, then the electric field intensity is in volts per meter; if x is in some other unit, E is the volts per that unit of distance. r_1 is the lowest value of x that has practical significance. This means that the maximum electric field intensity occurs at the surface of the inner conductor. The minimum electric field intensity occurs at the inner surface of the outer conductor, and we have

$$E_{\max} = \frac{V}{r_1 \ln r_2/r_1} \quad \text{and} \quad E_{\min} = \frac{V}{r_2 \ln r_2/r_1} \tag{2-39}$$

2-14 GRADED INSULATION

Some high-voltage cables use two or more different kinds of insulation (dielectric) in order to achieve a smaller difference in voltage gradient or electric field intensity throughout the insulation. Equation 2-38 shows that in coaxial cable, the electric field intensity varies inversely as the distance from the center of the inner conductor and also inversely as the relative dielectric constant. Therefore, in some types of high-voltage cable, next to the inner conductor a material is used that has a higher relative dielectric constant k_r than the material further out. It is then necessary to take the difference in the dielectrics into account in determining the capacitance of the cable. This is done as follows: in Fig. 2-16 the voltage across the dielectric, which has the relative dielectric constant kr_{12}, is expressed by

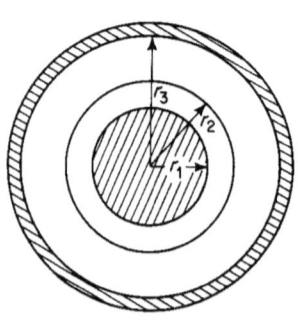

Figure 2-16. Coaxial cable with graded insulation.

$$V_{12} = \frac{q}{2\pi l \epsilon_0} \frac{1}{k_{r_{12}}} \ln \frac{r_2}{r_1} \tag{2-40}$$

and the voltage across the outer shell of the dielectric having a dielectric constant $k_{r_{23}}$ is expressed by

$$V_{23} = \frac{q}{2\pi l \epsilon_0} \frac{1}{k_{23}} \ln \frac{r_3}{r_2} \tag{2-41}$$

The total voltage from conductor to sheath is expressed by

$$V_{13} = V_{12} + V_{23} = \frac{q}{2\pi l \epsilon_0} \left[\frac{1}{k_{r_{12}}} \ln \frac{r_2}{r_1} + \frac{1}{k_{r_{23}}} \ln \frac{r_3}{r_2} \right] \tag{2-42}$$

and the capacitance is expressed by

$$C_{13} = \frac{q}{V_{13}} = \frac{2\pi l \epsilon_0 k_{r_{12}} k_{r_{23}}}{k_{r_{23}} \ln r_2/r_1 + k_{r_{12}} \ln r_3/r_2} \tag{2-43}$$

The electric field intensity at the surface of the inner conductor is

$$E_1 = \frac{q}{2\pi l \epsilon_0 k_{r_{12}} r_1} \tag{2-44}$$

and that inside the sheath is

$$E_3 = \frac{q}{2\pi l \epsilon_0 k_{r_{23}} r_3}$$

The ratio of these electric field intensities is

$$\frac{E_1}{E_3} = \frac{k_{r_{23}}}{k_{r_{12}}} \frac{r_3}{r_1} \tag{2-45}$$

whereas in the case of a homogeneous dielectric this ratio is

$$\frac{E_1}{E_3} = \frac{r_3}{r_1} \tag{2-46}$$

It can be seen, therefore, that if the dielectric $k_{r_{12}}$ is greater than $k_{r_{23}}$, the voltage gradient in the dielectric is more nearly uniform than it is in a homogeneous medium. Generally in high-voltage cables, in the absence of mechanical imperfections, the dielectric deteriorates more rapidly near the inner conductor because the electric field intensity and the temperature in the dielectric are the greatest at that location.

2-15 ENERGY STORED IN A CAPACITOR

The energy stored in a capacitor is said to reside in the dielectric permeated by the electric field. Figure 2-17 shows a simple capacitor in which a charge of q coulombs is stored in the electric field. The top plate carries a net charge of $+q$ coulombs and the bottom plate a net charge of $-q$ coulombs. The voltage between the top plate and bottom plate of the capacitor is v volts.

If a differential charge $+dq$ is carried from the bottom plate to the top plate, there will be a differential increase of energy dw.

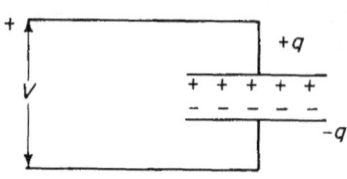

Figure 2-17. Capacitor with a charge.

Let $\mathbf{E}$ = electric field intensity at some point between the plates in volts per meter

Then dF = force on the differential charge dq expressed in newtons

$$\mathbf{dF} = dq\,\mathbf{E} \tag{2-47}$$

The differential energy required to carry the differential charge from the bottom plate to the top plate is given by the line integral along any path between the plates

$$dW = \int \mathbf{dF} \cdot \mathbf{ds} \tag{2-48}$$

When Eq. 2-47 is substituted in Eq. 2-48 the result is

$$dW = dq \int \mathbf{E} \cdot \mathbf{ds} \tag{2-49}$$

However, the line integral of the electric field intensity along any path from one plate to the other gives the voltage between plates; hence

$$v = \int \mathbf{E} \cdot \mathbf{ds} \tag{2-50}$$

Substitution of Eq. 2-50 in Eq. 2-49 yields

$$dW = v\,dq \tag{2-51}$$

In order to determine the energy required to store a charge q in the field of a capacitor when the capacitor is initially uncharged, Eq. 2-51 is integrated as follows

$$W = \int_{q=0}^{q=q} dW = \int_{q=0}^{q=q} v\,dq \tag{2-52}$$

but from Eq. 2-25 we get

$$v = \frac{q}{C} \qquad (2\text{-}53)$$

and from Eqs. 2-52 and 2-53 there results

$$W = \frac{1}{C}\int_0^q q\, dq = \frac{q^2}{2C} \qquad (2\text{-}54)$$

Making use of Eq. 2-53 in 2-54, we obtain the more commonly used expression for energy stored in a capacitor, i.e.

$$W = \frac{Cv^2}{2} \qquad (2\text{-}55)$$

2-16 ENERGY STORED IN A DIELECTRIC

The amount of energy that can be stored in a dielectric is theoretically limited by the electric field intensity that the material can withstand. For example air under standard conditions of temperature and barometric pressure has a dielectric strength of approximately 3 million v per m. If the electric field intensity exceeds this value, air breaks down. Energy stored in dielectrics plays an important part in the operation of such devices as oscillators and in the characteristics of high-voltage transmission lines.

Consider a section of area A in the parallel-plate capacitor of Fig. 2-13. The distance between plates, i.e., the thickness of dielectric, is d. The reason for selecting a section in this capacitor well away from the edges is to avoid the effects of fringing in this analysis. Hence, the electric flux density D is uniform throughout this section.

Let A = plate area of the section
d = distance between plates
k_r = relative dielectric constant of the material between plates
V = total voltage applied to the plates

Then

$$E = \frac{V}{d} = \frac{D}{k_r \epsilon_0} \quad \text{electric field intensity volts per meter} \qquad (2\text{-}56)$$

$$D = \frac{q}{A} \quad \text{electric flux density} \qquad (2\text{-}57)$$

where q = charge in coulombs in the section of area A. The incremental energy in the capacitor is $dW = v\,dq$ and $dq = C\,dv$ where C = capacitance of the section in farads. Then

$$dW = vC\,dv$$

and

$$W = C\int_0^v v\,dv = \frac{Cv^2}{2} = \frac{vq}{2} \tag{2-58}$$

Equations 2-56, 2-57, and 2-58 yield

$$W = \frac{EdAD}{2} = \text{Vol}\,\frac{ED}{2} \tag{2-59}$$

If E is the electric field intensity in any unit of length and D is the electric flux density in the same unit of area, then $ED/2$ is the energy per unit volume in those same dimensions. The quantity $ED/2$ is also known as the energy density of the electric field. Equation 2-59 can be rewritten as

$$W = \text{Vol}\,\frac{k_r\epsilon_0 E^2}{2} \tag{2-60}$$

or as

$$W = \text{Vol}\,\frac{D^2}{2k_r\epsilon_0} \tag{2-61}$$

Equations 2-60 and 2-61 are valid only for configurations in which the electric field intensity is uniform. Such is the case for capacitors with plane parallel plates in which the area is so large, relative to the separation, that fringing is negligible. Capacitors with liquid-impregnated paper or solid-impregnated paper or dielectric, and generally those with mica as a dielectric, are assumed to have uniform electric field intensity. Figure 2-18(a) shows a section of paper and foil for a liquid-impregnated capacitor. Figure 2-18(b) shows an assembly of 10 paper-foil sections in which there are two sets of five sections in parallel. The two sets are in series.

The plates in the capacitor sections in Fig. 2-18(a) and (b) are two long foils (usually aluminum), 0.00025 in. thick. The dielectric between foils is several layers of kraft paper ranging in thickness from 0.00020 in. to 0.001 in. Liquids commonly used for impregnating the paper are mineral oil with a relative dielectric constant $k_r = 2.2$, chlorinated diphenyl with a k_r of 4.9 and occasionally specially refined castor oil with a k_r of 4.7. The impregnant increases the capacitance and the dielectric strength.

Equation 2-60 shows that the amount of energy stored in a given volume of a certain dielectric is proportional to the square of the electric field intensity as long as the relative dielectric constant k_r does not vary with electric field intensity. The variation of k_r with electric field intensity is negligible for

Sec. 2-16 ENERGY STORED IN A DIELECTRIC

most dielectric materials in the normal operating range. We might thus have two parallel-plate capacitors of capacitances C_1 and C_2 rated at voltages V_1 and V_2 having thicknesses of identical dielectric material of d_1 and d_2

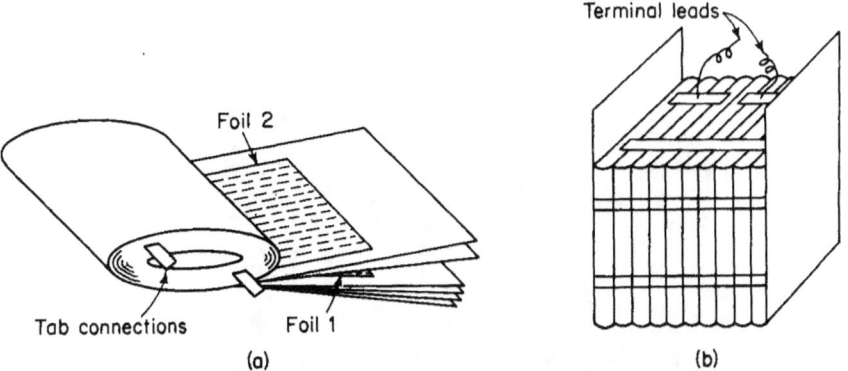

Figure 2-18. (a) Capacitor section of paper and foil; (b) assembly of paper-foil sections with series-parallel connections.

respectively. If both capacitors operate at the same value of electric field intensity of E v per m, then

$$\frac{V_1}{d_1} = \frac{V_2}{d_2} \qquad (2\text{-}62)$$

Also, from Eq. 2-55, if the stored energy is to be the same in both capacitors, there is the relation that

$$\frac{C_1 V_1^2}{2} = \frac{C_2 V_2^2}{2}$$

from which we get

$$\frac{C_1}{C_2} = \frac{V_2^2}{V_1^2} \qquad (2\text{-}63)$$

From Eqs. 2-33, 2-62 and 2-63 we get

$$A_1 d_1 = A_2 d_2 \qquad (2\text{-}64)$$

The product of area A and thickness d gives the volume of the dielectric. Equation 2-64 simply bears out Eq. 2-60 and shows that in capacitors equal volumes of dielectric will store the same amount of energy at the same value of electric field intensity regardless of the relative values of the capacitance.

In the case of large capacitors, such as are used on electric power, transmission, and distribution systems, it is generally not practical from the standpoint of manufacturing costs to build capacitors with widely differing

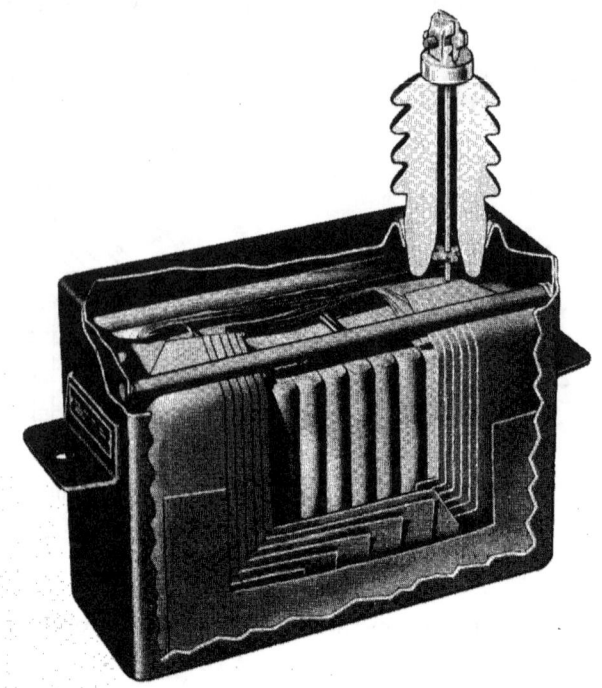

Figure 2-19. Power capacitor. (Courtesy of Line Material Industries.)

thicknesses of dielectric between the metal foils. Instead, the capacitors are made up of sections, as shown in Figs. 2-18(b) and 2-19, that are connected in series-parallel combinations to give the desired value of capacitance. There is also a practical lower limit to thickness of the dielectric. This means that below a certain value of operating voltage (about 460 v for capacitors used in distribution systems) the electric field intensity is reduced from values that can be used at the higher voltage ratings.

2-17 d-c ENERGY STORAGE CAPACITORS

Capacitors operating on direct current are used in a variety of applications, such as

1. d-c power supply filters.
    a. X-ray equipment.
    b. Radar equipment.
    c. Particle accelerators for nuclear research.

2. Intermittent duty pulse forming.
   a. Small surge generators.
   b. Linear accelerators.
3. Other energy storage applications.
   a. Capacitor discharge welding.
   b. Flash tube light sources.

Energy storage capacitors for rapid discharge involving relatively high amounts of energy (several thousand joules) are used in nuclear fission studies, hydraulic metal forming, and electric plasma research.*† In these applications energy is stored to be released subsequently in microseconds. Such rapid discharges involve large currents, and the construction of the capacitor must be strong enough to withstand the mechanical forces developed by these high currents. In addition it is necessary to minimize the internal inductance of the capacitor to permit rapid release of the energy. The effect of inductance in series with capacitance is discussed in Section 1-16.

Energy storage capacitors used in some of these operations are rated from 1 to 100 μf at voltages between 3 and 100 kv, usually storing between 1 and 3 kj (kilojoules); this means that the higher voltage ratings apply to the lower capacitance ratings. Since their internal inductance is less than 0.1 μh they are able to produce peak discharge currents of 100 ka (kiloamperes) routinely. Constant-current sources are preferred to constant-voltage sources because the efficiency of the former is considerably greater than that of the latter, which, according to Section 1-16, cannot exceed 50 percent.

2-18 KVA RATING OF CAPACITORS

Capacitors are used in electric transmission and distribution systems for power factor correction. Present loads on electric power systems are predominantly inductive and therefore take current that lags the voltage. Losses in transformers, transmission lines, and generators are lower for a given amount of real power when the power factor is near unity than when the current lags the voltage by an appreciable angle. The over-all power factor of a system that normally delivers current to an inductive load can be improved by installing capacitors at various locations in such a system.

---

* H. K. Jennings, "Charging Large Capacitor Banks in Thermonuclear Research," *Electrical Engineering*, June 1961, pp. 419–421.

† E. L. Kemp, "Elements of Energy Storage Capacitor Banks," *Electrical Engineering*, September 1962, pp. 681–685.

The amount of capacitance to be installed for a given improvement in the power factor is a function of the amount of reactive power taken by the system. Reactive power is measured in vars or kilovars and capacitors used for power factor correction are rated in kilovars.

Consider an ideal capacitor, i.e., one in which there are no energy losses.

Let C = the capacitance of the capacitor in farads
V = rms or effective a-c voltage ratings of the capacitor
f = frequency of the supply cycles per second

Under steady-state a-c operation the instantaneous voltage applied to the capacitor is expressed by

$$v = \sqrt{2} V \sin \omega t \qquad (2\text{-}65)$$

and the instantaneous current in the capacitor is

$$i = \frac{dq}{dt} = C \frac{dv}{dt} \qquad (2\text{-}66)$$

then substitution of Eq. 2-65 in 2-66 yields

$$i = \sqrt{2}\, \omega C V \cos \omega t \qquad (2\text{-}67)$$

where $\omega = 2\pi f$.

The maximum value of the instantaneous current expressed by Eq. 2-67 occurs when $\cos \omega t$ is unity, and we have

$$I_M = \sqrt{2}\,(2\pi f)\, CV \quad \text{amp} \qquad (2\text{-}68)$$

The rms or effective value of a sinusoidal current is maximum value divided by $\sqrt{2}$, and from Eq. 2-68 we get the rated current

$$I = \frac{I_M}{\sqrt{2}} = 2\pi f CV \quad \text{amp}$$

The kilovar rating of the capacitor is the product of the rated current at rated frequency and the rated voltage divided by 1,000, hence

$$\text{KVAR} = \frac{VI}{1{,}000} = \frac{2\pi f CV^2}{1{,}000} \qquad (2\text{-}69)$$

2-19 DIELECTRIC STRENGTH

The dielectric strength of a material is determined by the maximum electric field intensity it can support without electrical failure. If this value is exceeded the material becomes a conductor. The mechanism of breakdown is discussed in Section 2-25 under *Corona*. Values of dielectric strength of some common dielectric materials are listed in Table 2-2. The values of dielectric strength

TABLE 2-2. TYPICAL VALUES OF DIELECTRIC STRENGTH

Material	Dielectric strength volts per meter
Insulator porcelains	$10 \times 10^6 - 20 \times 10^6$
Glass	$20 \times 10^6 - 40 \times 10^6$
Vulcanized rubber	$16 \times 10^6 - 50 \times 10^6$
Transformer oils	$5 \times 10^6 - 15 \times 10^6$
Paraffined paper	$40 \times 10^6 - 60 \times 10^6$
Mica	$25 \times 10^6 - 200 \times 10^6$
Nitrocellulose plastics	$10 \times 10^6 - 40 \times 10^6$
Atmosphere dry air at 1 atmosphere	3×10^6

shown in Table 2-2 are not valid for continuous operation, nor are they valid for all configurations of the dielectric. If the materials listed were to be used for storing energy with constant d-c voltage or constant a-c voltage, only much lower values of voltage could be used with safety. To gain some measure of energy capacity, consider mica and its largest value of relative dielectric constant $k_r = 7.5$ and its largest value of dielectric strength 200×10^6 v per m, listed in Table 2-2. The energy that could be stored in a meter cube of mica on that basis would be

$$W = \frac{k_r \epsilon_0 E^2}{2} = \frac{7.5 \times 8.854 \times 10^{-12} \times (200)^2 \times 10^{12}}{2}$$

$$= 1.327 \times 10^6 \text{ j per cu m}$$

or 1.327 j per cu cm, which amounts to $1.327 \times 16.387 = 21.7$ j per cu in. This value of energy is unrealistically high and it is practically impossible to store this amount of energy in mica. This same amount of energy, however, can be stored magnetically in a cubic inch of air at a flux density of about 117.5 kilolines per sq in., a value that can be realized in practice without great difficulty. The amount of energy that can be stored magnetically in air is limited by the heating of the winding, which furnishes the magnetizing force, and in some cases by magnetic saturation of the iron, which is used to give the air gap its desired configuration.

2-20 TYPES OF CAPACITORS

Capacitors are generally classified according to the kind of dielectric used in them. They may be divided into the following four groups

1. Capacitors that use vacuum, air, or other gases.

2. Capacitors in which the dielectric is mineral oil or castor oil.

3. Capacitors in which the dielectric is a combination of solid and liquid dielectrics such as paper, films of synthetic materials, glass, mica, etc., and mineral oil, castor oil, silicone, nitrobenzene, chlorinated diphenyl, etc.

4. Capacitors with strictly solid dielectric such as glass, mica, titanium oxide, etc.

The first group of capacitors are used in applications where values of capacitance required do not have to be large, but where the energy loss in the dielectric must be very small. Energy losses in dielectrics are discussed in Section 2-22. The dielectric losses in vacuum, air, or other gases, are negligible. Some applications for these capacitors are in radio-frequency circuits and in low-frequency measuring circuits where great precision is required. A precision variable air capacitor is shown in Fig. 2-20.

Oil-insulated capacitors are used in applications where larger values of capacitance are required than can be obtained readily with the first group and where a small amount of dielectric power loss can be tolerated. Paper capacitors with liquid impregnation are used in applications where precise values of capacitance are unimportant but where large amounts of capacitance can be obtained in a relatively small volume. In these applications dielectric strength and long life are also factors. One application of this type of capacitor is correcting power factor in electric power distribution systems as discussed in Section 2-17.

Mica has excellent properties as a dielectric: it has a high dielectric constant and high insulation resistance, and is affected but little by time and temperature. The dielectric losses are relatively low for a solid material and there is very little drift in the dielectric constant. Capacitors with mica as a dielectric are extensively used in laboratories as standards because of their stability and the relatively high values of capacitance for a given volume.

There are other types of capacitors not included in the above groups. One of these is the electrolytic capacitor, which has an oxide film formed on aluminum. This film is the dielectric and is exceedingly thin, making for very large values of capacitance in relatively small volumes. This film may have a thickness as small as 10^{-5} cm. The aluminum serves as one of the plates and the electrolyte as the other plate. The electrolyte employed in wet electrolytic capacitors is an aqueous solution of boric acid. Dry

Figure 2-20. Precision variable air capacitor, type 1422.
(Courtesy of General Radio Company.)

electrolytic capacitors have electrolytes consisting of solutions in which the water content is small. Solutions of polyhydric alcohols such as ethylene glycol, boric acid, and small quantities of ammonium are electrolytes generally used in the dry electrolytic capacitors.

Electrolytic capacitors have certain limitations. The aluminum must be the positive electrode and the electrolyte the negative electrode. If the polarity is reversed the film becomes conducting. This arrangement is suitable for service where steady or pulsating unidirectional potentials are applied. However, dry electrolytic capacitors can be made to operate on alternating current. Such capacitors consist of two anode foils immersed in an electrolyte that is common to both. This is equivalent to operating two electrolyte capacitors in series back to back with the negative terminals connected together.

Capacitors that use ceramics for dielectrics have been classified by the Radio-Electronics-Television Manufacturers Association into Class I and Class II dielectrics as shown in Table 2-3.

TABLE 2-3. CLASSIFICATION OF CERAMIC DIELECTRICS FOR CAPACITORS

	Class I	Class II
K range (k' at room temperature)	6 to 500	500 to 10,000
Temperature coefficient of capacitance	P120 to N5600	
Power factor	0.04 to 0.4%	0.4 to 3%
Insulation resistance	10,000 to 100,000 megohms	10,000 to 100,000 megohms
Minimum capacitance decrease (-55 to $+85°$C)	0 to 40%	15 to 80%
Maximum capacitance increase (-55 to $+85°$C)	0 to 40%	0 to 25%
Aging Characteristics of k		Approximately 4% per decade time

RETMA Standard REC-107-A, Ceramic Dielectric Capacitors, MIL-C-11015A.

Mineral rutile TiO_2 and combinations of titanium oxide with other oxides are the ceramic materials generally used in these types of capacitors. The Class I dielectrics are used in resonant circuits and other applications where a low dissipation factor, which is a measure of the dielectric energy loss as explained in Section 2-22, or a high Q, which is the reciprocal of the dissipation factor, and good stability are required. These materials have dielectric constants that are highly sensitive to temperature and can be used as temperature compensators in electronic circuits.

The Class II dielectrics, because of their extremely high values of dielectric constant, have found their greatest application where general purpose capacitors of very small dimensions are needed. The applications, however, must be such that changes in capacitance with temperature are not too critical.

2-21 POLARIZATION AND DIELECTRIC CONSTANT

Equation 2-33 shows that dielectric materials increase the energy storage capability of a capacitor. This increase results from an increase in the charge on the plates of a capacitor due to charges that are induced by dielectric polarization, a phenomenon first recognized by Faraday.* Polarization is

---

* Michael Faraday, *Philosophical Transactions*, 1837–1838.

Sec. 2-21 POLARIZATION AND DIELECTRIC CONSTANT

produced by the orientation, in the direction of the electric field, of electric dipoles in the dielectric material. An electric dipole consists, in effect, of two equal charges of opposite polarity separated from each other by a small distance.

Consider the parallel conducting plates, in Fig. 2-21(a), separated from each other by a distance d in vacuum.* A potential difference of V volts is

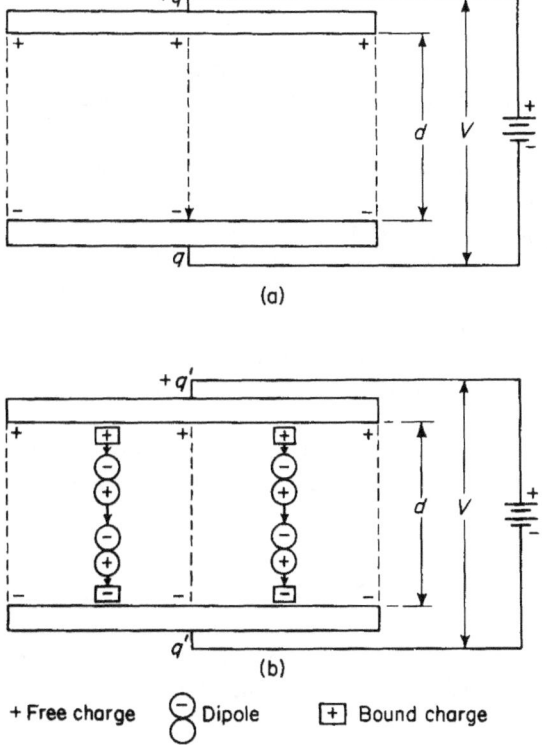

Figure 2-21. (a) Free charges on plates in vacuum; (b) free charges plus bound charge induced on plates by dipoles in dielectric.

maintained between the plates and the free charges indicated by the plus signs and minus signs are produced on the positive and negative plates respectively. The electric field is uniform and in the direction indicated in the region between plates away from the edges. At the edges there is a fringing of the electric flux.

---

* Thomas L. Martin, Jr., *Physical Basis for Electrical Engineering*. Englewood Cliffs, N.J.: Prentice-Hall, Inc., 1957.

The electric field intensity between plates away from the edges is

$$E = \frac{V}{d}$$

The electric flux density in this region is also uniform and is expressed by

$$D = \epsilon_0 E$$

Suppose that the voltage across the plates is maintained at its former value V and that the space between plates is filled with a dielectric material up to but not beyond the edges of the plates. The free charges on the plates will be the same as before since there is no change in the fringing of the field with the dielectric that fills the space between plates.

However, the electric dipoles will tend to align themselves in the direction of the electric field as indicated schematically in Fig. 2-21(b). With such an alignment the electric field intensity in the regions between the aligned dipoles is increased and coulomb forces produced by the dipoles nearest the plates induce the bound charge marked $\boxed{+}$ and $\boxed{-}$. Thus the total charge on the plates has increased from a value q in Fig. 2-21(a) to one of q' in Fig. 2-21(b).

This increased value of charge is the sum of the free charges and the charges induced by the dipoles. The flux density in the region between the plates has increased in the same proportion and can be expressed by

$$D' = \epsilon_0 E + P \tag{2-70}$$

The term P in Eq. 2-70 is defined as the dielectric polarization because it is the increase in flux density produced by the alignment of the dipoles in the dielectric material.

The dielectric constant ϵ is the ratio of electric flux density to electric field intensity. Thus, if Eq. 2-70 is divided by the electric field intensity E the expression for the dielectric constant is

$$\epsilon = \epsilon_0 + \frac{P}{E} \tag{2-71}$$

The expression for the relative dielectric constant is obtained by dividing Eq. 2-71 by ϵ_0 as follows

$$k_r = 1 + \frac{P}{\epsilon_0 E} = 1 + \frac{P}{D} \tag{2-72}$$

2-22 MECHANISM OF POLARIZATION

The mechanisms of polarization have been divided into four general kinds

1. Electronic polarization.
2. Atomic polarization.
3. Permanent dipole polarization.
4. Space-charge polarization.

Electronic polarization is a result of dipole moments induced by the electric field. Atoms are composed of heavy nuclei surrounded by negative electron clouds. In the absence of an external field the movements of the electrons are such that the resultant orbital paths are symmetrical about the nucleus as shown schematically in Fig. 2-22(a). These orbital paths become distorted by an external electric field because the paths of the electrons tend to shift against the direction of the applied field as shown in Fig. 2-22(b). A dipole is thus induced; this dipole is in effect positive on the left-hand side for the assumed direction of the field.

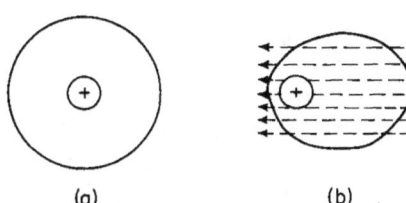

Figure 2-22. Electronic polarization; (a) no field; (b) electric field.

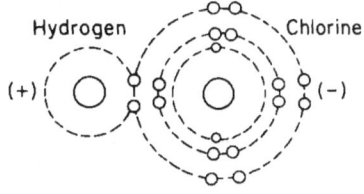

Figure 2-23. Molecular structure of HCl illustrating atomic dipoles.

Atomic polarization results from an unsymmetrical sharing of electrons when two different atoms combine to form a molecule. This type of polarization is illustrated in Fig. 2-23 for an HCL molecule in which the hydrogen atom and the chlorine atom are both polarized. In this case the electron of the hydrogen atom is attracted into the unfilled outer shell of the chlorine atom.

Permanent dipole polarization

The molecule illustrated in Fig. 2-23 is a permanent dipole because it exists as such even in the absence of an electric field. Such dipoles, however, are oriented at random when there is no resultant field in the dielectric material. However, upon the application of an electric field these dipoles will develop torque and will tend to align themselves in the direction of the electric field as indicated in Fig. 2-21(b).

Space-charge polarization is produced by charge carriers, such as free ions that can migrate for some distance through the dielectric. When the movement of such carriers is impeded by interfaces such as occur where two different dielectrics meet, or if the carriers are arrested in the material, space charges result. When there is no applied electric field the arrangement of these space charges is random and there is no net effect. However, in the presence of an electric field, the positive charge carriers will tend to take positions, relative to the negative charge carriers, in the direction of the field as shown in Fig. 2-24. Figure 2-24(a) shows the positions of the charge carriers when there is no field. On the one hand, a vertical line through the center of the charges in Fig. 2-24(a) will have zero net charge to the right and left of it. On the other hand, a vertical line through the center of Fig. 2-24(b) will have net positive charge to the left and net negative charge to the right.

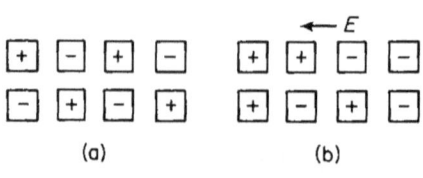

Figure 2-24. Space charge polarization; (a) no field; (b) electric field.

2-23 a-c CHARACTERISTICS OF DIELECTRICS

Equations 2-65 and 2-67 show that current in an ideal capacitor [Fig. 2-25(a)] leads the applied voltage, if sinusoidal, by an angle of 90° as shown in Fig. 2-25. In liquid dielectrics such as insulating oils and other liquids used in capacitors, transformers, and high-voltage cables, as well as in solid dielectrics, the current leads the voltage by an angle that is somewhat smaller than 90°. This departure from 90° results from dielectric energy losses that show up in the form of heat. In the case of a-c fields the dipoles undergo rotation, which is opposed, to some extent, by a sort of molecular friction, and as a result the current associated with the charges induced by the dipoles leads the voltage by an angle of less than 90°. Other losses result from conduction currents in the dielectric. The phasor diagram of Fig. 2-25(d) shows the relationships between the components of current resulting from the free charges, induced charges, and conduction in a capacitor with an imperfect dielectric.

In Fig. 2-25

let I_0 = charging current from the free charge
I_i = charging current from the induced charge
I_c = conduction current
I = total current
θ = angle by which the total current leads the applied voltage
$\delta = 90° - \theta$ = phase defect angle

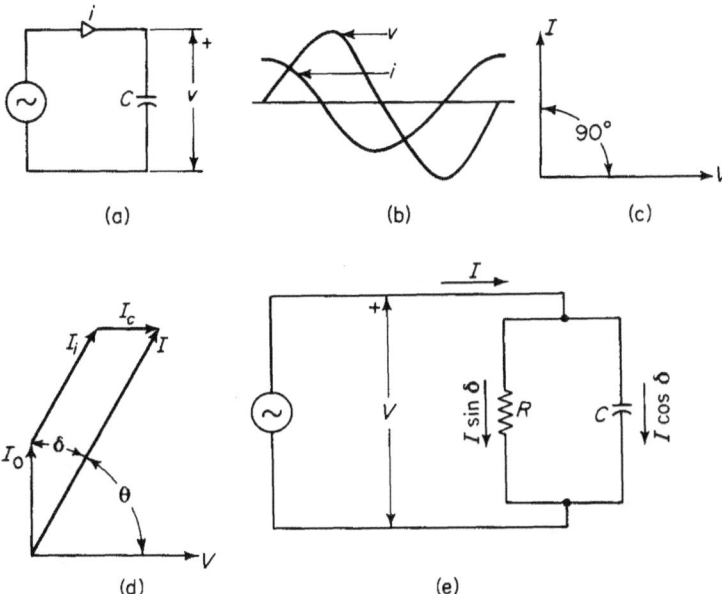

Figure 2-25. (a) Capacitor with applied a-c voltage; (b) instantaneous a-c voltage and current in ideal capacitor; (c) phasor diagram for ideal capacitor; (d) phasor diagram for capacitor with energy loss; (e) equivalent circuit of capacitor with energy loss.

The current I_0 is the current that would result if the dielectric were free space. The dielectric power factor is

$$\cos \theta = \sin \delta \qquad (2\text{-}73)$$

Dissipation factor is a term commonly used in connection with dielectric losses and the symbol for it is DF

$$\text{DF} = \tan \delta \qquad (2\text{-}74)$$

At small values of δ the difference between the power factor $\cos \theta$ or $\sin \delta$ and the dissipation factor $\tan \delta$ is negligible. For example when the power factor, i.e., $\sin \delta$ is 0.1000 the dissipation factor $\tan \delta$ is 0.1004. The value of dissipation factor is about 0.005 or less for dielectrics used in high-voltage cable, liquid-impregnated capacitors, and other applications where dielectric losses must be kept at low values. The angle δ shown in Fig. 2-25(d) is much larger than is normal for a good dielectric.

The quality factor Q is another term used in connection with capacitors; it is the reciprocal of the dissipation factor, i.e.

$$Q = \frac{1}{\text{DF}} = \cot \delta \qquad (2\text{-}75)$$

The equivalent circuit for an imperfect capacitor is shown in Fig. 2-25(e), which shows the loss component of current $I \tan \delta$ to flow through an equivalent resistance and the out-of-phase current $I \cot \delta$ to flow through an equivalent ideal capacitance.

This equivalent circuit greatly oversimplifies the actual conditions. In a given dielectric the dielectric constant and consequently the capacitance C of Fig. 2-25(e) varies with frequency and temperature. The dielectric losses are also functions of frequency and temperature so that the shunt resistance R is also a variable. Nevertheless, the equivalent circuit of Fig. 2-25(e) is quite useful for certain kinds of analyses.

2-24 COMPLEX DIELECTRIC CONSTANT

The phasor diagram of Fig. 2-25(c) shows the current in an ideal capacitor to lead the applied a-c voltage by an angle of 90°. The dielectric of an ideal capacitor is free space that has a relative dielectric constant of unity and is free of polarization and leakage. If the applied voltage and frequency are kept constant and a solid dielectric or a liquid dielectric, or a combination of both, is made to displace the free space the current will increase because of polarization. In addition the current will now lead the voltage by an angle less than 90° as shown in the phasor diagram of Fig. 2-25(d). For a given applied voltage and frequency, the relative dielectric constant of an imperfect dielectric is the ratio of the current phasor to that which would result by replacing the imperfect dielectric with free space. Because of the dielectric energy loss the relative dielectric constant is a complex quantity rather than a real number.

Let C_0 = capacitance of the capacitor when it has free space for its dielectric
V = applied sinusoidal voltage
$\omega = 2\pi f$ angular velocity in radians per second
f = frequency in cycles per second
I_0 = current in the capacitor when dielectric is free space
C = apparent capacitance when the dielectric is imperfect
I = current in the capacitor when the dielectric is imperfect

From Fig. 2-25(c)
$$I_0 = j\omega C_0 V \quad \text{amp} \tag{2-76}$$
From Fig. 2-25(e)
$$I = j\omega C V \epsilon^{-j\delta} \tag{2-77}$$

the dielectric constant is found by dividing Eq. 2-77 by Eq. 2-76. This yields

$$k^* = \frac{C\epsilon^{-j\delta}}{C_0} = k' - jk'' \tag{2-78}$$

The term expressed by Eq. 2-78 is also called the *complex relative permittivity* of the dielectric in which k'' is the *relative loss factor*.

Another term that is used is the *complex permittivity*. This is synonymous with absolute dielectric constant and is expressed as

$$\epsilon^* = \epsilon' - j\epsilon'' = \epsilon_0 k^* = \epsilon_0 k' - j\epsilon_0 k'' \qquad (2\text{-}79)$$

where ϵ_0 is the dielectric constant of free space, i.e., 8.854×10^{-12} farads per m.

The dissipation factor DF or loss tangent becomes

$$\mathrm{DF} = \tan \delta = \frac{k''}{k'} = \frac{\epsilon''}{\epsilon'} \qquad (2\text{-}80)$$

Typical values of relative dielectric constant and dissipation factor are shown in Table 2-4.*

Table 2-4 shows that the dielectric constant decreases with frequency. In the case of porcelain and Pyranol, the decrease is more pronounced than for Teflon and cable oil. This decrease in the dielectric constant is caused by the inability of some of the larger polar molecules to follow the reversals of the electric field. The dissipation factor $\tan \delta$ is quite small for all the four materials listed. In the case of liquid dielectrics appreciably larger values of dissipation factor are indications of deterioration or contamination of the dielectrics. In fact, dissipation factor or power factor tests have been used for years to determine the quality of capacitors, high-voltage cable, and dielectric materials themselves.

2-25 CORONA

An electric field modifies the random motion of the ions normally present in air. The components of motion produced by the electric field are such that negative ions are impelled toward the positive electrode and the positive ions toward the negative electrode. As the electric field intensity is increased the velocity of the ions increases accordingly. A field intensity of 3,000 kv per m will accelerate some of the ions in air, at atmospheric pressure, to velocities such that collisions between them and neutral molecules release an electron from the neutral molecule producing a pair of ions with opposite charges. This phenomenon is known as ionization by collision. Corona is thus produced. When the electric field intensity at the surface of the conductor is increased beyond the critical value, the formation of corona is accompanied by a hissing sound that is most intense in regions where the

---

* A. Von Hippel, *Dielectric Materials and Applications*, Cambridge, Mass.: The Technology Press of MIT, 1954.

TABLE 2-4. RELATIVE DIELECTRIC CONSTANTS AND DISSIPATION FACTORS

Solids		f cps	1×10^2	1×10^3	1×10^4	1×10^5	1×10^6	1×10^7	1×10^8	3×10^8	3×10^9	1×10^{10}
Porcelain dry process		ϵ'/ϵ_0	5.50	5.36	5.23	5.14	5.08	5.04	5.04	5.02		4.74
		$\tan \delta$	0.022	0.014	0.011	0.009	0.008	0.007	0.008	0.010		0.016
Teflon		ϵ'/ϵ_0	2.1	2.1	2.1	2.1	2.1	2.1	2.1	2.1	2.1	2.08
		$\tan \delta$	0.0005	0.0003	0.0003	0.0003	0.0002	0.0002	0.0002	0.00015	0.00015	0.0006
Liquids												
Cable oil		ϵ'/ϵ_0	2.25	2.25	2.25	2.25				2.24	2.22	2.22
		$\tan \delta$	0.0003	0.00004	0.00004	0.0001				0.0039	0.0018	0.0022
Pyranol		ϵ'/ϵ_0	4.42	4.40	4.40	4.40	4.40	4.40	4.08	3.19	2.84	2.62
		$\tan \delta$	0.0036	0.0003	0.0003	0.00036	0.0025	0.0260	0.15	0.15	0.12	0.074

conductor has sharp points or corners. At those regions the electric field intensity is the greatest. Corona accompanied by sound is known as audible corona. In a darkened room, corona formation produces a luminous glow around a conductor with streamers emanating from corners and points. If the electric field intensity is raised to a sufficiently high value the cumulative effect of the collisions of ions with neutral molecules will produce an ion avalanche and electrical breakdown. If the field is uniform, as between plane parallel-plate electrodes, complete breakdown occurs suddenly with the formation of corona.

Corona is undesirable in electrical equipment. In the case of high-voltage aerial transmission lines, the corona increases the losses and thus lowers the efficiency. In addition, it produces radio interference. Large conductors are used in high-voltage aerial lines to prevent the formation of corona. Some of these conductors are hollow and some are made of aluminum strands reinforced by a core of one or more steel strands. The insulation in high-voltage equipment such as underground cable and capacitors is thoroughly impregnated to prevent voids in which corona can form. Transformers in the higher voltage ratings are oil-insulated. The oil serves two purposes, one of which is to increase the dielectric strength, and the other to reduce the thermal resistance, thus making for good heat transfer. Another method of suppressing corona is to operate the insulation or dielectric under hydrostatic pressure. Some high voltage cables that use impregnated paper insulation are provided with channels where nitrogen gas is maintained under pressure.

Corona shortens the life of dielectric materials. In air, corona produces ozone O_3 and oxides of nitrogen. Although ozone is unstable, it is a strong oxidizing agent and attacks most organic materials. The oxides of nitrogen produce nitric and nitrous acids in the presence of moisture; these acids embrittle cellulose and corrode metal.

When corona starts in a dielectric, the dissipation factor increases with increasing electric field intensity. Dielectrics used in commercial insulation exhibit a practically constant dissipation factor at a given temperature for voltages well beyond the normal operating range as long as there are no voids. Corona loss increases the conduction current $I \tan \delta$ in Fig. 2-25.

2-26 RESISTANCE OF DIELECTRIC CONFIGURATIONS

The electrical resistance of most dielectric liquids and solids is a rather complex function of time, electrification, and in some cases of electric field intensity. The losses in a dielectric under alternating stresses are generally much greater than can be accounted for on the basis of d-c resistance. At least part of this difference is due to the polarization of the dielectric. In some kinds of dielectric a sudden application of d-c voltage is accompanied by a relatively large momentary current, which at first falls off rapidly and then decreases very slowly, sometimes still decreasing hours after the application of a constant potential.

However, if the resistivity of the dielectric is assumed constant, i.e., it is not affected by the electric field intensity or time of electrification, the resistance of the dielectric of a certain configuration is found by using the methods that were used for determining capacitance.

Consider two concentric spheres, the space between which is occupied by a dielectric that has an electrical resistivity of ρ ohms per meter cube. If Fig. 2-15 is applied to the case of concentric spheres instead of concentric cylinders, the resistance of the elemental spherical shell of thickness dx is expressed by

$$dR = \frac{\rho\, dx}{4\pi x^2} \quad \text{ohms}$$

The resistance between spheres is therefore

$$R = \frac{\rho}{4\pi} \int_{r_1}^{r_2} \frac{dx}{x^2} = \frac{\rho}{4\pi}\left(\frac{1}{r_1} - \frac{1}{r_2}\right) = \frac{\rho}{4\pi}\frac{(r_2 - r_1)}{r_1 r_2} \qquad (2\text{-}81)$$

It is interesting to compare Eq. 2-81 with Eq. 2-28 for the capacitance of concentric spheres. The product of these two quantities and of the relative dielectric constant k_r yields

$$RC = \rho k_r \epsilon_0 \qquad (2\text{-}82)$$

This product is valid for any configuration of homogeneous dielectric.

2-27 MECHANICAL ENERGY AND FORCE IN A CAPACITOR

Figure 2-26 shows a simple two-plate capacitor, the plates of which are in a liquid or gaseous medium and are free to move with respect to each other.

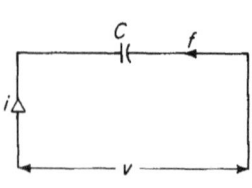

Figure 2-26. Movable plate capacitor.

If a voltage v is applied to the plates of the capacitor a positive charge is distributed on one plate and an equal charge of opposite polarity is distributed on the other plate. These charges produce a force of attraction between the plates, and if the left-hand plate is fixed while the right-hand plate is free to move, a force f is developed and tends to move the right-hand plate toward the left. As a result electrical energy is converted to mechanical energy. However, the conversion of electrical energy to mechanical energy is generally accompanied by a change in the amount of energy stored in the dielectric. If the dielectric is lossless then

Sec. 2-27 MECHANICAL ENERGY AND FORCE IN A CAPACITOR

all the energy absorbed by the dielectric is stored in reversible form and we have

$$dW_e = dW_{st} + dW_{mech} = v\,dq \qquad (2\text{-}83)$$

where dW_e = differential energy supplied by the electrical source in time dt
dW_{st} = differential energy stored in the dielectric in time dt
dW_{mech} = differential mechanical energy output of the capacitor in time dt

For a fixed configuration of the dielectric, and in this case for a fixed position of the right-hand plate relative to that of the left-hand plate, there can be no mechanical energy, and all of the energy supplied by the source is stored in the dielectric, i.e.

$$dW_e = dW_{st} + 0 \qquad (2\text{-}84)$$

but

$$dW_e = v\,dq \qquad (2\text{-}85)$$

Substitution of Eq. 2-85 in 2-84 gives

$$dW_{st} = v\,dq \qquad (2\text{-}86)$$

It must be remembered that Eq. 2-86 is valid only for a fixed configuration of dielectric. The energy absorbed by the dielectric in changing the charge from q_1 to q_2 is

$$\Delta W_{st} = \int_{q_1}^{q_2} v\,dq \qquad (2\text{-}87)$$

If the initial charge in the capacitor is zero, then the energy absorbed by the dielectric when the charge q is established is

$$W_{st} = \int_0^q v\,dq \qquad (2\text{-}88)$$

In these equations the charge q is a function of the voltage across the capacitor expressed by Eq. 2-25 as $q = vc$.

If the dielectric constant has a fixed value, then for a given configuration of the dielectric the capacitance C in Eq. 2-25 is constant, and the result of carrying out the integration in Eq. 2-88 is

$$W_{st} = \frac{v^2 C}{2} \qquad (2\text{-}89)$$

Equation 2-89 expresses the energy that is stored in the dielectric of a lossless capacitor of capacitance C farads. This is true whether C is constant or variable. W_{st} is the stored energy for the particular value of C at a given instant regardless of what the value of C was prior to that instant or the value it might attain at some later instant. From Eq. 2-89 it is evident that a change in the applied voltage v or a change in the capacitance C produces a change in the stored energy. Hence, in general, the differential energy stored in the dielectric in time dt is expressed by

$$dW_{st} = vC\,dv + \frac{v^2}{2}\,dC \qquad (2\text{-}90)$$

The differential energy supplied by the source is expressed by Eq. 2-86 whether the capacitance is fixed or variable, and if Eqs. 2-25, 2-86, and 2-90 are substituted in Eq. 2-83 the result is

$$v^2\,dC + vC\,dv = vC\,dv + \frac{v^2}{2}\,dC + dW_{mech} \qquad (2\text{-}91)$$

and the differential mechanical energy is

$$dW_{mech} = \frac{v^2}{2}\,dC \qquad (2\text{-}92)$$

Equations 2-90 and 2-92 show that for a constant applied voltage ($dv = 0$) the differential mechanical energy equals the differential stored energy. Substitution of Eq. 2-25 in 2-92 yields

$$dW_{mech} = \frac{q^2}{2C^2}\,dC \qquad (2\text{-}93)$$

When Eq. 2-25 is substituted in Eq. 2-90 the result is

$$dW_{st} = \frac{dq}{C} - \frac{q^2}{2C^2}\,dC \qquad (2\text{-}94)$$

If q is held constant (this can be done by disconnecting the capacitor from the electrical source) the differential stored energy is expressed by

$$dW_{st} = -\frac{q^2}{2C^2}\,dC \qquad (2\text{-}95)$$

This is to be expected because there is no differential electrical energy input, and the differential mechanical energy must therefore result from a change in the stored energy.

Equations 2-92 and 2-93 show that motor action results from an increase in the capacitance, which also follows from the fact that the coulomb forces act in such a direction as to increase the capacitance. The forces that are developed by an electric field in a dielectric are always in a direction that produces a dielectric configuration such that the capacitance becomes a maximum.

The mechanical power developed during the differential time according to Eqs. 2-92 and 2-93 is

$$p_{mech} = \frac{dW_{mech}}{dt} = \frac{v^2}{2}\frac{dC}{dt} = \frac{q^2}{2C^2}\frac{dC}{dt}$$

The force on the plates of the capacitor is found from

$$f\,dx = dW_{mech} = \frac{v^2}{2}dC = \frac{q^2}{2C^2}dC$$

to be

$$f = \frac{v^2}{2}\frac{dC}{dx} = \frac{q^2}{2C^2}\frac{dC}{dx} \quad (2\text{-}96)$$

If the arrangement is such that the developed forces produce rotation, the torque is expressed by

$$T = \frac{v^2}{2}\frac{dC}{d\theta} = \frac{q^2}{2C^2}\frac{dC}{d\theta} \quad (2\text{-}97)$$

A practical application of converting the energy stored in a dielectric to electrical energy by means of mechanical energy input is in the use of the capacitor microphone, also known as the condensor or electrostatic microphone. In this application, sound waves cause motion of a diaphragm that is one plate of an electrically energized capacitor in which the other plate is fixed.

2-28 ELECTROSTATIC SYNCHRONOUS MACHINE

Figure 2-27 shows a single-phase electrostatic machine. The rotor in this figure has two parallel plates connected electrically to each other and operating best at ground potential. The stator has stationary plates that are interleaved with the rotor plates. An a-c voltage $v_{a\text{-}c}$ is applied between the stator and ground while the stator is rotating at synchronous speed. Since the machine in this illustration has two poles [two lobes shown in Fig. 2-27(a)] the number of revolutions per second is the same as the applied frequency. A machine with p poles has a synchronous speed of

$$N_{syn} = 120\frac{f}{p}\text{ rpm} \quad (2\text{-}98)$$

where f = frequency in cycles per second.

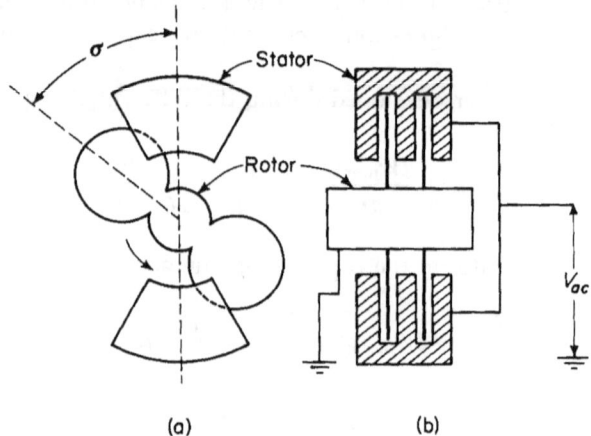

Figure 2-27. Electrostatic synchronous machine; (a) end view; (b) cross section.

Let the applied voltage be expressed by

$$v_{\text{a-c}} = \sqrt{2}V \cos(\omega t + \theta_0) \tag{2-99}$$

and let the capacitor plates be shaped so that the capacitance between the stator and rotor varies, as follows

$$C = \tfrac{1}{2}(C_{\max} + C_{\min}) + \tfrac{1}{2}(C_{\max} - C_{\min})\cos 2\sigma \tag{2-100}$$

From Eq. 2-96 the expression for the developed mechanical power is

$$p_{\text{mech}} = \frac{V^2}{2}\frac{dC}{dt} \tag{2-101}$$

Substitution of Eqs. 2-99 and 2-100 in Eq. 2-101 yields

$$p_{\text{mech}} = V^2 \cos^2(\omega t + \theta_0)(C_{\max} - C_{\min}) \sin 2\sigma \frac{d\sigma}{dt} \tag{2-102}$$

but

$$\sigma = \omega t + \sigma_0 \qquad \frac{d\sigma}{dt} = \omega$$

hence

$$p_{\text{mech}} = V^2(C_{\max} - C_{\min}) \cos^2(\omega t + \theta_0) \sin 2(\omega t + \sigma_0) \tag{2-103}$$

which, upon expanding, becomes

$$p_{\text{mech}} = \frac{\omega V^2}{2}(C_{\max} - C_{\min})[\sin 2(\omega t + \sigma_0) \\ + \tfrac{1}{2}\{\sin 2(2\omega t + \theta_0 + \sigma_0) + \sin 2(\sigma_0 - \theta_0)\}] \tag{2-104}$$

Equation 2-104 expresses the instantaneous *applied* power. The average *applied* power is

$$p_{mech} = \frac{1}{\pi} \int_{\omega t=0}^{\omega t=\pi} p_{mech} \, d\omega t = \frac{\omega V^2}{4} (C_{max} - C_{min}) \sin 2(\sigma_0 - \theta_0) \qquad (2\text{-}105)$$

Let $\sigma_0 - \theta_0 = \theta$, then the average applied power is

$$p_{mech} = \frac{\omega V^2}{4} (C_{max} - C_{min}) \sin 2\theta \qquad (2\text{-}106)$$

Electrostatic generators and motors operating on the principles outlined in the preceding pages have very little output for a given size, particularly if they are operating in air at atmospheric pressure because of the low value of dielectric strength at that pressure. However, when operating in a vacuum the rating can be increased considerably because the increased dielectric strength for vacuum allows the applied voltage V to be increased accordingly and the output is proportional to V^2. Nevertheless, the output of the electrostatic machine, even under ideal conditions, is very small in comparison with that of conventional electromagnetic motors and generators and therefore has limited applications. Since these machines operate in a vacuum they have practically no losses, and since there are no magnetic losses, efficiencies above 99 percent have been realized.*†

PROBLEMS

2-1 Given a charge of q coulombs in free space at the origin in a system of x, y, and z coordinates, determine in terms of q

  (a) The electric flux density.
  (b) The electric field intensity.
  (c) The voltage gradient at $x = 3$, $y = 4$, and $z = 12$ m.
  HINT: $r = \sqrt{x^2 + y^2 + z^2}$

2-2 In Problem 2-1, what is the voltage between two points at the following locations

  (a)
    (i) $x = 3$, $y = 4$, $z = 12$ m, and
    (ii) $x = 0$, $y = 5$, $z = 12$ m.
  (b)
    (i) $x = 1.5$, $y = 2.0$, $z = 6.0$ m, and
    (ii) $x = 4.0$, $y = 12.0$, $z = 3.0$ m.

---

* J. G. Trump, "Electrostatic Sources of Electric Power," *Electrical Engineering*, June 1947, pp. 525–534.
† Fitzgerald and Kingsley, *Electric Machinery*. New York: McGraw-Hill Book Company, 1952.

84 CAPACITANCE AND RELATED EFFECTS Chap. 2

2-3 Assume the earth to be a sphere having a radius of 6,380 km. Determine
   (a) The capacitance of the earth.
   (b) The potential of the earth if it were completely isolated in space and carried a uniformly distributed charge of 1 coulomb.
   (c) The energy stored in space due to the charge of 1 coulomb on the earth's surface.
   (d) The potential gradient at the earth's surface due to the 1-coulomb charge.

2-4 Given two concentric conducting spherical shells of negligible thickness having diameters of 15 cm and 10 cm. The medium between these shells and that external to them is free space. A surface integral of electric field intensity obtained between the shells gives a value of 1.13×10^6 m v, whereas a surface integral taken over the outer surface of the outer spherical shell also yields a value of 1.13×10^6 m v. In addition, the surface integral of electric field intensity taken over the inner surface of the inner shell has a value of zero. Determine
   (a) The charge within the inner shell.
   (b) The charge within the outer shell.
   (c) The charge on the inner shell.
   (d) The charge on the outer shell.

2-5 Under a different set of conditions the following values of surface integrals of electric field intensity in meter volts are obtained for the concentric spherical shells of Problem 2-4.
   (a) Over the outer surface of the outer shell, zero.
   (b) Over the outer surface of the inner shell, 1.13×10^5.
   (c) Over the inner surface of the inner shell, zero.

   Determine
   (i) The charge within the inner shell.
   (ii) The charge within the outer shell.
   (iii) The charge on the inner shell.
   (iv) The charge on the outer shell.

2-6 Under a third set of conditions the following values of surface integrals of electric field intensity in meter volts are obtained for the concentric spherical shells of Problem 2-4.
   (a) Over the outer surface of the outer shell, 1.13×10^5.
   (b) Over the inner surface of the outer shell, 1.13×10^5.
   (c) Over the inner surface of the inner shell, 1.13×10^5.

   Determine
   (i) The charge within the inner shell.
   (ii) The charge within the outer shell.
   (iii) The charge on the inner shell.
   (iv) The charge on the outer shell.

PROBLEMS

2-7 If the space between the concentric spherical shells is filled with transformer oil, the relative dielectric constant of which is 2.24, determine the values of the surface integrals if the charges are at the same locations as in Problem 2-6. The medium surrounding the outer sphere is free space and that enclosed by the inner sphere is free space.

2-8 Given a conducting sphere isolated in free space carrying a uniformly distributed charge q. The radius of the sphere is a meters. Determine the energy density of the electric field at

(a) The surface of the sphere.
(b) A point external to the surface and a distance r from the center of the sphere.

2-9 Determine the energy stored in the space between the equipotential surfaces at r_1 and r_2 m from the center of the sphere and external to the surface of the sphere of Problem 2-8.

2-10 Repeat Problem 2-9 for the condition that the space between r_1 and r_2 is occupied by a medium that has a relative dielectric constant k_r.

2-11 Determine the maximum energy density that can be supported by air on the basis that breakdown occurs at $E = 3$ million v per m. What is the charge density at that value of electric field intensity?

2-12 Determine the capacitance of two concentric spheres if the diameter of the inner sphere is 10 cm and that of the outer sphere is 11 cm if the dielectric is

(a) Air.
(b) Transformer oil.

2-13 Determine the charge on the spheres in Problem 2-12 when the voltage between them is 5,000 v and the dielectric is

(a) Air.
(b) Transformer oil.

2-14 What is the electric field intensity at the surface of the inner sphere and at the surface of the outer sphere in Problem 2-12, when the applied voltage is 5,000 v and the dielectric is

(a) Air.
(b) Transformer oil.

2-15 Given two concentric spheres the outer sphere of which has a fixed radius r_2. Determine r_1 the radius of the inner sphere in terms of r_2 such that the maximum electric field intensity in the dielectric is a minimum *for a given applied voltage between the spheres.*

2-16 Repeat Problem 2-15 for concentric cylinders.

2-17 The conductor of a 230-kv single-conductor oil-impregnated cable has a diameter of 1.835 in. and the insulation surrounding the conductor is 0.925 in. thick. A lead sheath 0.170 in. in thickness tightly surrounds the insulation.

The relative dielectric constant k_r of the insulation is 3.5. This cable is one of three operating at a 3-phase, 60-cycle voltage of 230 kv between the conductors of the cable. The voltage between the conductor and the sheath of each cable is therefore 132.5 kv. Determine for a length of one mile of cable

(a) The capacitance.
(b) The energy stored in the cable at a maximum emf of $\sqrt{2} \times 132.5$ kv.
(c) The capacitive reactance in ohms at 60 cycles.
(d) The charging current at 132.5 kv and 60 cycles.
(e) The electric field intensity in
  (i) Volts per meter.
  (ii) Volts per inch at
    (a') The surface of the conductor.
    (b') The inner surface of the lead sheath when the applied voltage is 2×132.5 kv.

2-18 A 2,400-volt capacitor is rated at 15 kva at 60 cps. The thickness of the dielectric is 0.008 in. and the relative dielectric constant is 5.20. Determine

(a) The area of the plates in this capacitor.
(b) The dielectric loss if the dissipation factor is 0.25 percent.
(c) The Q of the capacitor.

2-19 The relative dielectric constant of the dielectric in a ceramic capacitor is 1,000. Determine the ratio of area to thickness of the dielectric if the capacitor has a capacitance of 1,000 μf (a) with the meter as the unit of length (b) with the inch as the unit of length.

2-20 Measurements made with a capacitance bridge on a liquid dielectric test cell at a frequency of 1,000 cps gave the following results

(a) Cell empty

$$C = 110.0 \text{ mmf} \qquad DF = 0.0000$$

(b) Cell filled with liquid dielectric

$$C = 246.5 \text{ mmf} \qquad DF = 0.0030$$

  (i) Determine the equivalent parallel capacitance C and the equivalent parallel resistance R on the basis of the equivalent circuit of Fig. 2-25(e)
    (a') when the cell is empty
    (b') when the cell is filled with the dielectric liquid.
  (ii) Determine the quality factor Q for
    (a') The empty cell.
    (b') The filled cell.
  (iii) Determine the complex dielectric constant of the liquid dielectric.

2-21 Two parallel plane plates having an area of 20 sq ft each and a separation of 3 in. are charged to a potential difference of 100,000 v. The source of potential is removed and the separation of the plates is increased from 3 to 5 in. Determine

(a) The potential difference between plates when separated 5 in.
(b) The energy required to increase the separation from 3 to 5 in. (neglect fringing).

2-22 A constant potential difference of 100,000 v is maintained between the plates in Problem 2-21 while the spacing decreases from a value of 3.0 to one of 1.5 in. Neglect fringing and determine

(a) The mechanical energy.
(b) The change in the stored energy.
(c) The average electrical power input if this change in spacing takes place in 0.01 sec.
(d) The amount of the electric power in part (b) that is absorbed by the dielectric and the amount of electric power that is converted into mechanical power. (Is this motor power or generator power?)

2-23 A variable air capacitor has its rotor free to rotate completely. The minimum and maximum values of capacitance are 16×10^{-12} and 250×10^{-12} farads. Assume the capacitance to vary linearly with angular displacement and that in one-half revolution the capacitance goes from minimum to maximum and returns from maximum to minimum during the next half revolution.
A constant d-c potential of 700 v is applied to this capacitor. Assume the angular displacement to be zero when the capacitance is a minimum.

(a) Plot a graph of
  (i) Capacitance vs angular displacement.
  (ii) Torque vs angular displacement.
  (iii) Current vs angular displacement when the rotor is driven at a speed of 3,600 rpm.
(b) Determine the approximate power consumed in a resistor of one megohm connected in series with this capacitor when the rotor is driven at 3,600 rpm.
(c) How much of the power in part (b) is furnished
  (i) By the d-c source?
  (ii) By the device that drives the rotor of the capacitor?

2-24 The dielectric in the cable of Problem 2-17 has a resistance of 10^{15} ohms per cm cube at a certain temperature. Determine the resistance of the insulation in one mile length of this cable.

BIBLIOGRAPHY

Bloomquist, W. C. et al., *Capacitors for Industry*. New York: John Wiley & Sons, Inc., 1950.

Deeley, P. M., *Electrolytic Capacitors*. South Plainfield, New Jersey: The Cornell-Dubilier Electric Corp., 1938.

Jordan, E. C., *Electromagnetic Waves and Radiating Systems*. Englewood Cliffs, N.J.: Prentice-Hall, Inc., 1950.

Peek, F. W., Jr., *Dielectric Phenomena in High-Voltage Engineering*. New York: McGraw-Hill Book Company, 1929.

Schwaiger, A. and R. W. Sorensen, *Theory of Dielectrics*. New York: John Wiley & Sons, Inc., 1932.

Skilling, H. H., *Fundamentals of Electric Waves*. New York: John Wiley & Sons, Inc., 1948.

Timbie, W. H. *et al.*, *Principles of Electrical Engineering*. New York: John Wiley & Sons, Inc., 1951

Winch, R. P., *Electricity and Magnetism*. Englewood Cliffs, N.J.: Prentice-Hall, Inc., 1955.

MAGNETIC CIRCUITS

3

3-1 MAGNETISM

Electric charges in motion produce a magnetic field. Thus, when an electric charge q moves with a velocity **v** a magnetic field is produced as shown in Fig. 3-1, in which **v** shows the direction of motion and P indicates the point at which the magnetic field intensity is **H**. If the velocity vector **v** and the point P are in the plane of the paper, then for a positive charge the direction of **H** is away from the observer and into the plane of the paper. The magnitude of the magnetic field intensity at P is

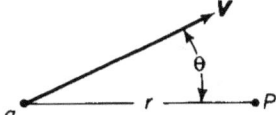

Figure 3-1. Magnetic field is produced in space surrounding a moving charge.

$$H = \frac{qv \sin \theta}{4\pi r^2} \quad (3\text{-}1)$$

However, both the magnitude of H and its direction are specified when this relationship is expressed in vector form as

$$\mathbf{H} = \frac{q}{4\pi r^2} \mathbf{v} \times \mathbf{i}_r \quad (3\text{-}2)$$

Where **v** is the vector that represents the velocity, and $\mathbf{i}_r$ the unit vector in the direction of the line joining the charge q with the point P. It is well to remember that the magnitude of the cross product is equal to the area of the parallelogram in which **v** and $\mathbf{i}_r$ are the adjacent sides. This corresponds to the quantity $v \sin \theta$. Also, the direction of the vector, resulting from the cross product, is at right angles to the plane containing the vectors **v** and $\mathbf{i}_r$ in accordance with the right-hand rule.

3-2 MAGNETIC FIELD ABOUT A STRAIGHT WIRE CARRYING CURRENT

An electric charge in motion constitutes an electric current. Then

$$i = qv \qquad (3\text{-}3)$$

where i = current in amperes
q = charge in coulombs
v = velocity of the charge in meters per second

Accordingly, an electric current is commonly described as a flow of electrons. Such a concept can be very misleading as it could easily lead to the erroneous conclusion that when the current in a conductor, as read by an ammeter, is zero all electron motion is arrested. This is far from the real condition. Actually, an enormous number of electrons swarm about in a conductor material with a random chaotic motion at velocities far greater than can be imparted to them by the application of an external voltage. The random current density or current crossing a unit surface in the conductor because of random electron velocities, found by taking into account only the electrons crossing in one direction and neglecting the equal numbers crossing in the other direction, is of the order of 10^{17} amp per sq m. An external voltage produces only a very minor modification of the random velocities,

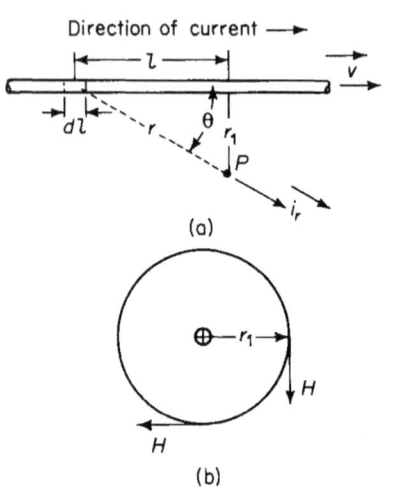

Figure 3-2. Magnetic field about a straight conductor (a) showing side view of conductor, and (b) showing end view.

causing a drift in the direction of the electric field resulting from the applied voltage. This drift gives rise to a net current density of the order of 1.5×10^6 amp per sq m for the values of electric field intensity normally encountered in metallic conductors such as are considered in the following.

The magnetic field produced by a filament of current will be treated first and the results then applied to long straight cylindrical conductors such as wires.

Figure 3-2 shows a straight filament carrying an electric current. Consider the point P, external to the wire, at which the magnetic field intensity

Sec. 3-2 MAGNETIC FIELD ABOUT A STRAIGHT WIRE

produced by the current in the wire is to be determined. Then for a differential length of wire dl

let ρ = net density of charge in coulombs per cubic meter drifting along the length of the wire
v = velocity of drift in meters per second
A = cross-sectional area of the filament in square meters
r = distance of the differential length dl from P in meters
r_1 = radial distance of P from the wire

The net electrical charge drifting along the filament and occupying the differential length dl of the wire is

$$dq = \rho \mathbf{A} \cdot \mathbf{dl} \quad \text{coulombs} \tag{3-4}$$

The vector that represents the area of a surface is normal to that surface. In the case of the wire under discussion the cross-sectional area A is at right angles to the length l, i.e., at right angles to the direction of drift. As a result there is no angular displacement between the vector $\mathbf{A}$ and the vector $\mathbf{dl}$, and their magnitudes can be multiplied directly so that we have

$$dq = A\, dl \tag{3-5}$$

The contribution that this incremental charge makes to the magnetic field intensity at P is expressed in vector form, according to Eq. 3-2, by

$$\mathbf{dH} = \frac{A\, \mathbf{dl}\, \mathbf{v} \times \mathbf{i}_r}{4\pi r^2} \tag{3-6}$$

From Eq. 3-1 the magnitude of the vector dH is found to be

$$dH = \frac{A\, dl\, v \sin \theta}{4\pi r^2} \tag{3-7}$$

The total magnetic field intensity H at P results from the current in the entire length of the filament and the magnitude of H is found by carrying out the following integration

$$H = \int_{l=-\infty}^{l=\infty} dH = \frac{Av}{4\pi} \int_{-\infty}^{\infty} \frac{\sin \theta}{r^2}\, dl \tag{3-8}$$

Since r and θ are functions of l the expressions in Eq. 3-8 can be simplified as follows

$$r = -\frac{r_1}{\sin \theta}$$

$$l = r \cos \theta = -r_1 \cot \theta$$

$$dl = \frac{r_1}{\sin^2 \theta} d\theta$$

also at $l = -\infty$ $\quad \theta = 0$ and

at $l = +\infty$ $\quad \theta = \pi$

When these relationships are used in Eq. 3-8 the result is

$$H = \frac{\rho A v}{4\pi r_1} \int_0^\pi \sin \theta \, d\theta = \frac{\rho A v}{2\pi r_1} \tag{3-9}$$

The magnetic field intensity H is most conveniently expressed as a function of current and is generally more useful in that form than in the form of Eq. 3-9. Accordingly

$$\rho A v = \rho A \frac{dl}{dt} \tag{3-10}$$

where $v = dl/dt$, i.e., the velocity at which the resultant charge is moving along the filament. A comparison of Eq. 3-10 with Eq. 3-5 shows that

$$\frac{dq}{dt} = \rho A \frac{dl}{dt} \tag{3-11}$$

However, an electric current is defined by

$$I = \frac{dq}{dt} \tag{3-12}$$

and Eq. 3-9 can be reduced to

$$H = \frac{I}{2\pi r_1} \quad \text{ampere turns per meter} \tag{3-13}$$

Although the current I is a scalar quantity it can be handled as a vector quantity when it is directed along a path and when the direction of the path

Sec. 3-3 MAGNETIC FLUX AND MAGNETIC LINES OF FORCE

is taken into account. In the case of the filament the current is directed along the length l of the filament and the magnetic field intensity at P can be expressed in vector form as follows

$$\mathbf{H} = \frac{\mathbf{I} \times \mathbf{i}_{r_1}}{2\pi r_1} \tag{3-14}$$

The vector notation of Eq. 3-14 shows that the direction of the magnetic field intensity at P is into the plane of the page when the direction of the current is shown as in Fig. 3-2. Equations 3-13 and 3-14 both show that the magnetic field intensity is constant at all points that are at a constant radial distance from the filament resulting in a magnetic field that is circular and concentric with the filament. The vector notation in Eq. 3-14 shows a cross-product that is merely a shorthand way of expressing the right-hand rule,

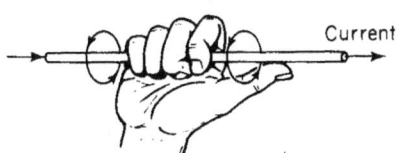

Figure 3-3. Right-hand rule.

which states that if the filament is grasped in the right hand, with the thumb pointing in the direction of the current as illustrated in Fig. 3-3, the fingers show the direction of the magnetic field.

The relationship expressed by Eq. 3-14 is valid for long straight conductors of any finite cross-section when the distance r_1 is large relative to the cross-sectional dimensions. It is valid for the magnetic field produced by a long straight cylindrical conductor at all points external to the conductor itself.

3-3 MAGNETIC FLUX AND MAGNETIC LINES OF FORCE

In Chapter 2 the concept of electric flux was used in connection with the electric field in which the electric flux density at a point in free space is directly proportional to the electric field intensity at that point. Analogous relations exist for the magnetic field in that the magnetic flux density at a point in free space is directly proportional to the magnetic field intensity at that point, as expressed by

$$B = \mu_0 H \tag{3-15}$$

where B = magnetic flux density and μ_0 = magnetic permeability of free space. In the rationalized MKS μ_0 has the value of $4\pi \times 10^{-7}$ henry per meter and B is expressed in webers per square meter.

The magnetic field is a field of force and requires energy for its production. A magnet when brought into the magnetic field produced by another magnet or by a current in a circuit experiences a force. Although a positive or negative charge can be isolated, a magnetic pole cannot exist by itself since every magnet has an equal number of poles of opposite polarity. The simplest magnet has one north pole and one south pole. However, if a north pole could exist by itself without the accompanying south pole, and

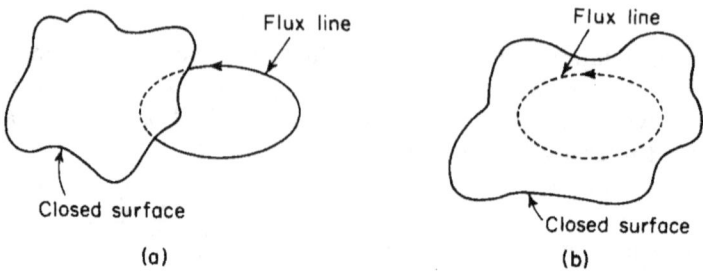

Figure 3-4. (a) Magnetic flux entering and leaving a closed surface; (b) line of force completely enclosed by closed surface.

this north pole were brought to within a distance of r_1 of the filament in Fig. 3-2, it would experience a force in a direction tangent to the circle of radius r_1 as determined by the right-hand rule. The locus of constant force would be a cylinder of radius r_1 concentric with the filament. The magnetic field can therefore be represented by lines of force.

It will be recalled that the flux lines in an electric field emanate from a positively charged body and terminate on a negatively charged body. The magnetic flux lines produced by the current in the straight conductor close upon themselves forming circles concentric with the conductor. In less simple arrangements of current-carrying conductors the magnetic flux lines close upon themselves although they do not necessarily form circles. Because of the property of the magnetic flux lines closing upon themselves (a) the number of magnetic flux lines that enter a closed surface in the field must also leave that surface or (b) the flux lines within a closed surface are completely enclosed by that surface as shown in Fig. 3-4(a) and (b). This law is expressed mathematically by

$$\oint \mathbf{B} \cdot d\mathbf{A} = 0 \qquad (3\text{-}16)$$

The surface integral in Eq. 3-16 merely states that the net magnetic flux emanating from a closed surface is zero.

Figure 3-5 shows an infinitesimal surface of area dA in a magnetic field. The flux density in this surface is B and the direction of the flux is at an angle θ with the normal. The magnetic flux passing through this infinitesimal area is given in vector notation by

$$d\phi = \mathbf{B} \cdot d\mathbf{A} = B \cos \theta \, dA \qquad (3\text{-}17)$$

The flux through a finite surface would be

$$\phi = \int_A \mathbf{B} \cdot d\mathbf{A} \qquad (3\text{-}18)$$

The difference between the integral of Eq. 3-16 and that of Eq. 3-17 is that the former is taken over a completely closed surface whereas that of

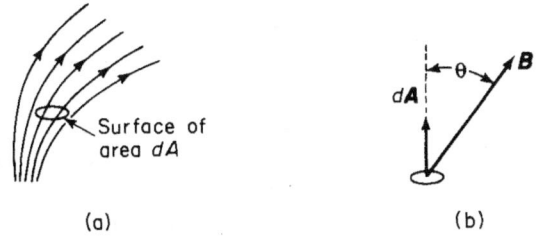

Figure 3-5. (a) Infinitesimal surface in a magnetic field; (b) vectors representing surface area and magnetic flux density.

Eq. 3-17 is taken over a surface that does not close upon itself if ϕ is to have a value other than zero.

3-4 THE UNIT MAGNET POLE

It is shown in Chapter 2 that electric charges produce an electric field and that forces between electric charges exist. The force between point charges isolated in free space is given by Eq. 2-1, which expresses Coulomb's law for point electrical charges. A similar law known as *Coulomb's law of force between point magnetic poles* applies to magnetic fields in free space and is given by the following formula

$$f = \frac{\mu_0 m_1 m_2}{4\pi r^2} \qquad (3\text{-}19)$$

where m_1 and m_2 are the strengths of two magnetic poles in terms of unit magnet poles. If the poles are of like polarity the force is one of repulsion and if the poles are of opposite polarity they are attracted toward each other.

A unit magnet pole is defined as one that exerts a force of one dyne (10^{-7} newton) on another unit magnet pole when poles are in free space and separated from each other by a distance of one cm. In the rationalized MKS system the force F is expressed in newtons and the distance r in meters. The term μ_0 expresses the magnetic permeability of free space having a value of $4\pi \times 10^{-7}$ h per m.

The concept of a magnet pole arises from the observation that the flux from a bar magnet emanates into surrounding space from a region within the magnet near one end and enters a region within the magnet located similarly with respect to the other end as shown in Fig. 3-6. Such observations are based on the pattern assumed by iron filings when placed in the space surrounding the magnet and on experiments in which the field is explored with a magnetic compass.

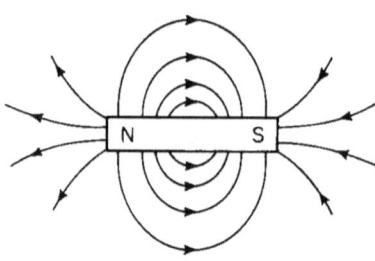

Figure 3-6. Lines of force of a bar magnet.

According to convention, the magnetic flux leaves the north (north-seeking) pole and enters the south pole, the lines of flux continuing on through the body of the magnet and closing upon themselves. Actually a north pole cannot be isolated from its companion south pole and vice versa. However, one can conceive of a slender magnet of such great length that when one of the two poles placed at a point in the magnetic field from some other source such as another bar magnet or an electric circuit, the other pole is so far removed from the point in the field that the force on it is negligible.

Permanent magnets produce a magnetic field even when there is no excitation by means of an electric current external to the magnets. In such magnets the flux is produced by currents within the magnets that result from uncompensated electron spins, a feature of ferromagnetic materials such as iron and cobalt.

The magnetic flux that emanates from a magnetic pole having a strength of m units is expressed by

$$\phi = \mu_0 m \quad \text{webers} \tag{3-20}$$

An imaginary magnetic pole of such strength concentrated to a point and isolated in free space would produce a magnetic flux density at a distance of r m of

$$\mathbf{B} = \frac{\mu_0 m}{4\pi r^2} \mathbf{i}_r \quad \text{webers per sq m} \tag{3-21}$$

A comparison of Eq. 3-21 with Eq. 3-15 shows that the magnetic field intensity in terms of unit poles is expressed by

$$\mathbf{H} = \frac{m}{4\pi r^2} \mathbf{i}_r \tag{3-22}$$

The force developed between unit magnetic poles for a given separation is very feeble in comparison with that developed between equal charges of one coulomb at the same separation. From Eq. 2-1 the force developed between two charges of one coulomb each is

$$f_e = \frac{1}{4\pi\epsilon_0 r^2} \tag{3-23}$$

and that developed between two unit magnetic poles at the same separation, according to Eq. 3-19, is expressed by

$$f_m = \frac{\mu_0}{4\pi r^2} \tag{3-24}$$

Division of Eq. 3-23 by Eq. 3-24 yields

$$\frac{f_e}{f_m} = \frac{1}{\mu_0 \epsilon_0} = \frac{1}{4\pi \times 10^{-7} \times 8.854 \times 10^{-12}}$$
$$= 8.99 \times 10^{16} = c^2 \tag{3-25}$$

where $c = 2.998 \times 10^8$ m per sec, which is the velocity of light or the velocity of an electromagnetic propagation in free space.

The concept of the unit magnet pole is useful for analytical exploration of the magnetic field since it brings out in a quantitative manner the fact that the magnetic field is a field of force and is comparable in some respects, although not entirely, to the electric field. Some of the relationships derived for the magnetic field, shown with their counterparts relating to the electric field, are shown in Table 3-1.

TABLE 3-1. MAGNETIC QUANTITIES IN TERMS OF THE UNIT POLE AND ELECTRIC QUANTITIES IN TERMS OF ELECTRIC CHARGE

Quantity	Magnetic		Electric	
	Formula	Unit	Formula	Unit
Field Intensity	$H = \dfrac{m}{4\pi r^2}$	amperes/meter	$E = \dfrac{q}{4\pi\epsilon_0 r^2}$	volts/meter
Flux Density	$B = \dfrac{\mu_0 m}{4\pi r^2}$	webers/square meter	$D = \dfrac{q}{4\pi r^2}$	coulombs/square meter
Flux	$\phi = \mu_0 m$	webers	$\psi = q$	coulombs

3-5 MAGNETOMOTIVE FORCE, MMF

The unit of magnetomotive force is the ampere turn. It was shown in Chapter 2 that the energy required to move a unit test charge from one point to another in an electric field is equal to the voltage or electromotive force between those two points. This relationship is expressed mathematically by Eq. 2-10. An analogous situation exists for the magnetic field in that the energy required to move a unit magnet pole from one point to another along a certain path in the field is proportional to the magnetomotive force between these two points *along the path taken by the unit pole*. It should be carefully noted that the magnetic units of measurement are such that the magnetomotive force or mmf expressed in ampere turns is *proportional* (not equal) to the energy as is the case of the electromotive force in relation to the electric field. Furthermore, the energy required to transport a test charge from one point to another in an electric field is independent of the path along which the test charge is moved. This is another way of saying that the line integral of the electric field intensity between two points in the field is independent of the path over which it is integrated. However, the amount of energy expended in transporting a magnet pole from one point to another in a magnetic field depends upon the path traced by the magnet pole. This is another way of saying that the line integral of magnetic field intensity between two points in the magnetic field depends upon the path of integration. The magnetic field produced by a current in a long straight filament affords a good illustration of the effect of the path of integration in the magnetic field.

Figure 3-7(a) shows the cross section of the filament (a cylindrical conductor) carrying an electric current away from the observer, i.e., into the page. Then, according to the right-hand rule, the direction of the magnetic field intensity resulting from this current is clockwise. Since the strength m of a unit pole is unity, the force required to hold a unit pole at a point where the magnetic field intensity is H can be evaluated from Eqs. 3-19 and 3-22. This force is found to be

$$\mathbf{f} = \mu_0 \mathbf{H} \tag{3-26}$$

The differential energy required to move the unit magnet pole against this force F through a differential distance ds is given by the following scalar product or dot product

$$dW = \mathbf{f} \cdot \mathbf{ds} = \mu_0 \mathbf{H} \cdot \mathbf{ds} \tag{3-27}$$

or simply from Fig. 3-7(a) by

$$dW = f \cos \theta \, ds \tag{3-28}$$

but

$$\cos \theta \, ds = r \, d\beta \tag{3-29}$$

When Eq. 3-29 is substituted in Eq. 3-27 the result is

$$dW = \mu_0 H r \, d\beta \tag{3-30}$$

Sec. 3-5 MAGNETOMOTIVE FORCE, MMF

The magnetic field intensity H at a point well away from the ends of a long straight filament and produced by the current in the filament is expressed by Eq. 3-13 which, upon substitution in Eq. 3-30, yields

$$dW = \frac{\mu_0 I}{2\pi} d\beta \tag{3-31}$$

where I is the current in the filament expressed in amperes.

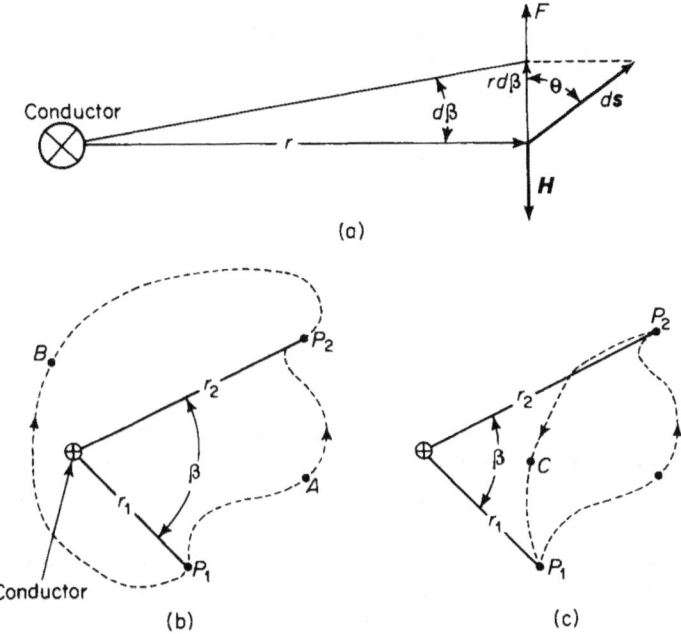

Figure 3-7. (a) Projection of differential distance on force vector in a magnetic field; (b) magnetic path around a straight wire; (c) magnetic path in a region of a straight conductor without linking the conductor.

If in Fig. 3-7(b) the unit magnet pole is carried from P_1 to P_2 along the path through A, the energy expended in carrying the pole from P_1 to P_2 is determined by integrating as follows

$$W = \int_0^\beta dW = \frac{\mu_0 I}{2\pi} \int_0^\beta d\beta = \frac{\mu_0 I \beta}{2\pi}$$

The magnetomotive force expressed in ampere turns is the energy divided by the permeability μ_0, thus

$$F_A = \frac{W}{\mu_0} = \frac{I\beta}{2\pi}$$

If on the other hand the unit magnet pole is carried from P_1 to P_2 in Fig. 3-7(b) along the path through B, the magnetomotive force expressed in amperes turns is evaluated as follows

$$F_B = \frac{1}{\mu_0} \int_0^{-(2\pi-\beta)} d\beta = \frac{I(\beta - 2\pi)}{2\pi}$$

For values of β other than π the magnitudes of F_A and F_B are unequal, but regardless of the value β, the signs of F_A and F_B will be opposite although the integration was carried out by progressing from P_1 to P_2 in both cases.

Suppose that the unit magnet pole is carried completely around the conductor along a closed path in a counterclockwise direction starting at P_1 following the path through A to P_2 then through B back to P_1. Then we have

$$F = \frac{1}{\mu_0} \oint dW = \frac{I}{2\pi} \int_0^{2\pi} d\beta = I \qquad (3\text{-}32)$$

Equation 3-32 shows that energy is required to carry a magnet pole around a closed path that links a current and that the energy is proportional to the current. If the magnet is carried against the direction of the magnetic field intensity the mechanical energy applied to the magnet pole is positive, and if carried in the direction of the magnetic field intensity the energy applied to the magnet pole is negative. Equation 2-16 shows that the energy expended in carrying an electric charge around a closed path in an electric field is always zero. This characterizes a conservative field. The magnetic field is nonconservative since the magnetic pole takes on energy each time it is carried around a closed path that links a current.

If the unit magnet pole is carried around the closed path [Fig. 3-7(c)] that does not link the current-carrying wire, i.e., from P_1 through A to P_2 and back through C to P_1, the mmf and the energy are found to be zero as follows

$$F = \frac{1}{\mu_0} \left[\int_{(P_1)}^{(P_2)} dW + \int_{(P_2)}^{(P_1)} dW \right]$$

$$= \frac{I}{2\pi} \left[\int_0^{\beta} d\beta + \int_{\beta}^{0} d\beta \right]$$

$$= 0 \qquad (3\text{-}33)$$

In Eq. 3-33, integral P_1 and P_2 are shown in parentheses because they are not limits of integration but are merely the end points of the paths of integration.

Although the integrations in Eqs. 3-32 and 3-33 are both taken around closed paths they do not lead to the same result. Equation 3-32 yields the current in the wire and Eq. 3-33 results in a value of zero.

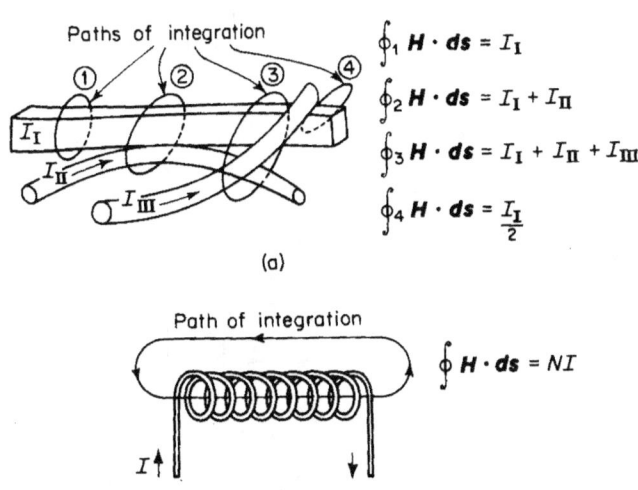

Figure 3-8. Line integral paths of H for (a) 3 conductors carrying different values of current; (b) N-turn solenoid carrying current.

The relationship expressed in Eq. 3-32 is very useful and is more commonly given in terms of the magnetic field intensity H and can be derived directly from Eq. 3-27 with the following result

$$F = \oint \mathbf{H} \cdot \mathbf{ds} = I \tag{3-34}$$

where I is the current linking the closed path of integration. When the closed path of integration is not linked by any current, as shown in Fig. 3-7(c), the value of the line integral $\oint \mathbf{H} \cdot \mathbf{ds}$ is zero.

Although Eq. 3-34 was obtained for a closed path that links a straight conductor, it is valid for any closed path linking one or more conductors of any configuration or even part of a conductor. This equation is considered a general law sometimes called Ampere's circuital law.

Paths of integration and the corresponding values of line integrals of H are shown in Fig. 3-8(a) and (b). In Fig. 3-8(a) path 4 links only one-half of the current in the rectangular conductor, path 1 links all the current in the rectangular conductor without linking any part of the round conductors, and path 3 links all three conductors in Fig. 3-8(a).

3-6 THE TOROID

Figure 3-9(a) shows a toroid with a uniformly distributed winding of N turns carrying a practically constant current of i amp. A toroid may be shaped like a doughnut or it may be in the form of a hollow cylinder such as is shown

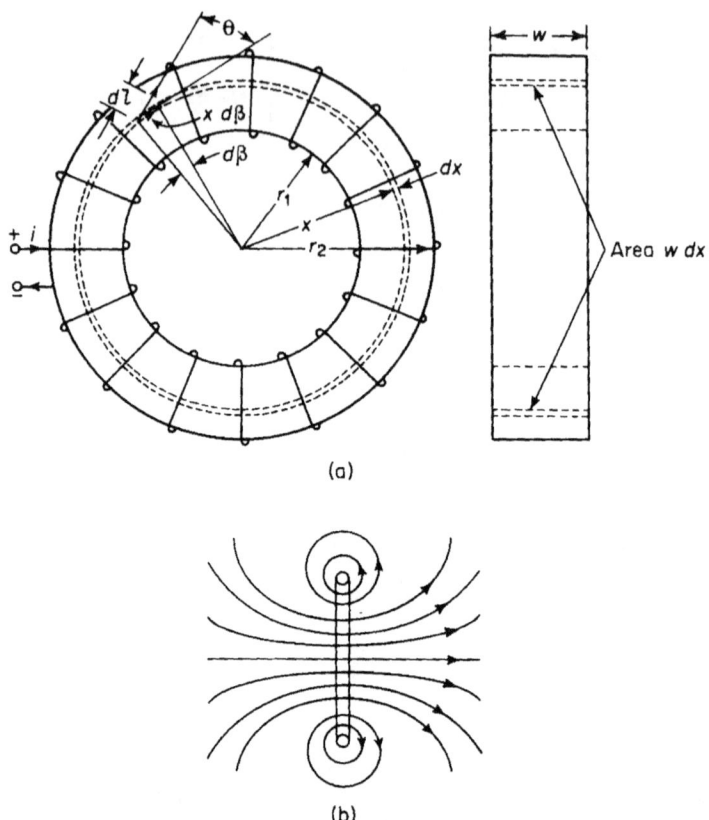

Figure 3-9. (a) Toroid with winding carrying current; (b) magnetic flux lines of a rectangular loop of current.

in Fig. 3-9(a). If the number of turns N is large, the current will produce magnetic lines of flux that are concentric circles confined to the toroid. This is evident from the direction of the flux lines through the plane of a rectangular loop carrying current as illustrated in Fig. 3-9(b). The magnetic flux is in a direction normal to that plane.

Each turn in the winding of Fig. 3-9(a) is such a loop, and if the number of turns or loops is large the flux will be normal to the radius everywhere in the toroid. This means that the flux lines will be concentric circles.

Consider the elemental flux path of thickness dx and radius x in the toroid. The width of this elemental path is the width of the toroid, namely w. Since this path closed upon itself, being a circle, Ampere's circuital law can be applied to express the magnetic field intensity in terms of the radius x. All the turns in the winding link the flux path, hence the total current that links the elemental path is the product of the current i and the number of turns N. Therefore, according to Eq. 3-34, we have

$$F = Ni = \oint \mathbf{H} \cdot \mathbf{dl} \tag{3-35}$$

From Fig. 3-9(a) it is apparent that in the scalar product $\mathbf{H} \cdot \mathbf{dl} = H \cos \theta \, dl$, the product $\cos \theta \, dl$ is the projection of dl on the circle of radius x and is expressed by

$$\cos \theta \, dl = d\beta \tag{3-36}$$

Since the flux lines are concentric circles and if the material in the toroid has constant magnetic permeability, as is the case for free space, air and most nonferrous materials, H is constant everywhere in the circular path of radius x. Under these conditions the magnetic field intensity H in the line integral of Eq. 3-35 is a constant multiplier when coupled with the relationship expressed in Eq. 3-36. As a result Eq. 3-35 can be reduced as follows for this simple circular path

$$F = Ni = H \oint x \, d\beta = Hx \int_{\beta=0}^{\beta=2} d\beta = 2\pi Hx \tag{3-37}$$

where F is the magnetomotive force or mmf expressed in ampere turns. Then from Eq. 3-37 we get

$$H = \frac{Ni}{2\pi x} \quad \text{amp turns per m} \tag{3-38}$$

Substitution of Eq. 3-15 in Eq. 3-38 yields the expression for the magnetic flux density in the circular path of radius x as follows

$$B = \mu_0 \frac{Ni}{2\pi x} \quad \text{webers per sq m} \tag{3-39}$$

The magnetic flux crossing the incremental area $dA = w \, dx$ in Fig. 3-9(a) is expressed by

$$d\phi = \mathbf{B} \cdot \mathbf{dA} = B \cos \theta \, w \, dx \tag{3-40}$$

where θ is the angle between the flux density vector $\mathbf{B}$ and the normal to the vector $\mathbf{dA}$. Since the vector associated with areas is perpendicular to the surface of the area being represented, the angle θ in this case is zero. The vector **B** is tangent to the circle and the surface of the area $w\,dx$ is radial. Then from Eqs. 3-39 and 3-40 we get

$$d\phi = B\,w\,dx = \frac{\mu_0 Niw}{2\pi}\frac{dx}{x}$$

The total flux within the toroid is found by integrating as follows

$$\phi = \int_{x=r_1}^{x=r_2} d\phi = \frac{\mu_0 Niw}{2\pi}\int_{r_1}^{r_2}\frac{dx}{x} = \frac{\mu_0 Niw}{2\pi}\ln\frac{r_2}{r_1} \quad \text{webers} \qquad (3\text{-}41)$$

Materials known as magnetic materials are ferrous materials and certain alloys of metals. They have a magnetic permeability that is much greater than that of free space. In the rationalized MKS system of units the permeability μ is taken as the product $\mu_r\mu_0$ where $\mu_0 = 4\pi \times 10^{-7}$ h per m, the magnetic permeability of free space and $\mu_r = $ the relative permeability of the toroid.

The relative permeability of ferromagnetic materials varies not only with the kind of material but also varies with the flux density in a given material. For example 24 gauge U.S.S. Electrical Sheet Steel has a relative permeability of about 1,300 at a value of H of about 24 amp turns per m, rising to a maximum of 5,800 at a value of H of about 120 amp turns, then decreasing to 1,300 at 960 amp turns per m.

If the toroid consists of a magnetic material that has a uniform relative permeability μ_r the total flux in the toroid is expressed by

$$\phi = \frac{\mu_r\mu_0 Niw}{2\pi}\ln\frac{r_2}{r_1} \quad \text{webers} \qquad (3\text{-}42)$$

In many applications it is sufficient to take H as the total ampere turns divided by the mean length of flux path. The value of B, resulting therefrom, multiplied by the area normal to the mean path is assumed to give the total flux. Thus for the toroid of Fig. 3-9(a) this approximation yields

$$H = \frac{NI}{2\pi(r_2 + r_1)/2}$$

and $A = (r_2 - r_1)w$, hence

$$\phi = \frac{\mu NI}{\pi}\,w\,\frac{r_2 - r_1}{r_2 + r_1} \quad \text{webers} \qquad (3\text{-}42a)$$

In general, if the magnetic circuit has a uniform cross-sectional area A normal to the direction of the magnetic flux and if the mean length of the

Sec. 3-7 THE MAGNETIC CIRCUIT AND THE ELECTRIC CIRCUIT

flux path is l, the steady flux or slowly varying flux can be expressed approximately as follows

$$\phi = NI\frac{\mu A}{l} \quad \text{webers} \tag{3-43}$$

3-7 COMPARISON OF THE MAGNETIC CIRCUIT WITH THE ELECTRIC CIRCUIT

The steady electric current in a simple electric circuit comprised of a conductor of length l and uniform cross-sectional area A is expressed by

$$I = V\frac{A}{\rho l} \quad \text{amp} \tag{3-44}$$

where $V =$ the voltage drop over the length l and $\rho =$ the resistivity of the conductor.

Equation 3-44 can be expressed in accordance with Ohm's law as

$$I = \frac{V}{R} \tag{3-45}$$

where $R = \rho l/A$, the electrical resistance expressed in ohms. Similarly Eq. 3-43 can be reduced to

$$\phi = \frac{F}{\mathcal{R}} \quad \text{webers} \tag{3-46}$$

where $F =$ mmf in ampere turns NI and $\mathcal{R} = l/\mu A$, the magnetic reluctance.

A comparison of Eq. 3-45 with Eq. 3-46 shows that the flux ϕ in the magnetic circuit corresponds to the current I in the electric circuit and that the mmf F or NI in the magnetic circuit corresponds to the emf V in the electric circuit. A similar comparison holds for the magnetic reluctance and the electrical resistance R.

There is a marked difference between the magnetic circuit and the electric circuit in an important respect. The current in an electric circuit can be confined very effectively to the desired path by insulating the conducting parts from each other. The conductivity of a good insulating material is of the order of 10^{-20}, that of a good electrical conducting material. This means that an electric current can be confined to the desired path with very little leakage. The magnetic circuit cannot be isolated nearly as effectively because the most practical magnetic insulator is air. There are no materials that have a substantially greater reluctivity than air. The permeability of air is generally greater than 10^{-4} and frequently only about 10^{-2} times that of the ferromagnetic material to which the flux is to be confined. Magnetic circuits generally have appreciable leakage that may become very pronounced if the magnetic material is interrupted by an air gap such that the major component of the flux crosses the gap. An air gap has relatively high reluctance and may have several times the reluctance of the remainder of the magnetic circuit although the length of the air gap may be quite short compared with the total circuit. An air gap is usually paralleled by surrounding air spaces. The flux divides between the air gap and the parallel air spaces, giving rise to relative high leakage, depending upon the length of the air gap.

3-8 OTHER COMMON SYSTEMS OF MAGNETIC UNITS

There are two other systems of magnetic units that are still in use. One of these is the cgs electromagnetic system, which has been used widely in scientific work for a number of years. In addition, there is the Mixed English System of Units, which has also been used for many years and is still used to a considerable extent in this country.

In the cgs system, the unit of flux density is the gauss or one maxwell per square centimeter. One maxwell is 10^{-8} weber. In the same system, the unit of magnetic field intensity is the oersted, which is $4\pi/10$ amp turns per cm. This choice of units results in a value of permeability for free space of unity. The unit of magnetomotive force is the gilbert and is equal to $0.4\pi NI$.

In the Mixed English System the unit of flux density is the maxwell per square inch and the unit of magnetic field intensity is the ampere turn per inch.

Actually the rationalized MKS system of units is one that has been accepted, though not universally, only in recent years.

Conversion factors are given in Table 3-2.

TABLE 3-2

Multiply	by	to obtain
F in ampere turns	0.4π	F in gilberts
H in ampere turns/inch	$\dfrac{0.4\pi}{2.54}$	H in oersteds
B in lines/square inch	$\dfrac{10^{-4}}{6.45}$	B in webers/square meter
B in lines/square inch	$\dfrac{1}{6.45}$	B in gausses

NOTE: 1 pragilbert = 1,000 gilberts
1 praoersted = 1,000 oersteds

3-9 MAGNETIC MATERIALS

In the atoms of nonmagnetic materials, the magnetic effect of electron spin in one direction is completely offset by equal electron spin in the opposite direction. Strictly speaking, all materials are magnetic in the sense that they can be magnetized. Materials that have a relative magnetic permeability μ_r not appreciably greater than unity are considered nonmagnetic. Similarly,

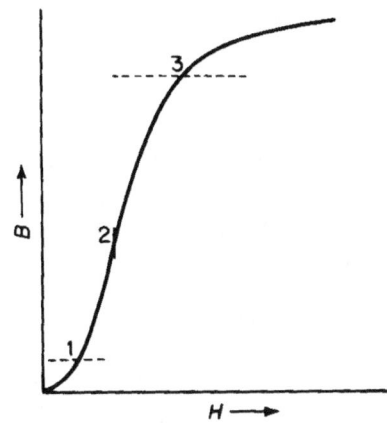

Figure 3-10. Magnetization curve showing three regions of domain behavior.

the composite effects of the orbital motions of charges cancel each other and there is no magnetic effect. However, in the case of magnetic materials, more properly called ferromagnetic materials, compensation of electron spin is not complete. As a result, tiny, completely magnetized regions called *domains* exist in the crystals of such materials. When ferromagnetic materials are in an unmagnetized state, the orientation of the domains is random and the magnetic moment of the whole crystal is zero. When subjected to low values of magnetic field intensity the domains undergo a boundary displacement.* As the magnetic field intensity is increased there is a sudden orientation of the domains toward the direction of the applied magnetic field. Further increases in magnetic field intensity are accompanied by a slower orientation of the domains. These three regions are shown in Fig. 3-10.

Magnetic materials form an indispensable part of electric generators, power transformers, audiotransformers, telephone receivers, relays, loud speakers, magnetic recorders, and numerous other devices. Magnetic materials include certain forms of iron and its alloys in combination with

---

* Richard M. Bozorth, *Ferromagnetism*. Princeton, N.J.: D. Van Nostrand Co., Inc., 1951.

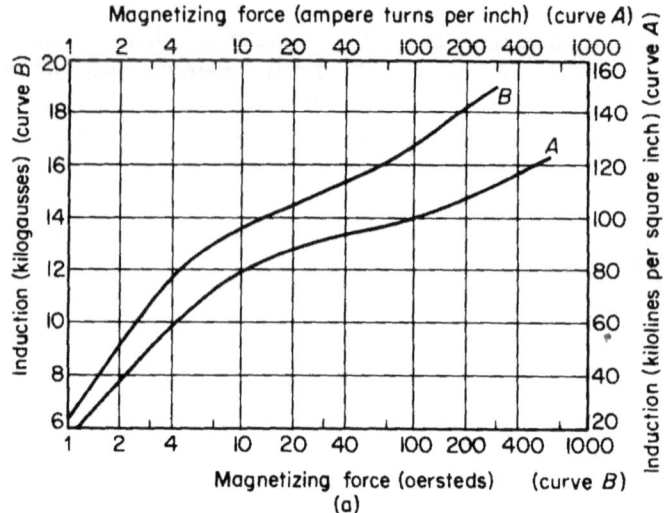

(a)

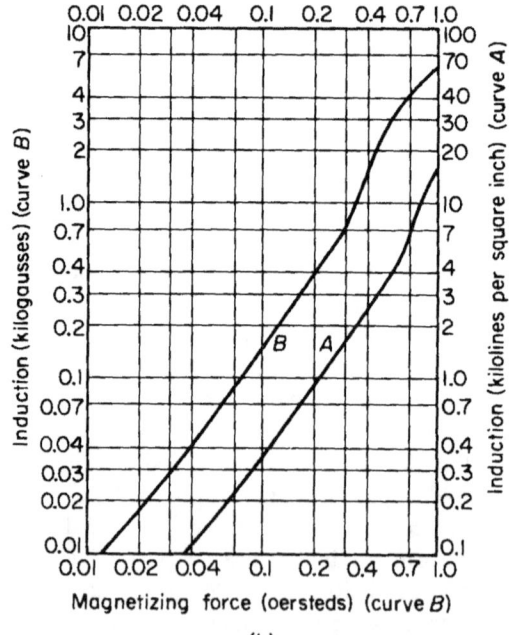

(b)

Figure 3-11. d-c magnetization curves U.S.S. Transformer 72, 29 Gage (a) nominal induction, (b) low induction. (Courtesy of United States Steel Corporation.)

cobalt, nickel, aluminum, and tungstens and are known as ferromagnetic materials. Such materials are easy to magnetize and if these were to be

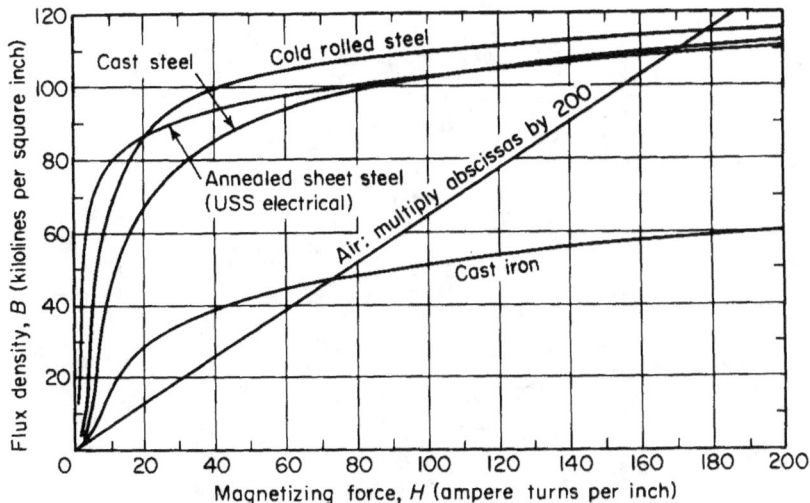

Figure 3-12. Normal magnetization curves for common magnetic materials.

replaced by nonmagnetic materials, the size of most electromagnetic apparatus would become prohibitive if it did indeed permit functioning.

The magnetic permeability of ferromagnetic materials is not constant. At low values of flux density the permeability is low and as the flux density increases, the permeability reaches a maximum until magnetic saturation is approached, upon which the permeability falls off with further increases in flux density. This is illustrated by the curves in Figs. 3-11 and 3-12. In addition the permeability is generally different for increasing flux density than for decreasing flux density at the same value of magnetic field intensity, giving rise to a hysteresis loop when the magnetization is carried through a complete cycle. Figure 3-13 shows a hysteresis loop.

Ferrite cores are molded from a mixture of metallic oxide powders and are used in many high-frequency operations. Their physical properties and manufacturing processes are similar to those of ceramic materials; their d-c

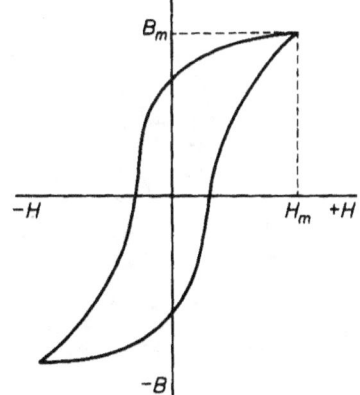

Figure 3-13. Magnetic hysteresis loop.

Figure 3-14. Shapes of steel laminations used in magnetic circuits. (Courtesy of United States Steel Corporation.)

resistivities, however, are more than 10^6 times greater than ceramics, a property which makes them particularly suitable for high-frequency applications. Although such materials have relative permeabilities as high as 5,000 their use is somewhat limited because they become saturated at less than one-half the flux densities at which annealed sheet steel saturates.

3-10 CALCULATION OF MAGNETIC CIRCUITS WITHOUT AIR GAPS

Magnetic circuits are constructed in a variety of shapes. In a-c equipment and in devices in which rapid response is required the magnetic core is usually built up of laminations having coated surfaces to provide interlaminar resistance. If the magnetic structure in a-c magnets were solid, large objectionable currents, which lead to excessive losses and heating, would be induced in the iron. In such d-c magnets where rapid response is required the currents induced in the magnetic cores would, in accordance with Lenz's Law, oppose changes in the magnetic flux and thus retard the building up of the flux. A few typical shapes of laminations used in motors, generators, relays, and transformers are shown in Fig. 3-14. Some of these shapes are somewhat complex and the magnetic calculations involved are far from simple since the flux density is not uniform throughout the structure. Thus, some parts of the structure may be highly saturated while others carry only moderate flux densities. In addition there is fringing of flux at the sides of air gaps as well as flux leakage across air spaces. Figure 3-15 shows small magnetic cores built by Arnold Engineering. Several thousand of the smallest of these may be used in a single computer or data-processing machine. Such cores are also applicable to high-frequency magnetic amplifiers where high gain is needed.

EXAMPLE 3-1: Consider the simple core shown in Fig. 3-16 comprised of U.S.S. Transformer 72, 29 (0.0140") Gage Steel. Determine the current necessary to produce a flux of 25,000 maxwells (lines) in the core.

Solution:

Gross area of core $= \frac{1}{2} \times \frac{1}{2} = 0.25$ sq in.

Stacking factor to correct for nonmagnetic coating on laminations $= 0.95$

Net area of core $= 0.25 \times 0.95$

Flux density $B = \dfrac{\phi}{A} = \dfrac{25{,}000}{0.25 \times 0.95} = 105{,}300$ lines sq in.

Mean length of flux path $= 3 + 2\frac{1}{2} + 2 + 1\frac{1}{2}$
$= 9.0$ in.

112   MAGNETIC CIRCUITS   Chap. 3

Figure 3-15. Collection of magnetic cores. (Courtesy of The Arnold Engineering Co.)

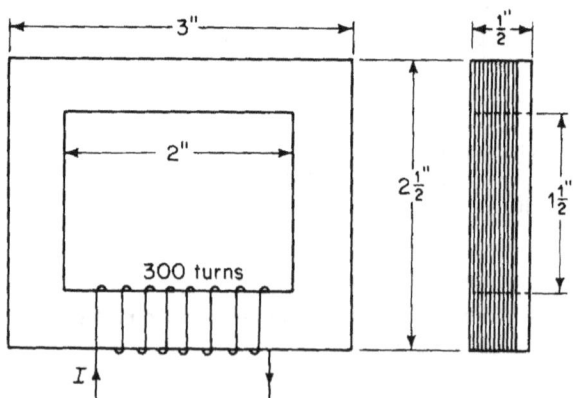

Figure 3-16. Laminated electromagnet.

Sec. 3-10 MAGNETIC CIRCUITS WITHOUT AIR GAPS 113

From Fig. 3-12

$H = 124$ amp turns per in., magnetic field intensity

$F = NI = Hl = 124 \times 9.0 = 1{,}116$ amp turns, mmf

$$I = \frac{Hl}{N} = \frac{1{,}116}{300} = 3.72 \text{ amp}$$

The shape of the core shown in Fig. 3-17 is that used in the core-type transformer. It is also used in certain kinds of reactors or chokes. In the

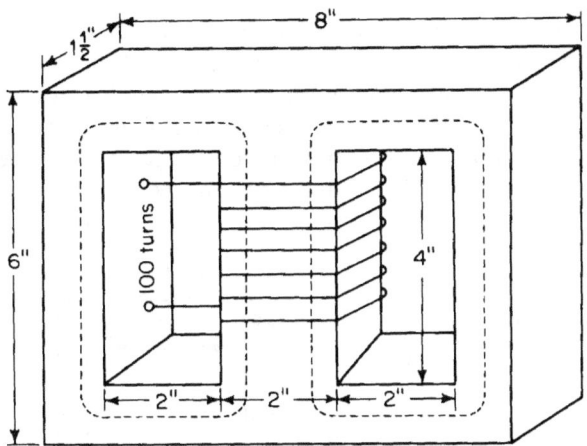

Figure 3-17. Three-legged core and winding.

shell-type transformer and in some reactors the shape of the core is as shown in Fig. 3-17. Reactors are used in circuits as current-limiting devices and to smooth out ripples in direct current.

EXAMPLE 3-2: Figure 3-17 shows a three-legged core comprised of U.S.S. Transformer 72, 29 Gage Steel. The winding has 100 turns and carries a current of 0.64 amp. Determine the flux in the center leg and in each outer leg. Assume a stacking factor of 0.95 for the core.

Solution: Mean length of flux path as indicated by the broken lines in Fig. 3-17 is 16 in., and the magnetizing force or magnetic field intensity is

$$H = \frac{NI}{l} = \frac{100 \times 0.64}{16} = 4.0 \text{ amp turns per in.}$$

From Fig. 3-12 the flux density corresponding to a magnetizing force of 4.0 amp turns per in. is 60,000 lines per sq in. Hence $B = 60{,}000$

Gross area of center leg of core $= 1\frac{1}{2} \times 2 = 3.00$ sq in.

Then if the stacking factor is 0.95

    Net area of center leg of core = 3.00 × 0.95 = 2.85 sq in.

Since the outer legs are one-half as wide as the center leg

    Net area of each outer leg = 1.50 × 0.95 = 1.425 sq in.

    Flux in center leg = BA

                            = 60,000 × 2.85 = 171,000 lines

    Flux in each outer leg = 60,000 × 1.425 = 85,500 lines

3-11 MAGNETIC LEAKAGE

Figure 3-18(a) shows a laminated steel core with an air gap of length g. If there were no magnetic leakage the magnetic flux in all parts of the iron in the air gap would have the same value. The broken line in Fig. 3-18 indicates the mean path of the flux when there is no leakage. The leakage may be kept low by placing the exciting winding on the leg of the core that contains the air gap as shown in Fig. 3-18(b) and by keeping the flux density at

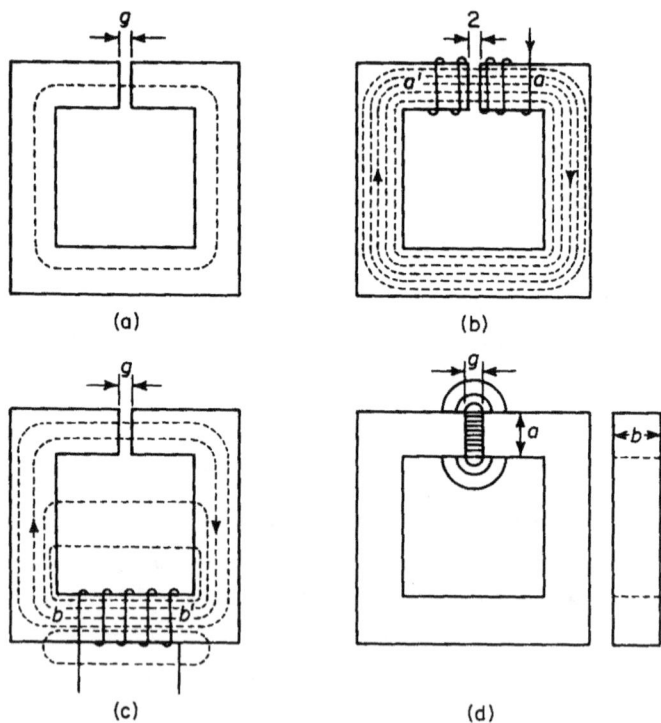

Figure 3-18. (a) Ferromagnetic core with air gap; (b) exciting winding surrounding leg with air gap; (c) exciting winding on leg not containing air gap; (d) fringing at air gap. Note: The effect of fringing is not shown in (a), (b), and (c).

Sec. 3-12 CORRECTION FOR FRINGING AT SHORT AIR GAPS

a low enough value so that the iron does not saturate. However, if the exciting winding is placed on a leg of the core that does not contain the air gap as shown in Fig. 3-18(c) for example, then appreciable leakage flux will result. In Fig. 3-18(b) the mmf required to maintain the flux along the path from a to a' in a clockwise direction is small and if the core had infinite permeability this mmf would be zero. As a result there would be no mmf to drive flux across the window in the core. If on the other hand the exciting winding is placed on the leg of the core opposite the leg containing the air gap as shown in Fig. 3-18(c), the mmf required to maintain the flux along the path from b to b' would be substantial because it would need to be sufficient to overcome the reluctance of the air gap, a condition that would exist even if the iron had infinite permeability. This mmf acts upon the air spaces that parallel the air gap and produces flux in these parallel air spaces, as for example the window in the core of Fig. 3-18(c). Although the electromagnets of Figs. 3-18(b) and 3-18(c) are identical the magnetic leakage in Fig. 3-18(b) is negligible, whereas that in Fig. 3-18(c) is appreciable even if the exciting ampere turns in both cases are the same.

There are arrangements, however, in which it is not possible to have the exciting winding cover the air gap. Here the magnetic leakage may be appreciable. Appreciable leakage fluxes may also be present in magnetic structures with and without air gaps when the iron is saturated magnetically. In precise magnetic calculations of such structures the effect of leakage must be taken into account. Such calculations are seldom straightforward and they are therefore not within the scope of this textbook.*

3-12 CORRECTION FOR FRINGING AT SHORT AIR GAPS

The magnetic field intensity across an air gap is much greater than that across an equivalent length of iron, particularly below the region of saturation. This relatively high magnetic field intensity drives appreciable flux through the air near the sides of the air gap. This effect is known as fringing and in the case of short air gaps has the effect of increasing the effective area of the air gap. Therefore an empirical correction is made. Figure 3-18(d) shows a reactor core with an air gap of length g. The cross-sectional dimensions of the core are a and b. The corrected area, when fringing is taken into account, is found by adding the length of the gap to each dimension. Thus

$$A = (a + g)(b + g)$$

---

* Methods of taking magnetic leakage fluxes into account are shown in Herbert C. Roter, *Electromagnetic Devices*. New York: John Wiley & Sons Inc., 1941.

3-13 IRON AND AIR

Many magnetic structures include one or more air gaps. There are air gaps between the rotor and stator of motors and generators. Relays depend upon a change in the length of air gap as do solenoids of the plunger and clapper types. In such circuits the exciting winding must develop sufficient mmf to overcome not only the reluctance of the iron but also that of the air gap.

EXAMPLE 3-3: Consider the simple magnetic circuit shown in Fig. 3-19 which consists of a laminated core of U.S.S. Transformer 72, 29 Gage Steel with an air gap g and an exciting winding of 300 turns. If the air gap

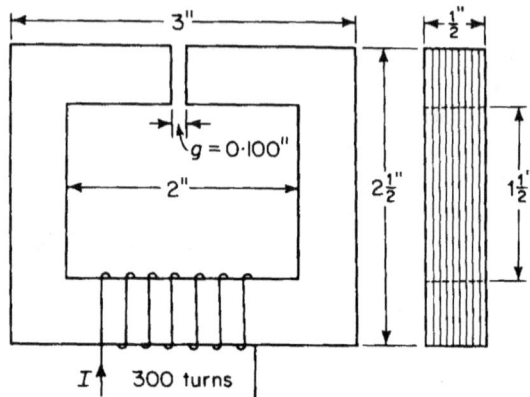

Figure 3-19. Laminated electromagnet with air gap.

has a length $g = 0.100$ in., determine the current to produce a flux of 20,000 lines in the core. Neglect leakage and fringing. Assume a stacking factor of 0.95 for the core.

Solution:

    Net area of core $= 0.50 \times 0.50 \times 0.95 = 0.238$ sq in.
    Length of iron circuit $= 2(2\frac{1}{2} + 2) = 9.00$ in.

$$B = \frac{\phi}{A} = \frac{20,000}{0.238} = 84,200 \text{ lines per sq in.}$$

    H (for iron, from Fig. 3-11) $= 18$ amp turns per in.
The mmf for the iron is

$$F_{\text{iron}} = H_{\text{iron}} l_{\text{iron}} = NI_{\text{iron}}$$

$$NI_{\text{iron}} = 18 \times 9 = 162 \text{ amp turns}$$

    Length of air gap $= 0.10$ in.

The mmf for air is

$$B_{\text{air}} = 3.19 \frac{NI_{\text{air}}}{l_{\text{air}}}$$

$$B_{\text{air}} = \frac{\phi}{A_{\text{air}}}$$

The area of the air gap is $0.50 \times 0.50 = 0.25$ sq in.
It should be noted that the stacking factor applies only to laminated iron and not to the air gap.

$$B_{\text{air}} = \frac{20{,}000}{0.25} = 80{,}000 \text{ lines per sq in.}$$

$$80{,}000 = 3.19 \frac{NI_{\text{air}}}{0.10}$$

$$NI_{\text{air}} = \frac{8{,}000}{3.19} = 2{,}510 \text{ amp turns}$$

The air gap is in series with the iron as the same amount of flux traverses both, hence the total mmf is

$$NI_t = NI_{\text{iron}} + NI_{\text{air}}$$
$$= 162 + 2{,}510 = 2{,}672 \text{ amp turns}$$
$$I = 2{,}672 \div 300 = 8.91 \text{ amp}$$

Air gaps are undesirable in magnetic circuits where low mmf are required for given values of flux. This is true of power transformers where a low value of exciting current is desirable. However, the cost of keeping air gaps short or entirely eliminated must not be so great as to outweigh other factors such as cost and difficulty of construction. Air gaps can be eliminated by using punchings in the form of hollow discs, as for example in the toroidal core of Fig. 3-9(a). This construction is practical for small cores, but the cost of dies becomes prohibitive beyond a certain limited size. Figure 3-15 shows a group of small magnetic cores in toroidal form. A construction frequently used for reasons of low cost has the core made up of E and I laminations as shown in Fig. 3-20.

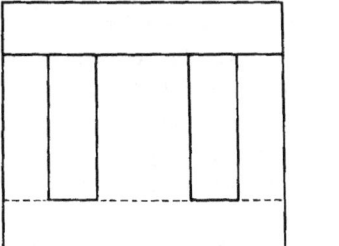

Figure 3-20. Core comprised of E and I laminations with overlapping butt joints.

The least laborious method for assembling such a core is to stack all the I laminations together to the desired thickness and all the E laminations together and then pushing the two stacks together. This, however, produces three butt joints where the I stack meets the three legs of the E stack and a small air gap results at each joint. The effect of these joints can be minimized by going to lap joints and assembling the I pieces one at a time with the E pieces in such a fashion that for successive layers the position of the I pieces alternates between top position and bottom position, and the corresponding E pieces between bottom position and top position. Thus, the gap in each layer is bridged by a solid piece of lamination on each side of the gap, as shown in Fig. 3-20. Lap joints can also be achieved by using L-shaped laminations for making up a two-legged core. Some two-legged cores are constructed from straight I-shaped laminations and using lap joints.

There are applications, however, in which an air gap is desired in a core to give a smaller variation of inductance with current or to prevent saturation of transformer cores. An air gap is also desired in the cores of chokes when direct current is present.

3-14 GRAPHICAL SOLUTION FOR SIMPLE MAGNETIC CIRCUIT WITH SHORT AIR GAP

In simple structures with an air gap, such as that shown in Fig. 3-19, it is a relatively easy matter to determine the mmf for a given value of magnetic flux when leakage can be neglected. For the condition of negligible leakage the flux in the air gap is the same as that in the iron. It is therefore necessary only to compute the mmf for the iron for the given value of flux and the mmf for the air gap at the same value of flux. The sum of these mmf gives the total. However, in the case of a given magnetic structure of iron and air, if the total mmf is given and it is required to determine the magnetic flux, the procedure is not quite as straightforward, since the characteristic of the iron is nonlinear. Nevertheless, such calculations can be facilitated by graphical methods. In one of these the flux vs magnetomotive characteristic is plotted for the iron and the negative air-gap line of flux vs mmf is plotted on the same graph. This is illustrated in Example 3-4.

> EXAMPLE 3-4: A magnetic structure similar to that shown in Fig. 3-19 is comprised of a steel core with a 0.050-in. air gap. The core is made of laminations 0.014 in. thick. The characteristics of the iron are shown in Fig. 3-11. The core is stacked to a thickness of $b = 1.25$ in., a width of $a = 1.00$ in., and the mean length of path $= 12$ in. There are 600 turns of #14 A.W.G. copper wire; the current is 2.5 amp. Assume a stacking factor of 0.90 for the steel and determine the flux.

Sec. 3-14 SIMPLE MAGNETIC CIRCUIT WITH SHORT AIR GAP

Solution: The flux vs mmf (ϕ vs Ni) characteristic is plotted for the iron in Fig. 3-21 on the basis of the magnetization curve of Fig. 3-12 and
Mean length of iron = 12.0 in.
Ampere turns for iron = $12.0 \, H_{\text{iron}}$
Net area of iron = $(1.00 \times 1.25 \times 0.90) = 1.125$ sq in.
Flux in iron = $1.125 \, B_{\text{iron}}$

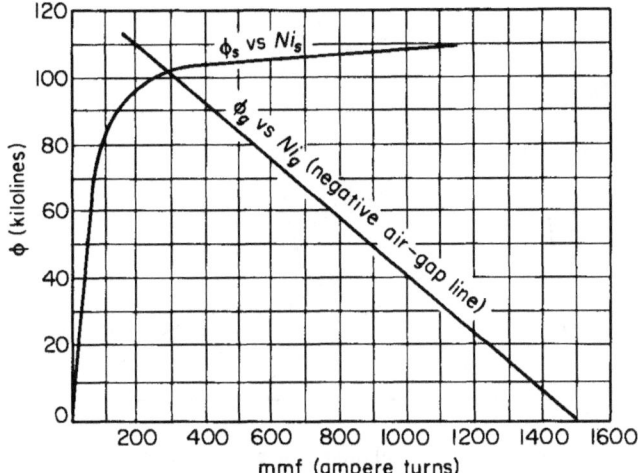

Figure 3-21. Construction for graphical solution of steel core with air gap.

Area of air gap corrected for fringing

$$A_g = (1.00 + 0.05) \times (1.25 + 0.05) = 1.365 \text{ sq in.}$$

$$B_g = \frac{3.19 \, Ni_g}{l_g} = \frac{\phi}{A_g} \quad \text{or} \quad Ni_g = \frac{\phi l_g}{3.19 \, A_g}$$

Let $Ni_g = 1,000$ amp turns, then

$$\phi = \frac{Ni_g \times 3.19 \, A_g}{l_g} = \frac{1,000 \times 3.19 \times 1.365}{0.05} = 87,000 \text{ lines}$$

Two points are now available to fix the *negative* air-gap line in Fig. 3-21. The first point is selected at $Ni = 1,500$ (the total mmf) and $\phi = 0$, whereas the second has been established at $Ni = 1,500 - 1,000 = 500$ and $\phi = 87,000$. The other point at $Ni = 500$ and $\phi = 87,000$. The air-gap line in passing through these two points intersects the iron curve at a flux of 101,000 lines.

The mmf for the iron then is 330 amp turns; for the air it is 1,500 − 330 or 1,170 amp turns.

The flux for other values of total mmf can be obtained simply by shifting the air-gap line parallel to itself so that it intersects the desired value of mmf on the abscissa as shown in Fig. 3-22.

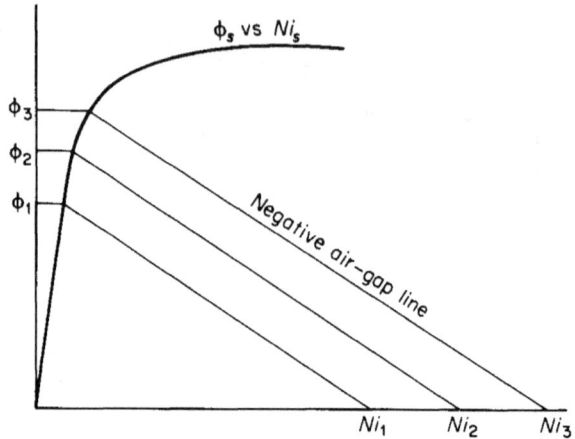

Figure 3-22. Construction for graphical solution of steel core with air gap and various mmfs.

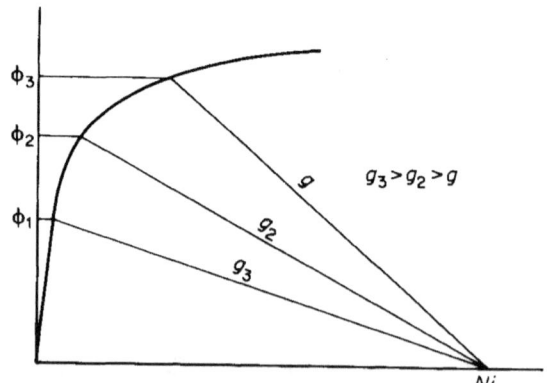

Figure 3-23. Construction for graphical solution of steel core with constant mmf for various air gap lengths.

On the other hand, if it is desired to determine the flux for a given value of mmf but for various lengths of air gap, the different air-gap lines are drawn from the intersection of the mmf on the abscissa as shown in Fig. 3-23.

Figures 3-22 and 3-23 apply only to fairly simple structures. As magnetic structures become more complex the calculations accordingly become more involved.*

3-15 FLUX LINKAGES

It was stated in Section 3-3 that magnetic lines of flux close upon themselves. As a result it is possible for magnetic flux lines to link electric circuits in much the same manner as flux lines link the windings on the cores in Figs. 3-16 and 3-17 and the manner in which the flux in the toroid of Fig. 3-9(a) links the turns in the winding. Another example of flux linking a circuit is the turn shown in Fig. 3-9(b).

In the case of a winding having N turns and the same value of flux ϕ links all the turns, the flux linkages are expressed by

$$\lambda = N\phi \quad \text{weber turns} \tag{3-47}$$

There are many cases, however, in which the different turns of a winding serve to link different amounts of flux. Electric motors and generators usually have windings in which the turns are distributed among slots, an arrangement which makes for differing amounts of flux linking the turns. Figure 3-24 shows the stator of a 3-phase synchronous motor. The coil ends of the stator winding are visible on the right-hand side of Fig. 3-24.

Consider a winding in which the fluxes $\phi_1, \phi_2, \phi_3, \ldots, \phi_n$ link $N_1, N_2, N_3, \ldots, N_n$ turns respectively where N_1, N_2, N_3, etc. may each be one or more turns, then from Eq. 3-54 we have

$$\lambda_1 = N_1\phi, \lambda_2 = N_2\phi_2, \lambda_3 = N_3\phi_3 \cdots \lambda_n = N_n\phi_n$$

and the total flux linkages will be

$$\lambda = \lambda_1 + \lambda_2 + \lambda_3 + \cdots + \lambda_n$$

or

$$\lambda = \sum_{i=1}^{i=n} N_i \phi_i \tag{3-48}$$

---

* Examples of magnetic calculations in more complex circuits are shown in: MIT Staff, *Magnetic Circuits and Transformers*. New York: John Wiley & Sons, Inc., 1943; Herbert C. Roters, *Electromagnetic Devices*. New York: John Wiley & Sons, Inc., 1941; and Alexander S. Langsdorf, *Principles of Direct-Current Machines*, 5th Ed. New York: McGraw-Hill Book Company, 1940.

Figure 3-24. Stator of a 3-phase synchronous motor. (Courtesy of Westinghouse Electric Corporation.)

3-16 INDUCED emf. LENZ'S LAW

Lenz's law states that whenever the magnetic flux linking an electric circuit undergoes a change an emf is induced in that circuit which is proportional to the time rate at which the flux changes, i.e.

$$e = \frac{d\lambda}{dt} \tag{3-49}$$

The polarity of this induced voltage is such that if the circuit were closed a current would result and flow in a direction that would oppose any *change* in the flux.

For example, if the current in the toroid of Fig. 3-9(a) is in the direction shown and is increasing, then the flux would be increasing also. According to Lenz's law the induced emf would be in a direction such as to prevent the flux from increasing, and if the induced emf were acting by itself it would *produce* a current in a direction opposite to that shown in Fig. 3-9(a). This would make the upper terminal positive as marked.

Sec. 3-17 ENERGY STORED IN MAGNETIC CIRCUITS

If we were to pass a current i through the winding on the toroid of Fig. 3-9(a) the applied voltage v must be sufficient not only to overcome the resistance drop of the winding but also to overcome the induced emf, and we would have

$$v = Ri + e = Ri + \frac{d\lambda}{dt} \qquad (3\text{-}50)$$

where R is the resistance of the winding.

In the case of the toroid, if the winding has a large number of turns and the turns are uniformly distributed, then all the turns are linked by the same value of flux, ϕ, and Eq. 3-50 may be written as

$$v = Ri + N\frac{d\phi}{dt} \qquad (3\text{-}51)$$

where N is the number of turns in the winding.

3-17 ENERGY STORED IN MAGNETIC CIRCUITS

Several examples of energy storage were discussed in Chapter 1. One of these is the R-L circuit for which it was shown that, in building up a current in such a circuit, energy equal to $Li^2/2$ is stored in the inductance. Self-inductance is a property of magnetic circuits and the energy stored in a constant self-inductance is the energy delivered to the magnetic field of the circuit.

The power input to the winding on the toroid is obtained by multiplying both sides of Eq. 3-51 by the current i, and we have

$$p = vi = Ri^2 + Ni\frac{d\phi}{dt} \qquad (3\text{-}52)$$

where Ri^2 = power converted into heat

$Ni\dfrac{d\phi}{dt}$ = power that stores energy in the field in the core

In problems dealing with magnetic forces and core losses, i.e., hysteresis and eddy-current losses, it is usually convenient to determine these quantities in terms of the flux density B.

If in the toroid of Fig. 3-9(a), the ratio of R_2/R_1 does not differ too much from unity, i.e., if the radial thickness of the toroid is small in comparison with the mean radius $(R_1 + R_2)/2$ then the flux density B is nearly uniform over the cross section of the core. Then

$$\phi = AB \quad \text{webers} \qquad (3\text{-}53)$$

where A is the cross-sectional area of the core. Also

$$H = \frac{Ni}{l} \quad \text{amp turns per m} \qquad (3\text{-}54)$$

or

$$Ni = Hl \quad \text{amp turns} \qquad (3\text{-}55)$$

where $l = \pi(R_1 + R_2)/2$, the mean length of the core. The power that stores energy is expressed by

$$p_\phi = Ni \frac{d\phi}{dt} \qquad (3\text{-}56)$$

and from Eq. 3-53

$$\frac{d\phi}{dt} = A \frac{dB}{dt} \qquad (3\text{-}57)$$

From Eqs. 3-55, 3-56, and 3-57, there results

$$p_\phi = HlA \frac{dB}{dt} = \text{Vol } H \frac{dB}{dt} \qquad (3\text{-}58)$$

where Vol = volume of the core in cubic meters.

The energy input into the magnetic field during the differential time interval dt is

$$dW_\phi = p_\phi \, dt = \text{Vol } H \, dB \qquad (3\text{-}59)$$

The energy put into the field as the flux density is changed from a value B_1 to any other value, B_2 is

$$W_\phi = \text{Vol} \int_{B_1}^{B_2} H \, dB \qquad (3\text{-}60)$$

If the relationship between B and H is linear, i.e., μ is constant

$$H = \frac{B}{\mu}$$

and Eq. 3-60 becomes

$$W_\phi = \frac{\text{Vol}}{\mu} \int_{B_1}^{B_2} B \, dB = \text{Vol} \frac{(B_2^2 - B_1^2)}{2\mu} \tag{3-61}$$

If the flux density B is increased from zero the energy stored in the field is

$$W_\phi = \text{Vol} \frac{B^2}{2\mu} \tag{3-62}$$

Equation 3-62 can also be expressed as

$$W_\phi = \text{Vol} \, \mu \frac{H^2}{2} \tag{3-63}$$

or

$$W_\phi = \text{Vol} \frac{BH}{2} \tag{3-64}$$

It should be noted that *if H is expressed in the ampere turns per unit length in any system and B is in webers per unit area in the same system, and if the volume is calculated in the same units, Eq. 3-64 expresses the energy in joules for that volume.* In Eq. 3-64 the quantity $BH/2$ is known as the magnetic energy density.

In nonlinear magnetic circuits, i.e., those in which the relative permeability μ_r is not a constant, the simple relationship for the energy stored in the field expressed by Eq. 3-64 is not valid. For the nonlinear case the energy is expressed by

$$W_\phi = \text{Vol} \int_0^B H \, dB \tag{3-65}$$

Since it is not practical in most cases to express B as an analytic function of H in nonlinear magnetic circuits the integral in Eq. 3-65 is evaluated by graphical means as discussed in Section 3-25 on magnetic hysteresis.

From Eqs. 3-61 and 3-62 it is evident that the lower the value of the permeability μ, the greater is the energy stored in the field for a given value of B. Thus, in a magnetic structure with an air gap, even if quite short, the energy stored in the air gap may be quite large compared with that stored in the iron. Thus, consider the magnetic structure of Example 3-3, in which the mean length of path in iron is 9 in. and the length of air gap is only 0.100 in. From Fig. 3-12 it is found that at a flux density of 80,000 lines per sq in. the relative permeability is approximately 2,200. If fringing is neglected the volume of the air gap is proportional to its length and the energy stored in the gap is

$$W_g \sim \frac{0.100}{1}$$

whereas the energy stored in the iron is

$$W_s \sim \frac{9}{2,200}$$

and the ratio of the energy stored in air to that in iron is

$$\frac{W_g}{W_s} = \frac{0.100 \times 2,200}{9} = 24.4$$

Although the air gap has a volume of only 1.1 percent that of the iron, it stores 24.4 times the energy stored in the iron. However, as the iron becomes saturated its permeability decreases and the ratio of the energy in the air gap to that in the iron also decreases.

3-18 MAGNETIC FORCE IN TERMS OF FLUX DENSITY

The magnetic flux that crosses the air gap in a structure of magnetic material produces a force of attraction between the faces of the air gap. This force,

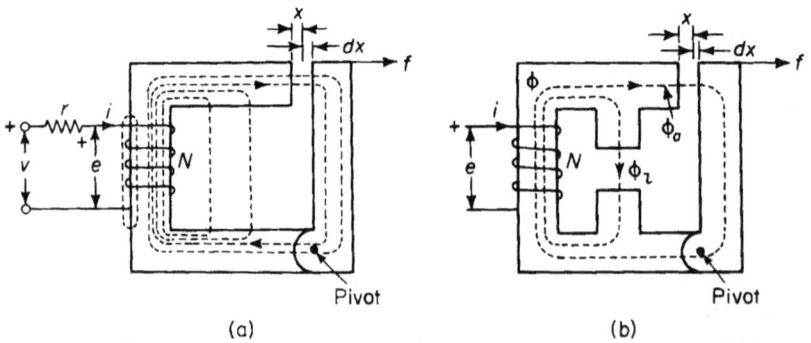

Figure 3-25. Electromagnet (a) approximate flux path; (b) simplified flux path.

which can be exerted only on materials with a relative permeability μ_r greater than unity, is a function of the magnetic flux density B_a in the air gap as shown in the following derivation.

A simple iron structure with an air gap of length x, and with plane parallel faces is shown in Fig. 3-25(a). Fringing at the air gap can be

Sec. 3-18   MAGNETIC FORCE IN TERMS OF FLUX DENSITY   127

neglected for short air gaps. A pivot, shown in the bottom of the structure, makes it possible to vary the length of the air gap.

Assume a current of i amp, in the N turns of the exciting winding, producing a magnetic flux distributed approximately as shown in Fig. 3-25(a). Part of this flux crosses the air gap, the remainder is leakage flux. A further simplification of the flux paths is indicated in Fig. 3-25(b) where ϕ_a is the flux that crosses the air gap, ϕ_l is the equivalent leakage flux, and $\phi = \phi_a + \phi_l$ is the equivalent flux linking all N turns of the winding. Then for a given air gap length x and a given current i there is a certain amount of energy stored in the magnetic field.

Now if a force f is applied to the movable member of the structure causing the air gap to increase its length by an amount equal to the differential distance dx, a differential amount of mechanical energy is put into the field. The differential amount of energy may also be accompanied by a positive or negative contribution of energy to the field from the electrical source. The energy input to the field corresponds to "the gain in reversible energy" in Eq. 1-3, which for the case under discussion can also be written as

$$\begin{bmatrix} \text{all the energy} \\ \text{that enters} \\ \text{the magnet} \end{bmatrix} = \begin{bmatrix} \text{the electrical} \\ \text{energy input} \end{bmatrix} + \begin{bmatrix} \text{the mechanical} \\ \text{energy input} \end{bmatrix} \quad (3\text{-}66)$$

The work done by the force f in increasing the air gap by the differential distance dx is the mechanical differential energy input

$$dW_{\text{mech}} = f \, dx \quad (3\text{-}67)$$

whereas the electrical differential energy input is

$$dW_e = p_e \, dt = R_{\text{eff}} i^2 \, dt + i \, d\lambda \quad (3\text{-}68)$$

where dt = the differential time in which the displacement dx occurs
R_{eff} = the effective resistance that includes the ohmic resistance of the winding and takes into account the energy dissipated in the form of heat due to hysteresis and eddy currents in the iron resulting from time variations in the flux
λ = flux linkage with the N-turn winding

The total differential energy input to the electromagnet in Fig. 3-25 is the sum of the mechanical and electrical input, or

$$dW = dW_{\text{mech}} + dW_e = f \, dx + R_{\text{eff}} i^2 \, dt + i \, d\lambda \quad (3\text{-}69)$$

The quantity $R_{\text{eff}} i^2 \, dt$ represents the electrical energy converted into heat and accounts for two terms in Eq. 1-3 as follows

$$R_{\text{eff}} i^2 \, dt = \begin{bmatrix} \text{all the differential} \\ \text{energy that leaves} \\ \text{the magnet} \end{bmatrix} + \begin{bmatrix} \text{the differential} \\ \text{gain in irreversible} \\ \text{energy} \end{bmatrix}$$

In this case the energy that leaves the magnet is in the form of heat transmitted to the surrounding medium, and the gain in irreversible energy is that evidenced by the increase in the temperature of the various parts of the electromagnet. These components of differential energy make no contribution to the energy stored in the magnetic field.

To account for the term remaining in Eq. 1-3 the following must hold

$$f \, dx + i \, d\lambda = \begin{bmatrix} \text{the differential} \\ \text{gain in reversible} \\ \text{energy} \end{bmatrix} = dW_\phi$$

and in which the reversible energy is the energy stored in the magnetic field.

The quantity $i \, d\lambda$ is the electromagnetic differential energy dW_{em} supplied by the source of voltage and is necessary to produce any change in flux linkage. The flux linkage can be divided into two components on the basis of Fig. 3-25(b). These are

$$\lambda = \lambda_a + \lambda_l$$

where

$$\lambda = N(\phi_a + \phi_l) = N\phi_a + N\phi_l$$

with ϕ_a being the flux that crosses the air gap and ϕ_l the equivalent flux in the leakage path.

$$dW_{\text{em}} = i \, d\lambda_a + i \, d\lambda_l = Ni \, d\phi_a + Ni \, d\phi_l$$

Suppose that the current i in the winding is adjusted so that the flux ϕ_a in the air gap remains constant while the displacement dx transpires. Then $d\phi_a$ is zero and the electromagnetic differential energy supplied by the source is

$$dW_{\text{em}} = Ni \, d\phi_l$$

which means that the only contribution made by the voltage source to the reversible energy is that stored in the *leakage field only*. The voltage source makes no contribution whatsoever to the reversible energy in the air gap. It follows, therefore, that as long as the flux in the air gap remains constant, any gain in the energy stored in the field of the air gap must equal the mechanical energy input $f \, dx$.

Sec. 3-18 MAGNETIC FORCE IN TERMS OF FLUX DENSITY

The permeability μ_0 of the air gap is independent of B, i.e., is constant, and Eq. 3-62 therefore applies to the energy stored in the air gap since the faces of the gap are parallel planes making the flux density B_a uniform throughout since fringing is neglected. Hence

$$W_a = \text{Vol}_a \frac{B_a^2}{2\mu_0}$$

and if the flux density B_a is constant, the differential air gap energy is

$$dW_a = \frac{B_a^2}{2\mu_0} d\,\text{Vol}_a = \frac{B_a^2}{2\mu_0} A_a\, dx \qquad (3\text{-}70)$$

the volume of the air gap being $\text{Vol}_a = A_a x$.

Equation 3-70 must also represent the mechanical input as long as B_a is constant. This leads to the relationship expressed by

$$f\, dx = \frac{B_a^2}{2\mu_0} A_a\, dx$$

and the force is therefore

$$f = \frac{B_a^2 A_a}{2\mu_0} \quad \text{newtons} \qquad (3\text{-}71)$$

where B_a is in webers per square meter, A_a is in square meters, and $\mu_0 = 4\pi \times 10^{-7}$ h per m.

When the flux density is expressed in kilolines per square inch and the area of the air gap in square inches, the force, expressed in pounds, is

$$f = 0.0139\, A_a B_a^2 \quad \text{lb} \qquad (3\text{-}72)$$

It should be remembered that no magnetic force is exerted on a material that has a relative permeability μ_r of unity. If the toroid of Fig. 3-9 had a wooden core interrupted by an air gap at right angles to the flux path there would be no force on the faces of that gap regardless of the current in the winding, whereas an iron core would experience a force in accordance with Eqs. 3-71 and 3-72. In any electromagnet the force on the steel or magnetic faces of the air gap or pole pieces is only that due to the flux that exists over and above the flux that would exist if the pole pieces were replaced by air. Therefore the actual force on the faces of the air gap is proportional to the difference between the effect of iron and the air, so that

$$F_{\text{net}} = \frac{B^2 A_a}{2\mu_0}\left(1 - \frac{1}{\mu_r}\right) \qquad (3\text{-}73)$$

The correction term $1/\mu_r$ is generally negligible, but it may be appreciable in the case of high saturations or slightly magnetic materials.

If the air gap were replaced by a magnetic material having a constant relative permeability of μ_r the amount of energy stored in that material for a given value of B_a would be $1/\mu_r$, that of the air gap. The force therefore would be reduced in the same proportion and Eq. 3-71 would be modified to

$$f = \frac{B^2 A_a}{2\mu_0 \mu_r} \qquad (3\text{-}74)$$

3-19 HYSTERESIS LOOP

The nonlinear relationship between the magnetic field intensity, also known as the magnetizing force, and the magnetic flux density in ferromagnetic materials is of great importance in engineering applications. In some cases this nonlinearity is undesirable since it may produce distortion of currents and voltages in a-c circuits. There is also the hysteresis loss, which is proportional to the area of the loop, (see Fig. 3-13) and the number of cycles that the flux undergoes in a given period of time. However, there are other applications in which this nonlinear characteristic is highly desirable if not altogether indispensable. Permanent magnets would not exist if the flux density B would return to zero when the magnetizing force H is removed. Self-excited d-c generators require operation of their magnetic circuits in the saturated region. Magnetic amplifiers, peaking transformers, and other devices depend for their operation upon nonlinear characteristics of magnetic materials.

Consider the case of the toroid in Fig. 3-9(a) as a completely unmagnetized ferromagnetic material. Let a current be passed through the winding in the direction shown. A magnetic flux will then be set up in the core, let us say in the positive direction. As this current is increased further the flux increases, rapidly at first, but beyond a certain value of current or H, the increase is less and less for a given increase in H. The curve *oab* of Fig. 3-26(a) indicates the approximate relation between the flux density and the magnetizing force. Suppose that the current is increased to a certain maximum value for which the corresponding value of the magnetizing force is $+H_{max}$, then the flux density reaches a maximum value *eb* in Fig. 3-26(a). If the current is now decreased to zero, the flux density does not drop to zero but to a value of 0*f* shown in Fig. 3-26(a). Reversing the current and with the same max value it had before produces the magnetizing force $-H_{max}$, and the flux density *e'b'*. If the current is reversed and the same maximum value is applied the flux density will not return to its former positive maximum value of *eb* but to a somewhat lower value *ec*. Similarly the succeeding negative value will be

$e'c'$. After applying alternate equal values of $+H_{max}$ and $-H_{max}$ a number of times, a process known as cycling, the magnitude of the flux density $+B_{max}$ will equal that of $-B_{max}$ and the loop known as the hysteresis loop will be symmetrical. A family of three hysteresis loops is shown in Fig. 3-26(b). If the ferromagnetic material is worked through a number of smaller loops the

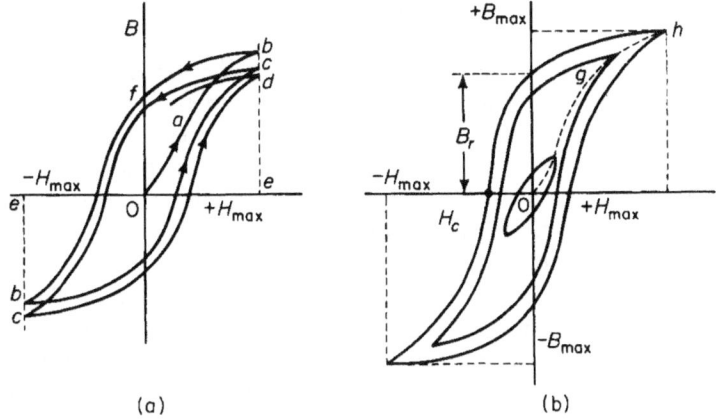

Figure 3-26. (a) B-H characteristic of initially unmagnetized iron; (b) hysteresis loops.

positive tips of the loops will lie on the curve ogh in Fig. 3-26(b). The curve ogh is called the *normal magnetization curve*.

The flux density B_r, which is still present after the magnetizing force is reduced to zero, is called the *residual flux density*. The magnetizing force H_c required to reduce the flux density from the residual value B_r to zero is called the *coercive force*. These two quantities, i.e., residual flux density and coercive force are of great importance in permanent magnets.

3-20 PERMANENT MAGNETS

There are two general kinds of magnetic materials, namely those that are magnetically soft and those that are magnetically hard. The magnetically soft materials are easy to magnetize and generally have low residual flux density, i.e., they retain a relatively small amount of magnetic flux when the applied magnetizing force is removed. Such materials are not suitable for permanent magnets. The magnetically hard materials are more difficult to magnetize and have high residual flux density and are capable of maintaining appreciable flux in the presence of relatively strong external demagnetizing forces.

Permanent magnets have many applications. In some applications the magnet exerts a force on other magnetic members; in other applications the field that the magnet maintains external to itself is used to influence the path of electrons flowing in a vacuum, a gas, or in a conductor, thus converting electrical energy into mechanical energy. The following are typical examples of permanent magnet applications: magnetic clutches and couplings, loud speakers, generators, television focusing units, magnetrons, measuring instruments, relays, information storage in computers and in video recording, etc.

Methods of magnetizing magnets depend upon the shape of the magnet. Magnets in the form of a straight bar are usually magnetized by placing them between the poles of a powerful electromagnet. When the magnets are U-shaped or circular they may be magnetized individually or in groups by placing a conductor through their centers and applying a current surge of several thousand amperes to the conductor. During this procedure the air gap of each magnet is short-circuited magnetically by bridging it with a soft iron bar so that practically all of the applied mmf is expended on the reluctance of the permanent magnet material. Removal of the bridging piece of iron causes a reduction in the flux of the magnet. If the air gap is again closed there is an increase in the flux but not to the value that it had before the bridging iron was initially removed. No appreciable permanent change occurs in the flux upon subsequent opening and reclosing of the air gap.

Carbon steel, chrome, tungsten magnet, and cobalt magnet steels have been used in permanent magnets for many years and are still used to some extent. These materials have now been largely replaced by carbon-free magnet alloys.* The most widely used alloy of this type is of the nickel-aluminum-iron group known as *Alnico*. Although these latter materials are difficult to machine, they come in a variety of sizes and shapes. In many cases the most economical design is one in which the permanent magnet material is arranged in a simple shape so as to require a minimum of machining (or rather grinding) and to provide the magnet with soft iron pole pieces that can be readily machined to the desired shape. The magnetic circuit of an indicating instrument that makes use of Alnico in combination with soft iron parts is shown in Fig. 3-27. Common shapes of Alnico magnets are rods, bars, and U-shapes, although there are numerous other shapes as well. Figure 3-28 shows a U-shaped magnet. Since the magnet is generally required to produce flux through an air gap, the space between the vertical sections is in parallel with the air gap and provides a leakage path. As a result the flux must be greater in the part of the magnet away from the poles (at the open ends) than it is in the vicinity of the poles. In order to provide

---

* For an extensive treatment see R. J. Parker and R. J. Studders, *Permanent Magnets and Their Applications*. New York: John Wiley & Sons, Inc., 1962.

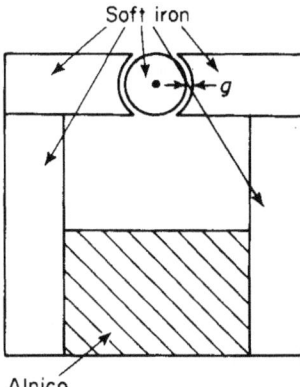

Figure 3-27. Instrument magnet circuit.

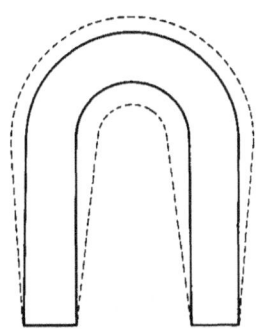

Figure 3-28. U-shaped magnet. Solid lines indicate size and shape required if there were no leakage. Broken lines show size and shape to maintain constant flux density when there is leakage flux.

for more nearly uniform flux density in the magnet material, the area carrying the greater amount of flux is built up as shown by the broken lines in Fig. 3-28.

3-21 DEMAGNETIZATION CURVE

Permanent magnets operate in that region of the hysteresis loop known as the demagnetization curve. The greater the area under the demagnetization curve the more effective is the material in the permanent magnet. Figure 3-29 shows a U-shaped magnet with a soft iron bar, sometimes called a keeper, across its open ends. The magnet is provided with an exciting winding that is usually removed after the magnet has been magnetized. It is common practice to use only one turn with a correspondingly heavy current, which is applied for only a fraction of a second.

Let $N = $ the number turns in the winding
  $l = $ the mean length of the permanent magnet (the U-shaped portion only)

Suppose that for the condition depicted in Fig. 3-29(a) a current I is applied to the winding such that the magnetizing force is H_{max} in Fig. 3-30, which shows the curve for the magnetic material. Then $H_{max} = NI/l$ if the reluctance of the horizontal soft iron piece is neglected.

The corresponding flux density throughout the permanent magnet is B_{max} if leakage is neglected. If the current is reduced to zero, the magnetizing force becomes zero and the flux density drops from the value B_{max} to B_r, the retentivity. In order to reduce the flux density to zero it is necessary to apply

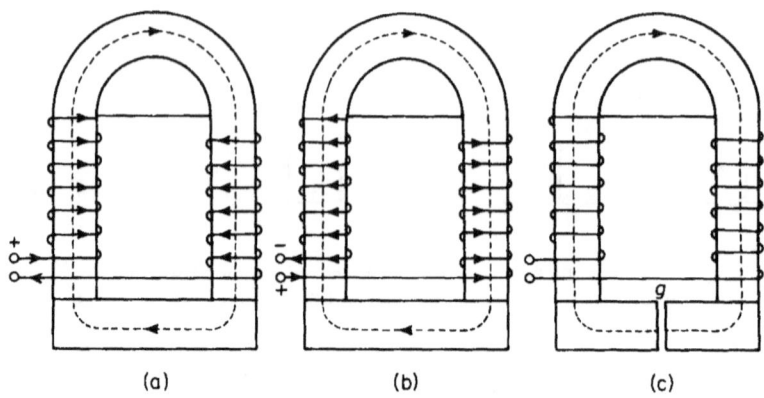

Figure 3-29. Magnetized U-shaped magnet (a) without air gap and with current in magnetizing direction, (b) without air gap and with current in demagnetizing direction, (c) with air gap and without current.

a current in the reverse direction as shown in Fig. 3-29(b) of such a value as to produce a magnetizing force equal to the coercivity H_c in Fig. 3-30. The portion $B_r P H_c$ of the curve is known as the demagnetization curve; it represents the region of interest in the operation of permanent magnets.

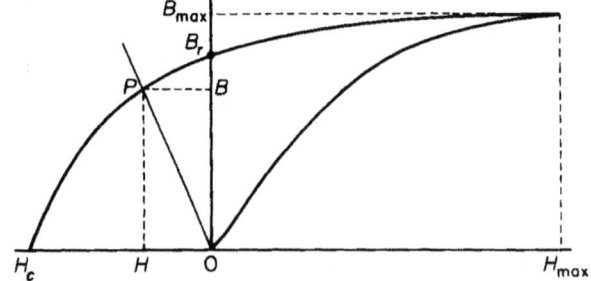

Figure 3-30. Curve of a permanent magnet material.

Suppose that the magnetizing force H_{max} has been applied to the magnet with the current in the direction shown in Fig. 3-29(a). Then if the current is reduced to zero, without in the meantime reversing its direction, the flux density will not drop to zero but rather to the value B_r as mentioned previously and its direction will remain unchanged. If now a current of such a value as to produce a magnetizing force H in Fig. 3-30 is passed through the winding in the reverse direction as shown in Fig. 3-29(b), the flux density drops from its value of B_r to B. It is important to note that although the direction of the current in Fig. 3-29(b) is opposite that in Fig. 3-29(a) the direction of the flux density is the same in both cases.

The same magnetic state, i.e., a reduction in the flux density from the value B_r to that of B can be produced when there is no current in the winding by introducing an air gap g of the proper length into the magnetic circuit as shown in Fig. 3-29(c), assuming that leakage can be neglected. This can be shown as follows

Let l_m = the mean length of the permanent magnet
$\quad A_m$ = the cross-sectional area of the magnet
$\quad H$ = the magnetizing force for the permanent magnet material (see Fig. 3-30)
$\quad B$ = the magnetic flux density in the permanent magnet material (see Fig. 3-30)
$\quad g$ = the length of the air gap
$\quad A_a$ = the cross-sectional area of the air gap, which does not need to be the same as the cross-sectional area A_m of the magnet
$\quad H_a$ = the magnetizing force for the air gap

The reluctance of the soft iron pieces on both sides of the air gap is neglected. The mmf for the air gap is

$$F_a = H_a g \qquad (3\text{-}75)$$

and the mmf for the permanent magnet is

$$F_m = H_m l_m \qquad (3\text{-}76)$$

The total mmf F_t around the closed flux path must be the sum of these two mmfs, i.e.

$$F_t = F_m + F_a \qquad (3\text{-}77)$$

Since for the condition represented by Fig. 3-29(c) there is no current in the exciting winding, the total mmf F_t must be zero in accordance with Eq. 3-34. Hence

$$F_t = \oint \mathbf{H} \cdot \mathbf{ds} = 0 \qquad (3\text{-}78)$$

Then from Eqs. 3-77 and 3-78 there results

$$F_m = -F_a \qquad (3\text{-}79)$$

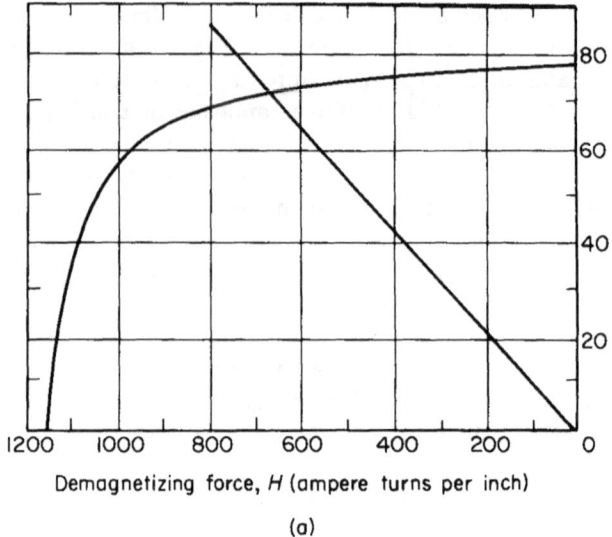

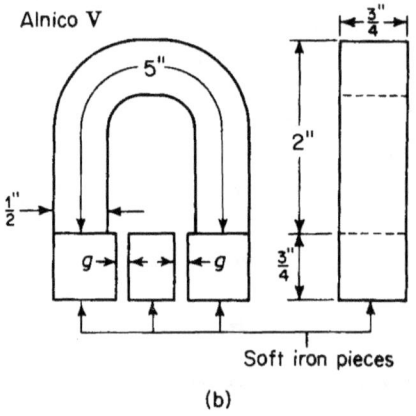

Figure 3-31. (a) Demagnetization curve for Alnico V; (b) alnico magnet with soft iron pieces and air gaps.

The magnetizing force H for the magnet is obtained in terms of H_a, the magnetizing force for the air gap, by comparing Eqs. 3-75 and 3-76 with Eq. 3-79 and is found to be

$$H = \frac{-H_a g}{l_m} \tag{3-80}$$

which can be expressed by

$$H = \frac{B}{\mu_0} \frac{A_m}{A_a} \frac{g}{l_m} \tag{3-81}$$

Sec. 3-22   ENERGY PRODUCT   137

This relationship is represented graphically by the straight line OP in Fig. 3-30. The values of the magnetizing force H and the flux density B for the permanent magnet are determined by the intersection of the line OP with the demagnetization curve. On the basis of no magnetic leakage the flux must be the same in all parts of the magnetic circuit, i.e., in the permanent magnet, in the soft iron pieces, and in the air gap. The flux is expressed by

$$\phi = BA_m = B_a A_a \tag{3-82}$$

EXAMPLE 3-5: Figure 3-31(a) shows the demagnetization curve for the Alnico magnet of Figure 3-31(b). The length of each of the two air gaps in Fig. 3-31(b) is $g = 0.10$ in. Neglect leakage but allow for fringing at the air gaps and determine the flux in the air gaps.

Solution: The flux density in the permanent magnet is determined by making use of the graphical construction illustrated in Fig. 3-30. The values to be used in Eq. 3-81 are

$l_m = 5$ in., mean length of the permanent magnet
$A_m = \frac{1}{2} \times \frac{3}{4} = \frac{3}{8}$ sq in., cross-sectional area of the magnet
$2_g = 2 \times 0.10 = 0.20$ in. total air gap length
$A_a = (0.75 + 0.10)(0.50 + 0.10) = 0.51$ sq in.
    cross-sectional area of the air gap
$\mu_0 = 3.19$, permeability of free space in the Mixed English System

$$H = \frac{-BA_m(2g)}{\mu_0 A_a l_m} = \frac{-B \times 0.375 \times 0.20}{3.19 \times 0.51 \times 5}$$
$$= -0.00922 B$$

This is the equation of a straight line, which when plotted in Fig. 3-31(a) intersects the demagnetization curve at approximately $B = 72$ kilolines per sq in. and $H = 660$ amp turns per in. Hence

$$\phi = BA_m = 72{,}000 \times 0.375 = 27{,}000 \quad \text{maxwells}$$

3-22 ENERGY PRODUCT

The energy product for the magnet material operating at the point P is the product BH at that point. When a magnet supplies an air gap with flux, as is its usual function, it supplies the air gap with energy. If the flux density B_a in the air gap and the magnetizing force H_a in the air gap are both uniform, then the magnetic energy stored in the air gap is according to Eq. 3-64.

$$W_a = \text{Vol}_a \frac{B_a H_a}{2} = A_a g \frac{B_a H_a}{2} \tag{3-83}$$

where Vol_a is the volume of the air gap. Substitution of Eqs. 3-80 and 3-82 in Eq. 3-83 yields

$$W_a = A_m\, l_m\, \frac{BH}{2} = \text{Vol}_m\, \frac{BH}{2} \qquad (3\text{-}84)$$

Then from Eqs. 3-83 and 3-84 we have

$$\text{Vol}_m\, BH = \text{Vol}_a B_a H_a \qquad (3\text{-}85)$$

where Vol_m is the volume of the permanent magnet.

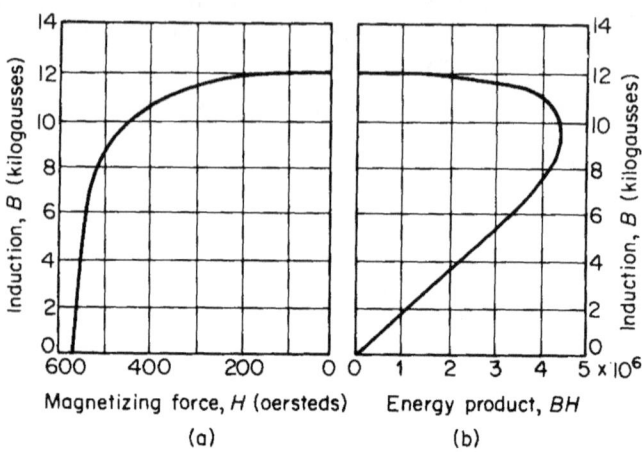

Figure 3-32. (a) Demagnetization curve, and (b) energy product curve of Alnico V.

It is generally the aim of designers, unless other considerations such as available space or configuration are prohibitive, to design the magnet with its air gap such that the energy product is a maximum. It is evident that if the length of air gap were zero then the point P in Fig. 3-30 would coincide with the point B_r and H would be zero and, of course, the energy product would be zero as well. If, on the other hand, the air gap were made very large, then the value of B would become accordingly low, resulting in a correspondingly low energy product.

A typical demagnetization curve and the corresponding energy product curve for Alnico V are shown in Fig. 3-32. In that figure the flux density B is expressed in kilogausses (kilolines or kilomaxwells per square centimeter) and the magnetizing force H is expressed in oerstedts ($4\pi/10\, Ni$ per centimeter). These units are generally used in connection with the design of permanent magnets and are included in this textbook for that reason.

The maximum energy product occurs at about $B = 9.3$ kilogausses and has a value of approximately 4.4×10^6.

If the effects of leakage are neglected, the relationship between the size of the magnet and that of the air gap is obtained from Eq. 3-85 and found to be

$$\text{Vol}_m = \text{Vol}_a \frac{B_a H_a}{BH} \tag{3-86}$$

The volume Vol_m of the magnet is a minimum for a given air gap volume when the energy product BH is a maximum.

Unless the length of the air gap is very small in relation to the length of the magnet there will be appreciable leakage, and a correction factor $\mathscr{F}$ is applied. The value of $\mathscr{F}$ lies in a range of from about 2 to 15. The length

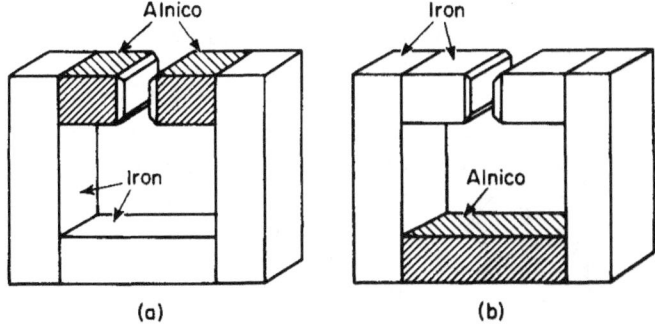

Figure 3-33. Location of magnet material. (a) Good design, low leakage; (b) poor design, high leakage.

of the air gap may not be uniform and the simple corrections for fringing discussed in Section 3-11 are not applicable. A factor f ranging from a value of 1.1 to about 1.5 is therefore applied. When these factors are applied the expression for the volume of the magnet is

$$\text{Vol}_m = \text{Vol}_a \mathscr{F} f \frac{B_a H_a}{BH} \tag{3-87}$$

The precise evaluation of the constants $\mathscr{F}$ and f is not straightforward, and requires judgment based on experience as well as on a thorough understanding of the application in which the magnet is put to use.*

The leakage may be minimized by locating the air gap in the magnet as shown in Fig. 3-33(a). With such an arrangement the mmf across the

---

* For more complete treatment of design considerations see R. J. Parker and R. J. Studders, *Permanent Magnets and Their Applications* (New York: John Wiley & Sons, Inc., 1962), Chaps. 4, 5; *Indiana Permanent Magnet Manual Number 6*. Valparaiso, Indiana: The Indiana Steel Products Company; and Earl M. Underhill, *Permanent Magnet Handbook* (Pittsburgh, Pa.: Crucible Steel Company of America, 1952), Section 2.

leakage path is due to the relatively low reluctance of the iron pieces. This situation corresponds to placing the exciting winding in an electromagnet over the leg of the core that contains the air gap. Figure 3-33(b) shows an arrangement in which the air gap is located in the iron pieces and which therefore makes for high leakage. The high mmf across the leakage path in this case is substantially equal to that across the air gap. The flux in the air gap of the structure in Fig. 3-33(a) would be considerably lower than that in the air gap of arrangement in Fig. 3-33(b) for the same size structure.

3-23 OPERATING CHARACTERISTICS OF PERMANENT MAGNETS

Variations in the flux of permanent magnets are caused by certain operating conditions. Many devices using permanent magnets have fixed air gaps, but there are others, such as magnetos and permanent-magnet generators and motors, in which flux variation may be produced by external demagnetizing forces resulting from the generator or motor current or external magnetic fields. Variations in the length of air gap for various positions of the rotor produce variations in the flux of permanent magnets.

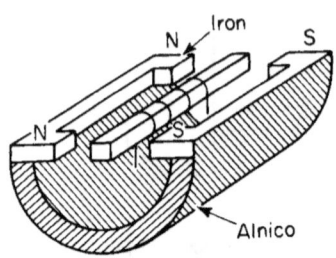

Figure 3-34. Polarized device used on loud speakers, phonograph recorders, relays, etc.

Figure 3-34 shows an arrangement consisting of a permanent magnet, usually alnico, with suitable pole pieces of iron and an iron armature pivoted at its center so that it is free to oscillate in the plane of the pole pieces. The iron armature carries a winding in which a current produces opposite poles at the ends of the armature. If this current is alternating, the armature oscillates at the frequency of the current. As a result there are variations in the reluctance of the air gap. The magnetic circuit of a permanent-magnet generator is shown in Fig. 3-35. The length of the air gap undergoes variations as the rotor occupies different angular positions.

The effect of a demagnetizing force on the flux of a permanent magnet is demonstrated by means of the demagnetization curve for a permanent magnet in Fig. 3-36. Suppose that the air gap of this permanent magnet is short-circuited by means of soft iron and a very high magnetizing force is applied. This is accomplished by passing a very heavy current, for a relatively short time period, through one or more turns of winding, linking the magnetic circuit. After the current has fallen to zero, the flux has dropped to the value ϕ_0 as shown in Fig. 3-36. If the soft iron short circuit is removed from

Sec. 3-23   CHARACTERISTICS OF PERMANENT MAGNETS   141

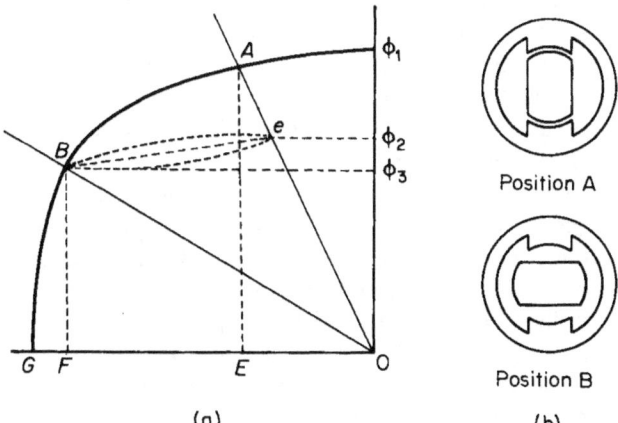

Figure 3-35. (a) Variations in the flux of a permanent magnet produced by changes in the air gap; (b) magnetic circuit of a permanent magnet generator.

the air gap, the flux drops from ϕ_0 to ϕ_1 as determined by the point A on the intersection of the demagnetization curve with the air-gap line. If a demagnetization force of OC amp turns is now applied to the magnetic circuit, the flux falls from ϕ_1 to ϕ_2 as determined by the point on the intersection of the

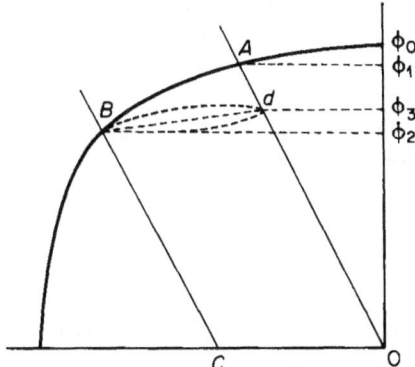

Figure 3-36. Change of flux in a permanent magnet produced by a demagnetization mmf.

demagnetization curve when the air-gap line is shifted to the left by an amount equal to OC. If the demagnetization force OC is now removed, the flux will not rise from ϕ_2 back to ϕ_1 but to some lower value ϕ_3. The point B will not follow the original characteristic back to point A, but will follow along the lower portion of the minor hysteresis loop to point d. Use is made of this effect to stabilize permanent magnets by applying an opposing mmf OC to the magnet.

In the original application of the mmf OC, the flux dropped from a value of ϕ_1 to ϕ_2, whereas successive applications of the same value of mmf will produce a change in the flux only from ϕ_3 to ϕ_2. In dealing with a permanent magnet that has been stabilized, it may be assumed with reasonable accuracy that the locus of the flux is along a straight line Bd, which has the same slope as the demagnetization curve at ϕ_0.

Consider now the effect of the variation of air-gap length. Figure 3-35(a) shows the demagnetization curve of a permanent-magnet generator, the magnetic circuit that is shown in Fig. 3-35(b). When the rotor of the generator is in position A the effective length of the air gap is a minimum. With the rotor in position B the effective length of the air gap is a maximum. The rotor in this generator would preferably be the permanent magnet and the armature winding would be placed on the stator. Such an arrangement would obviate the use of slip rings. The locus of operation on the demagnetization curve of Fig. 3-35(a) is the minor hysteresis loop between points B and e, and the flux oscillates between the values ϕ_2 and ϕ_3 twice for each revolution of the rotor.

3-24 CORE LOSSES

When ferromagnetic materials are subjected to magnetic fluxes, which vary with respect to time, such materials experience energy losses which show up in the form of heat. No such losses occur in the magnetic circuits of devices in which the magnetic flux is constant. However, the cores of transformers, iron-core choke coils, and the armature of d-c and a-c machines carry alternating fluxes that give rise to core losses or iron losses. The core losses consist of hysteresis and eddy-current losses.

3-25 HYSTERESIS LOSS

The hysteresis loss results from the B–H characteristic following a different path for decreasing values of H than for increasing values of H. If H is carried through a complete cycle from $+H_{\max}$ to $-H_{\max}$ and back to $+H_{\max}$, the B–H characteristic becomes a loop known as a hysteresis loop. Figure 3-37(a) shows a typical hysteresis loop.

Consider a unit volume of core material. Starting from point 1, Fig. 3-37(b), H is zero and is increased to $H_{\max}$; the energy absorbed by the unit volume is

$$W_1 = \int_{-Br}^{B_{\max}} H\, dB \tag{3-88}$$

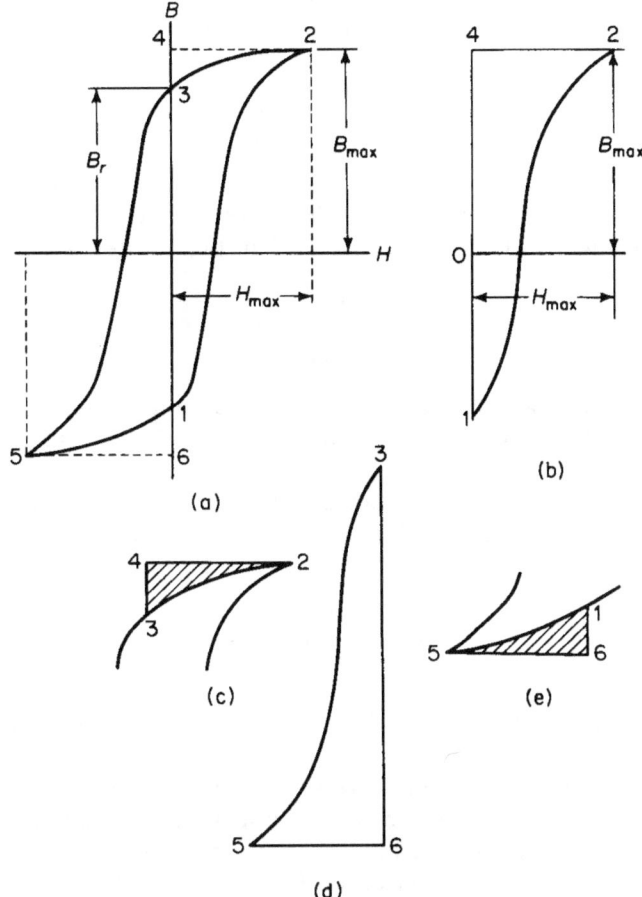

Figure 3-37. (a) Hysteresis loop. (b) and (d) show energy absorbed; (c) and (e) show energy given up by steel.

This amount of energy is represented by the area 1-2-4 in Fig. 3-37(b). If H is now decreased from $H_{\max}$ to 0, the path taken by the B–H characteristic is from point 2 to point 3. The energy is represented by the shaded portion 2-3-4 in Fig. 3-37(c), which is

$$W_2 = \int_{B_{\max}}^{B_r} H\, dB \qquad (3\text{-}89)$$

This energy input is negative since H is positive, but dB is negative, hence W_2 represents the energy given up by a unit volume of the core. If H is now taken from zero to $-H_{\max}$, the path 3-5 is traced. The energy put into the core material is represented by the area 3-5-6 in Fig. 3-37(d). This area is positive and represents energy absorbed by the core, and

$$W_3 = \int_{B_r}^{-B_{max}} H \, dB \qquad (3\text{-}90)$$

In Eq. 3-90, H is negative and dB is negative and the energy is therefore positive. If H is now increased from $-H_{max}$, the core gives up an amount of energy represented by the shaded area 5-6-1 in Fig. 3-37(e), and

$$W_4 = \int_{-B_{max}}^{-B_r} H \, dB \qquad (3\text{-}91)$$

In Eq. 3-91, H is negative and dB is positive, hence the absorbed energy is negative. This means that the core gives up energy. If the sum of the four values of energy is taken with due regard for the signs, whether positive or negative, the total energy is represented by the area of the hysteresis loop in Fig. 3-37(a). Thus, for a core having a volume Vol, and a uniform flux density B throughout the entire volume, the energy loss is

$$W_h = \text{Vol} \oint H \, dB \qquad (3\text{-}92)$$

where the line integral represents the area of the hysteresis loop. If H is in ampere turns per unit length and B is in webers per unit area and the volume is in the same system of units, Eq. 3-92 is valid for any system of dimensions. Thus, if H is in ampere turns per inch and B is in webers per square inch, then if the volume Vol is in cubic inches, W_h expresses the energy loss in joules for the entire volume. This is true also if the unit of length is the meter and the volume is in cubic meters.

In the case of the flux undergoing a cyclic variation at a frequency f cps there are f hysteresis loops per second, so to speak, and the power is

$$P_h = fW_h \quad \text{j per sec or w} \qquad (3\text{-}93)$$

or

$$P_h = \text{Vol} f \times (\text{area of loop}) \qquad (3\text{-}94)$$

In order to use the area of the loop in Eq. 3-93 it is necessary to take into account the scale to which the loop is plotted on the graph.

Let p = the number of units of H per inch of graph
$\quad\;\; q$ = the number of units of B per inch of graph
then $W = pq \times$ (area of loop in square inches)

The area of the loop may be determined from planimeter measurements or by counting squares. Then

$$P_h = \text{Vol} \, fpq \times (\text{area of loop in square inches}) \qquad (3\text{-}95)$$

The hysteresis loss, in a volume Vol in which the flux density is uniform and varies cyclically at a frequency of f cps, is expressed empirically as

$$P_h = \eta \, \text{Vol} \, f B_{\max}^n \qquad (3\text{-}96)$$

The nature of the magnetic material determines the values of η and n. The exponent may vary from a value of 1.5 to 2.5 for different materials and is

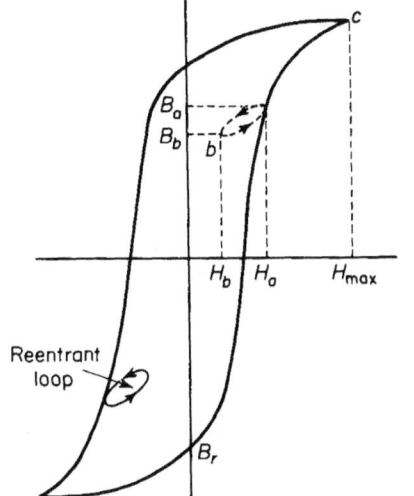

Figure 3-38. Hysteresis loop with reentrant loops.

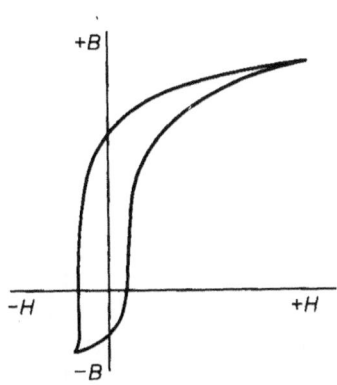

Figure 3-39. Unsymmetrical hysteresis loop.

usually not even constant for the same core. Steinmetz found in 1892 that the exponent n had a value of 1.6 for a range of flux densities from about 1,000 to 12,000 gausses for the magnetic materials then commonly in use. However, this exponent no longer applies to all present day materials. The hysteresis loss can, however, be computed from the area of the hysteresis loop for a given material if there are no re-entrant loops. If, on the other hand, the hysteresis loop has re-entrant loops, their areas must be added to that of the main loop. Re-entrant loops result from a reduction in a positive value of H and then an increase before $H_{\max}$ is reached. This is shown in Fig. 3-38 where H, after having increased from 0 to a value H_a, is decreased to a value H_b and then increased to $H_{\max}$. The locus then is from $H = 0$ at $-B_r$, to H_a

at B_a, back to H_b at B_b, back along the minor loop to H_a at B_a, then along the major loop from a to c at which we have $B_{\max}$ and $H_{\max}$. A similar situation exists in this case in the bottom half of the loop.

Equation 3-96 is not valid for unsymmetrical loops such as occur when there is a d-c component of flux in the presence of a-c flux, a situation that exists in some choke coils and transformers in vacuum tube circuits. An unsymmetrical hysteresis loop is shown in Fig. 3-39.

3-26 ROTATIONAL HYSTERESIS LOSS

The cores of transformers, choke coils, and a-c electromagnets are subject to magnetic flux that oscillates along a fixed axis under steady conditions. However, the rotors of rotary electromagnetic devices, such as generators

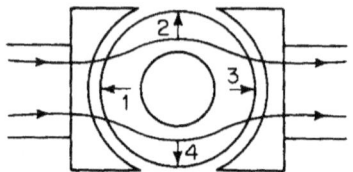

Figure 3-40. Pole cores and smooth armature core.

and motors, are subjected to rotating fluxes. Thus, in Fig. 3-40, when the arrow on the rotating armature core is in position 1, it is aligned opposite the direction of the magnetic flux, and in positions 2 and 4, the arrow is at right angles to the flux. In position 3, the arrow is in the direction of the flux. Thus, in one revolution the arrow has rotated through 360° relative to the flux. This would also be true if the armature core were stationary and the field poles rotated about the armature core. At low magnetization the rotating hysteresis loss is greater than the corresponding alternating hysteresis loss, and at very high densities the rotating hysteresis loss actually decreases, becoming quite low at very high flux densities.

3-27 EDDY-CURRENT LOSS

From earlier considerations it is evident that a variation in the flux linking a circuit generates an emf in the circuit. This is true also for the ferromagnetic circuit since the flux in the iron must naturally link various parts of the iron. Hence emfs are induced in the iron portions, which in turn produce currents, known as eddy currents, that circulate in the iron. In order to keep the eddy currents within practical values, the iron in the magnetic circuits is

Sec. 3-27 EDDY-CURRENT LOSS 147

laminated, although for some high-frequency applications molded ferrite materials are used in solid form. The laminations are coated with an insulating oxide or, in the case of large transformers, with a coating of insulating enamel. Laminating the core, with the plane of the laminations parallel to the direction of the flux, confines the eddy currents to paths of relatively

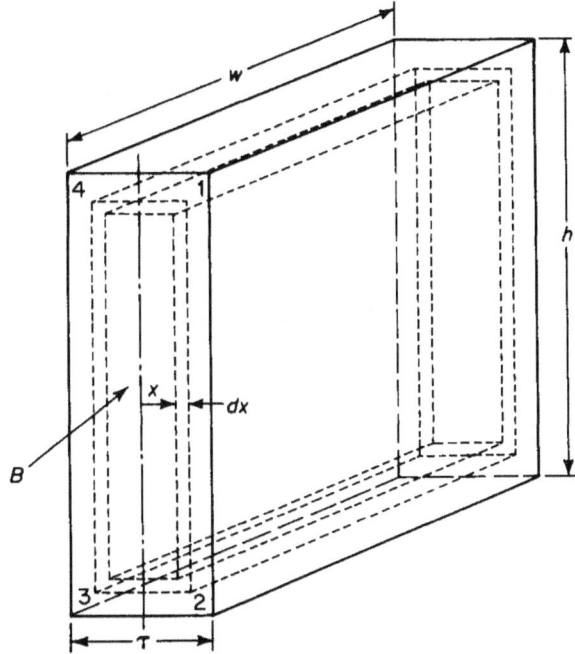

Figure 3-41. Lamination. Direction of flux is perpendicular to the current path 1-2-3-4-1.

small cross section and consequently high resistance. Shapes of typical laminations for small and medium size magnetic circuits are shown in Fig. 3-14.

Consider the piece of lamination shown in Fig. 3-41. The thickness of the lamination is τ and the height and width are h and w respectively. In Fig. 3-41 the thickness τ is shown much greater in proportion to h and w than is normally the case. The direction of the magnetic flux is at right angles to the thickness. Let $B =$ the magnetic flux density, considered uniform throughout the lamination.

If the flux density varies with respect to time, an emf is induced in the elemental path 1-2-3-4 of Fig. 3-41 by the flux linking this path, and

$$e = \frac{d\phi}{dt} = A\frac{dB}{dt} \tag{3-97}$$

where $A = 2\,hx$, the area of the path 1-2-3-4. This approximation is justified because τ is small in relation of h. In the case of most transformers and iron-core chokes the variation of flux with respect to time is sinusoidal or of such a nature that the flux can be expressed as a function of time in a Fourier series, which in general consists of a series of sine and cosine terms. However, let

$$B = B_m \sin \omega t \qquad (3\text{-}98)$$

where $\omega = 2\pi f$ and f is the frequency in cycles per second, and the voltage induced in the elemental path is expressed by

$$e = AB_m \frac{d}{dt}\sin \omega t = 2\omega hx B_m \cos \omega t \qquad (3\text{-}99)$$

This voltage produces a current in the elemental path that is practically in phase with the voltage because of the relatively high resistance of the path.

Let ρ = resistivity of the lamination; then the resistance of the path is

$$R = \frac{2h\rho}{w\,dx} \qquad (3\text{-}100)$$

from

$$R = \rho \frac{l}{A}$$

where $l = 2h$, the length of current path, and $A = 2w\,dx$, the cross-sectional area of the current path.

Here again the length of the path can be approximated to the value $2h$ because the thickness τ is so small. The current in the differential path is obtained by dividing the induced voltage in Eq. 3-99 by the resistance of the current path as given by Eq. 3-100, which yields

$$di = \frac{\omega B_m w x \cos \omega t\, dx}{\rho} \qquad (3\text{-}101)$$

The instantaneous power converted into heat in the elemental path is, by Eqs. 3-99 and 101

$$dp_e = e\,di = \frac{2\omega^2 B_m^2 h w x^2 \cos^2 \omega t\, dx}{\rho}$$

Sec. 3-27 EDDY-CURRENT LOSS

and the instantaneous loss through the entire lamination is

$$p_e = \frac{2\omega^2 B_m^2 hw \cos^2 \omega t}{\rho} \int_{x=0}^{x=\tau/2} x^2 \, dx$$

$$= \frac{2\omega^2 B_m^2 hw \cos^2 \omega t}{\rho} \frac{\tau^3}{24} \qquad (3\text{-}102)$$

$$= \frac{\omega^2 B_m^2 hw\tau^3 \cos^2 \omega t}{12\rho}$$

The quantity $hw\tau$ in Eq. 3-102 is the volume of the lamination, hence Eq. 3-102 can be rewritten as

$$p_e = \frac{\text{Vol } \omega^2 B_m^2 \tau^2 \cos^2 \omega t}{12\rho} \qquad (3\text{-}103)$$

Where $\text{Vol} = hw\tau$. Equation 3-103 expresses the instantaneous power. The power read by an ordinary wattmeter is the average power, the quantity of interest here, which is

$$P_e = \frac{1}{\pi} \int_{\omega t=0}^{\omega t=\pi} p_e \, d\omega t \qquad (3\text{-}104)$$

Substitution of Eq. 3-103 in 3-104 yields

$$P_e = \frac{\text{Vol } \omega^2 B_m^2 \tau^2}{12\rho\pi} \int_{\omega t=0}^{\omega t=\pi} \cos^2 \omega t \, dt$$

$$= \frac{\text{Vol } \omega^2 B_m^2 \tau^2}{12\rho\pi} \left[\frac{1}{2} \int_{\omega t=0}^{\omega t=\pi} (1 - \cos 2\omega t) \, d\omega t \right] \qquad (3\text{-}105)$$

$$= \frac{\text{Vol } \omega^2 B_m^2 \tau^2}{12\rho\pi} \frac{\pi}{2}$$

Substituting $2\pi f$ for ω in Eq. 3-105 gives

$$P_e = \frac{\text{Vol } \omega^2 f^2 \tau^2 B_m^2}{6\rho} \qquad (3\text{-}106)$$

The eddy-current loss can, in some cases, be expressed in terms of the voltage applied to the exciting winding. If the winding that is energized from the a-c source has N turns and all the flux in the core links all of these turns, then the induced voltage is expressed by

$$e = \frac{N\, d\phi}{dt} \tag{3-107}$$

and if the flux density in the core varies according to Eq. 3-100, then the flux in the core is expressed by

$$\phi = \phi_m \sin \omega t \tag{3-108}$$

Equations 3-107 and 3-108 yield

$$e = \omega N \phi_m \cos \omega t$$

in which the maximum instantaneous voltage is

$$\sqrt{2}E = \omega N \phi_m = 2\pi f N \phi_m$$

and the effective value of the voltage is

$$E = \frac{2\pi f N \phi_m}{\sqrt{2}} = 4.44\, f N \phi_m \tag{3-109}$$

If A_c is the cross-sectional area of the core, then Eq. 3-109 can be rewritten as

$$E = 4.44\, f B_m N A_c \tag{3-110}$$

Equation 3-110 shows that, for a sinusoidal emf or flux variation, the product fB_m is directly proportional to the induced emf E. Hence, the eddy-current losses are directly proportional to E^2 as can be seen if we substitute Eq. 3-110 in Eq. 3-106. This results in

$$P_e = \frac{\mathrm{Vol}\, \pi^2 \left(\dfrac{E}{4.44\, NA_c}\right)^2 \tau^2}{6\rho} \tag{3-111}$$

Generally the resistance drop in large transformers and large chokes is so small that the induced emf is very nearly equal to the applied emf V and the eddy-current loss in a given core and winding can be expressed approximately by

$$P_e = k_e V^2 \tag{3-112}$$

3-28 FACTORS INFLUENCING CORE LOSS

The hysteresis loss, as has been shown previously, is proportional to the area of the hysteresis loop, which in turn is affected by the heat treatment applied to the sheet steel. The hysteresis loss is thus kept low by proper annealing methods. The hysteresis losses are also related to grain sizes. The larger the grain size, the lower the hysteresis loss. The eddy-current losses vary as the square of the "lamination" thickness, and in order to keep the eddy-current losses low, the thickness of laminations must be kept as low as is economically feasible. The eddy-current losses also vary as the square of the frequency. A common lamination thickness for 60-cycle operation is 0.014 in. (29 gage), although thicker laminations are not unusual at that frequency. Laminations of 0.001 to 0.003 in. and cores of powdered iron or powdered nickel-iron alloys (known as permalloy) are used for radio-frequency applications.

Improved characteristics, not only in the sense of reduced core losses but also in increased permeability, are achieved in grain-oriented steel.[*] This is accomplished in the manufacture of the steel sheets or steel strips by cold rolling in reducing the material to the desired thickness. The grain-oriented materials are characterized by the property that the core loss is much lower and the permeability much higher when the flux path is parallel to the direction of rolling than when it is at right angles to it. This is taken into account in the manufacture of the most efficient power and distribution transformers.[†‡]

Generator laminations made of this material are segmented, being of such design that the major flux path does not deviate more than about 15° from the rolling direction.

Eddy current-losses determined by test on silicon steel have losses about 50 percent in excess of those computed from Eq. 3-106 or 3-107. The reason this excess is so great in the case of silicon steel is that the grain size for this type of iron is especially large. Generally the larger the grain size the greater the eddy-current loss for a given resistivity. Other factors tending to increase the eddy-current losses over and above the calculated values are low values of interlaminar resistance, bolts or rivets through the laminations to hold them together, etc. If the frequency is too high or the laminations are too thick, the flux will not completely penetrate the laminations and the eddy-current losses will actually be greater than normal.

---

[*] N. P. Goss, "New Development in Electrical Strip Steels," *Trans. Amer. Soc. Metals*, 23 (1935), 511.

[†] E. D. Treanor, "The Wound-Core Distribution Transformer," *A.I.E.E. Trans.*, 57 (November 1938), 622–625.

[‡] J. K. Hodnette and C. C. Horstman, "Hipersil, a New Magnetic Steel and Its Use in Transformers," *The Westinghouse Engineer* (August 1941), 52–56.

Nevertheless, Eqs. 3-106 and 3-107 show the relative effects of the various factors that influence the eddy current losses and are therefore quite useful for estimating purposes.

3-29 MAGNETIC CIRCUITS IN SERIES AND IN PARALLEL

Although the magnetic circuit is similar in many aspects to the electric circuit, calculations of magnetic circuits are generally more complex because of magnetic leakage and because of the nonlinearity of magnetic materials. Equation 3-43 expresses the relation between the magnetic flux and the mmf in a magnetic circuit of uniform cross section A and length l. In that equation the quantity $\mu A/l$ is known as the permeance $\mathscr{P}$, which is the reciprocal of the reluctance $\mathscr{R}$. If we had a single magnetic circuit with n components in series for which the reluctances would be

$$\mathscr{R}_1 = \frac{l_1}{\mu_1 A_1}, \quad \mathscr{R}_2 = \frac{l_2}{\mu_2 A_2}, \quad \mathscr{R}_n = \frac{l_n}{\mu_n A_n}$$

we could get the total reluctance simply as follows, provided that there is no leakage

$$\mathscr{R}_T = \mathscr{R}_1 + \mathscr{R}_2 + \cdots + \mathscr{R}_n \tag{3-113}$$

If the total mmf is NI then the flux in such a circuit is found to be

$$\phi = \frac{NI}{\mathscr{R}_T} = \frac{NI}{\mathscr{R}_1 + \mathscr{R}_2 + \cdots + \mathscr{R}_n} \tag{3-114}$$

However, since the values of the various permeabilities $\mu_1, \mu_2, \ldots, \mu_n$ depend upon the degree of magnetization, which is generally different for a given value of ϕ since the n components that the relationship suggested in Eqs. 3-113 and 3-114 are not particularly useful. It is, therefore, generally more convenient to compute the mmfs for each component for a given value of flux ϕ on the basis of the magnetization curve for that component.

However, in the cases of magnetic circuits that have air gaps in series and in which the iron is unsaturated so that its reluctance is small in comparison with that of the air gaps, useful approximations can be made by computing the reluctance of each of the air gaps and taking their sum as the total reluctance.

EXAMPLE 3-6: An electromagnet with a cylindrical plunger is shown in Fig. 3-42. Neglect the reluctance of the iron, leakage, and fringing and determine the flux in the magnet when the 800-turn coil carries a current of 3.5 amp.

Sec 3-29 MAGNETIC CIRCUITS IN SERIES AND IN PARALLEL

Solution: There are two air gaps in series, namely the 0.10-inch gap within the coil and the concentric air gap 0.01 in. in length between the throat and the plunger.

Since all dimensions are given in inches the mixed English system of units will be used for the calculations.

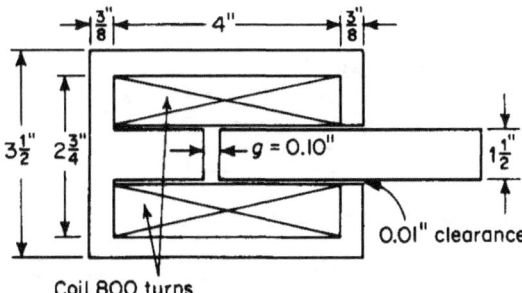

Figure 3-42. Plunger-type electromagnet.

The magnetic reluctance of 0.10-in. air gap is

$$\mathscr{R}_1 = \frac{l_1}{\mu_0 A_1} = \frac{0.10}{3.19\pi(1.5/2)^2} = 0.01772$$

and that of the 0.01-in. air gap between the plunger and throat of the magnet is

$$\mathscr{R}_2 = \frac{l_2}{\mu_0 A_2} = \frac{0.01}{3.19\pi(1.50 + 0.01)0.375} = 0.00176$$

The total reluctance is the sum of the two reluctances and

$$\phi = \frac{F}{\mathscr{R}_T} = \frac{800 \times 3.5}{0.0195} = 143{,}500 \quad \text{maxwells}$$

The reasoning that applies to the series magnetic circuit also holds for the parallel magnetic circuit and for rigorous calculations the fluxes in the various parallel branches are computed for a given value of mmf using the magnetization curves. Again if the iron is unsaturated and there are air gaps in parallel of such lengths that the reluctance of the iron is negligible the permeance, which is the reciprocal of reluctance, of the magnetic circuit is the sum of the permeances of the various air gaps provided that leakage is negligible. When the leakage is appreciable the permeance of the leakage paths must be added to those of the air gaps to obtain the total permeance.

EXAMPLE 3-7: Figure 3-43 shows an electromagnet with two air gaps in parallel. Neglect leakage and the reluctance of the iron but correct for fringing and determine the flux in each air gap and the total flux when the current in the 900-turn winding is 2.0 amp.

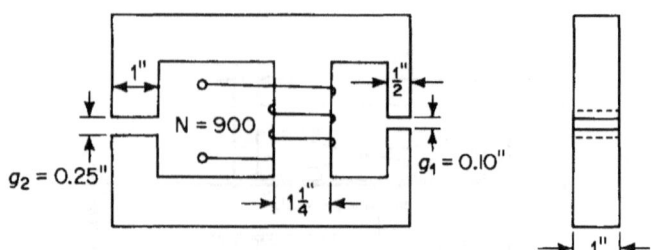

Figure 3-43. Electromagnet with two air gaps in parallel.

Solution: Since there are two air gaps in parallel their combined permeance is the sum of their permeances.

The permeance of the 0.10-in. air gap g_1 is

$$\mathcal{P}_1 = \frac{3.19 A_1}{l_1} = \frac{3.19(0.50 + 0.10)(1.00 + 0.10)}{0.10}$$
$$= 21.05$$

Since the reluctance of the iron is negligible and the two air gaps are in parallel the entire mmf is expended across each air gap, and the flux in the 0.10-in. air gap is

$$\phi_1 = \frac{F}{\mathcal{R}_1} = F\mathcal{P}_1 = NI\mathcal{P}_1 = 900 \times 2.0 \times 21.05 = 37,900 \quad \text{maxwells}$$

The permeance of the 0.25-in. air gap g_2 is

$$\mathcal{P}_2 = \frac{3.19 A_2}{l_2} = \frac{3.19(1.00 + 0.25)(1.00 + 0.25)}{0.25} = 19.95$$

and the flux in the 0.25-in. air gap is

$$\phi_2 = NI\mathcal{P}_2 = 900 \times 2.0 \times 19.95 = 35,900 \quad \text{maxwells}$$

The total flux, which is the flux in the middle leg, is

$$\phi_T = \phi_1 + \phi_2 = 37,900 + 35,900 = 73,800 \quad \text{maxwells}$$

The total flux can also be found from the total permeance and the mmf as follows

$$\mathcal{P}_T = \mathcal{P}_1 + \mathcal{P}_2 = 21.05 + 19.95 = 41.00$$
$$\phi_T = F\mathcal{P}_T = 1{,}800 \times 41.00 = 73{,}800 \quad \text{maxwells}$$

PROBLEMS

3-1 Given a long straight cylindrical conductor of radius a m, in free space, carrying a constant current I distributed uniformly over its cross section. Plot the magnetic field intensity H vs distance from the center of the conductor in terms of the current I for values of distance varying between 0 and $4a$. Assume there are no magnetic fields other than the one produced by the conductor in this problem.

3-2 Figure 3-44 shows a cross-sectional view of a straight cylindrical conductor of radius a m carrying a constant current of I amp distributed uniformly over its cross section. Assume the conductor to be in free space.

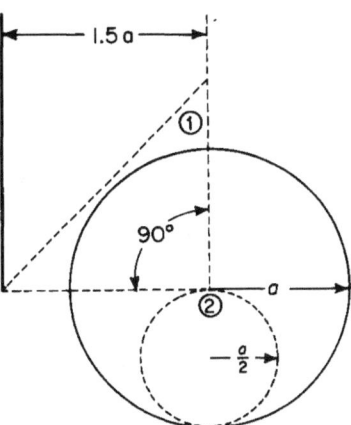

Figure 3-44. Cross section of a cylindrical conductor for Problem 3-2.

(a) Determine the value of the line integral of the magnetic field intensity H taken (1) around the triangular path 1 and (2) around the circular path 2.

(b) How would the values of part (a) above be affected if the current I in the conductor remained unchanged but the conductor were surrounded by, and electrically insulated from, a medium having a relative magnetic permeability of μ_r?

(c) Another identical conductor runs parallel to the conductor in Fig. 3-44, the spacing being $5a$ between centers. The second conductor carries a current of $10a$. Repeat part (a) if the current in the conductors is flowing (1) in the same direction (2) in opposite directions.

3-3 A magnetic circuit linking a winding of two coils having N_1 and N_2 turns respectively is shown in Fig. 3-45. The two coils are connected in series aiding, i.e., the direction of the current in each coil is such as to produce flux in the same direction around the core. The current in both coils is I amp.

(a) Determine the value of the line integral of the magnetic field intensity H taken around paths 1, 2, 3, 4, and 5 in Fig. 3-45.

(b) Repeat part (a) above but with the polarity of the lower coil (the one with N_2 turns) reversed.

3-4 A toroid as shown in Fig. 3-9 has a ratio of $R_2/R_1 = 1.5$ and has a uniform magnetic permeability throughout. Determine the percent error in using Eq. 3-42(a) instead of 3-42 to determine the flux in the core.

3-5 (a) Determine the current in the exciting winding of the magnetic circuit shown in Fig. 3-16 when the flux in the core is 19,000 maxwells. The core is comprised of U.S.S. Transformer 72, 29 Gage Steel. Stacking factor = 0.95.

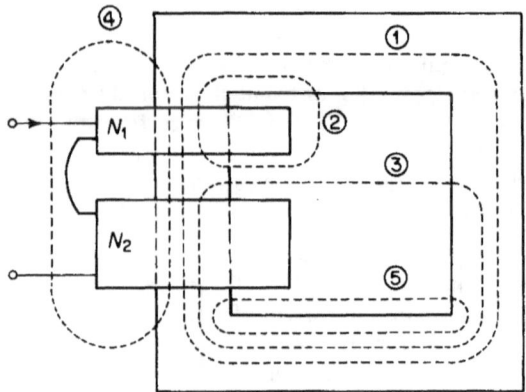

Figure 3-45. Magnetic core with a winding of 2 coils for Problem 3-3.

(b) What is the ratio of this value of current to the one in Example 3-1 for the same magnetic structure?

(c) How does this ratio compare with the corresponding ratio of fluxes?

3-6 The three-legged core in Fig. 3-17 is stacked to a depth of 1.50 in. The laminations are of 0.0140-in. U.S.S. Electrical Sheet Steel. Use Fig. 3-11 to determine the flux in the center leg and in the outer legs for a current of

(a) 10.0 amp.
(b) 3.0 amp. Assume a stacking factor of 0.90.

3-7 Plot in rectangular coordinates on the same graph curves of flux vs ampere turns for the magnetic circuit of Fig. 3-19, using the curve in Fig. 3-11 if the iron is U.S.S. Electrical Sheet Steel, for

(a) The iron alone.
(b) The air gap alone for $g = 0.100$ in.
(c) The combination of iron and air for $g = 0.100$ in. for values of flux up to 25,000 lines using the circuit of Fig. 3-19 and assuming a stacking factor of 0.90. Correct for fringing but neglect leakage.

3-8 (a) In Problem 3-7 what is the mmf when the flux is 12,500 lines, required by

(i) The iron?
(ii) The air gap?
(iii) The entire magnetic circuit?

(b) If the permeability of the iron were doubled for the conditions above at 12,500 lines, what would be the mmf required by

(i) The iron?
(ii) The air gap?
(iii) The entire magnetic circuit?

3-9 Determine the mmf for a flux of 10,000 maxwells in the core of Fig. 3-19 for an air gap 0.100 in. long. Neglect leakage flux, but correct for fringing.

3-10 Using the graphical construction shown in Fig. 3-22 determine the values of flux in the magnetic structure of Example 3-4 and Fig. 3-19 for the following values of mmf: 200, 400, 600, 800, and 1,200 amp turns. Plot, in rectangular coordinates, these values of flux for the various mmfs above.

3-11 Determine the flux in the magnetic structure of Problem 3-10 if the air gap has a length of 0.100 in. and the current in the 600-turn exciting winding is 2.5 amp.

3-12 The core of a large 60-cycle power transformer weighs 100 tons. The core material is comprised of 4.25 percent silicon steel laminations 0.014 in. thick. The specific weight of the core material is 465 lb per cu ft. The magnetic characteristics of this material are shown graphically in Fig. 3-11. Determine:

(a) The energy stored magnetically in a cubic inch of the core when the flux density is 80 kilolines per sq in. (Assume a linear relationship between B and H.)
(b) The energy stored in the entire core when the flux density is 80 kilolines per sq in.
(c) The energy in joules per kilogram for a flux density of 80 kilolines per sq in.

3-13 If the permeance, i.e., the ratio of flux to mmf, of the core in Problem 3-12 above were reduced 20 percent, what would be the energy stored in the entire magnetic circuit when the flux density is 80 kilolines per sq in.? (Assume a linear relationship between B and H.)

3-14 The electromagnet in Fig. 3-46 is made up of laminations stacked to a height of 1 in. The winding consists of 8,000 turns of No. 28 enamel wire. Neglect

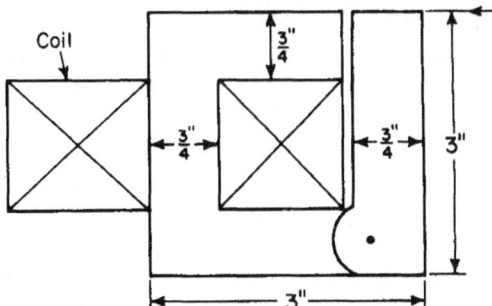

Figure 3-46. Electromagnet for Problem 3-14.

the reluctance of the iron, the effect of leakage, and fringing and determine the current required to develop a force of 37.5 lb when

(a) The air gap length is 0.05 in.
(b) The air gap length is 0.10 in.

3-15 In Problem 3-14 how much energy is stored in the air gap when its length is 0.10 in. and the force is 37.5 lb?

3-16 Figure 3-47 shows an Alnico V permanent magnet with soft iron pole shoes and an air gap g. The demagnetization curve* for this material is given below.

H oersteds	0	100	200	300	400
B kilogauss	12.5	12.3	12.0	11.5	10.7
H oersteds	500	550	600		
B kilogauss	8.8	6.6	0		

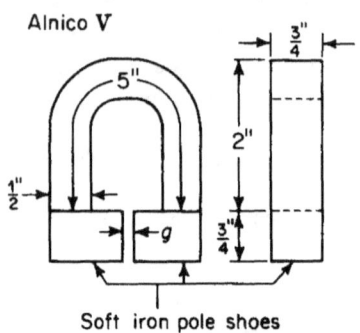

Figure 3-47. Alnico magnet with soft iron pole shoes and air gap for Problem 3-16.

Plot a demagnetization curve similar to that shown in Fig. 3-31(a) and determine the flux for an air gap length.
(a) $g = 0.125$ in.
(b) $g = 0.250$ in.
(c) $g = 0.060$ in.
Neglect leakage but correct for fringing at the air gap.

3-17 If the reluctance of the iron can be neglected and the alnico in the magnet of Problem 3-16 were to be replaced by silicon steel with negligible residual flux, how many ampere turns would need to be furnished to produce the same values of flux in the respective air gaps of Problem 3-16?

3-18 A permanent magnet-moving coil instrument has a magnetic circuit similar in configuration to that shown in Fig. 3-27. The volume of the double air gap (the two air gaps in series) is 0.24 cu in. The magnet flux in the structure is furnished by an alnico magnet that has the demagnetization characteristic shown in Fig. 3-31(a). Neglect leakage, fringing, and the reluctance of the soft iron portions and determine the smallest volume of alnico that will furnish the air gap with a flux density of 70 kilolines per sq in.

3-19 The rotor of a 2-pole permanent-magnet generator consists of Alnico V material and has an effective length of 1.00 in. (see Fig. 3-35(b) for rotor shape). The effective cross-sectional area is $\frac{1}{8}$ sq in. ($\frac{1}{4} \times \frac{1}{2}$). The minimum double air gap has an equivalent length of 0.020 in. and a maximum equivalent gap length of 0.065 in. The characteristic of the Alnico V material is listed in the table of Problem 3-16. Determine

(a) The flux when the rotor position is such that the air gap is a maximum.
(b) The flux when the rotor position is such that the air gap is minimum after having occupied a position of maximum air gap.

* Courtesy of Indiana Steel Products Co.

PROBLEMS

3-20 The following data define the upper half of the hysteresis loop for a sample of U.S.S. Transformer 72, 29 Gage Steel.

B Kilolines per square inch	0	10	20	30	40	45	50	55	60	64.3
H ampere turns per inch	0.85	1.00	1.17	1.45	1.87	2.14	2.50	2.97	3.65	4.45

B Kilolines per square inch	60	55	50	45	40	30	20	10	0
H ampere turns per inch	2.56	1.47	0.80	0.37	0.05	−0.35	−0.60	−0.73	−0.85

Plot the complete hysteresis loop and determine

(a) The residual flux density in
 (i) Kilolines per square inch.
 (ii) Gausses.
(b) The coercive force in
 (i) Ampere turns per inch.
 (ii) Oersteds.
(c) The energy product in joules for a flux density of $B = +24$ kilolines per square inch on the decreasing portion of the loop.

3-21 On the basis of the hysteresis loop of Problem 3-20 determine the hysteresis loss in joules per cycle per cubic inch of iron at a maximum flux density of 64.3 kilolines per sq in.

3-22 A transformer core has a mean length of 50 in. and a uniform gross cross-sectional area of 20 sq in. The core material is the same as that of Problem 3-20.

(a) Determine the hysteresis loss in watts for a frequency of 60 cps and a maximum flux density of 64.3 kilolines per sq in. The stacking factor is 0.90.
(b) What is the hysteresis loss in watts per pound if the density is 0.272 lb per cu in. ?

3-23 The thickness of 29 gage sheets is 0.0140 in. and the resistivity of the core material in Problem 3-22 is 22 microhms per cu in. Determine the eddy-current loss in the core of Problem 3-22 for a frequency of 60 cps and a maximum flux density of 64.3 kilolines per sq in.

3-24 Determine the total core loss for a transformer core of Problem 3-22 if the flux density is 64.3 kilolines per sq in. and the frequency is 120 cps.

3-25 Assume the hysteresis loss in the core of Problem 3-22 to vary in accordance with the equation

$$P_h = kfB_m^{1.6}$$

and determine the total core loss for a frequency of 60 cps and a flux density of 90,000 maxwells per sq in.

3-26 The core loss in an iron-core reactor is 548 w of which 402 w is hysteresis loss when the applied voltage is 240 v and the frequency is 30 cps. What is the core loss when the voltage and the frequency are both doubled? Neglect the resistance of the winding.

3-27 The core loss in a transformer is 500 w of which 100 w is eddy-current loss. If the number of turns in the exciting winding were to remain unchanged but the volume of the core were doubled by increasing the cross-section of the iron, what would be the core loss for the same applied voltage and frequency? Neglect winding resistance and assume $P_h = kfB_{max}^{1.6}$ for a given volume.

3-28 The dimensions of the core in a core-type transformer are as follows: the outside dimensions of the core parallel to the plane of the window are $8\frac{3}{4}$ in. × 10 in.; cross section of iron = $2\frac{5}{8}$ in. × $4\frac{1}{2}$ in. The laminations are 26 ($\tau = 0.0185$ in.) gage and are stacked to a depth of $4\frac{1}{2}$ in. with a stacking factor of 0.93. The primary winding has two coils of 87 turns each. The rated primary voltage when the two coils are connected in series is 440 v at 60 cps. The density of the iron is 0.272 lb per cu in. The following no-load tests were made by energizing the primary winding.

Volts	Frequency	Watts
440	60	103.5
220	30	45.2

Determine

(a) The 60-cycle hysteresis loss at rated voltage and rated frequency.
(b) The 60-cycle eddy-current loss at rated voltage and rated frequency.
(c) The core loss per pound at rated voltage and rated frequency.
(d) The core loss per pound if the same iron were used but if the laminations were 24 gage ($\tau = 0.0250$ in.).

3-29 Determine the magnetic flux in each of the three legs of the electromagnet shown in Fig. 3-43 if the air gaps in the outer legs are as indicated and an air gap of 0.05 in. is inserted in the middle leg and when the current in the 900-turn winding is 3.5 amp. Neglect the reluctance of the iron, and magnetic leakage but correct for fringing.

3-30 Determine the forces on each of the faces of the air gaps in the electromagnet of Problem 3-29.

BIBLIOGRAPHY

Bozorth, Richard M., *Ferromagnetism*. Princeton, N.J.: D. Van Nostrand Co., Inc., 1951.

Crucible Steel Company of America, *Permanent Magnet Handbook*. Pittsburgh, Pa., 1957.

Jordan, Edward C., *Electromagnetic Waves and Radiating Systems*. Englewood Cliffs, N.J.: Prentice-Hall, Inc., 1950.

Kuhlmann, John H., *Design of Electrical Apparatus*. New York: John Wiley & Sons, Inc., 1950.

Langsdorf, Alexander S., *Principles of Direct-Current Machines*. New York: McGraw-Hill Book Company, 1931.

MIT Staff, *Magnetic Circuits and Transformers*. New York: John Wiley & Sons, Inc., 1943.

Roters, Herbert C., *Electromagnetic Devices*. New York: John Wiley & Sons, Inc., 1941.

Spooner, Thomas, *Properties and Testing of Magnetic Materials*. New York: McGraw-Hill Book Company, 1927.

United States Steel Corporation, *Electrical* Steel Sheets. Pittsburgh, Pa., 1955.

Winch, Ralph P., *Electricity and Magnetism*. Englewood Cliffs, N.J.: Prentice-Hall, Inc., 1955.

INDUCTANCE-ELECTROMAGNETIC ENERGY CONVERSION

4

4-1 INDUCTIVE CIRCUITS

An inductive circuit may be defined as an *electric* circuit in which a current produces appreciable magnetic flux linkage. Windings in electromagnetic devices are inductive, as for example those in choke coils, relays, loud speakers, transformers, motors, generators, transmission lines, and antennas. It is important to note that inductance is associated with magnetic circuits, although such circuits do not necessarily have cores of magnetic materials since the path of the magnetic flux may be confined to air and to the current-carrying portions of the circuit. Such is the case with air-core chokes and air-core transformers having windings of nonmagnetic material, and with transmission lines having nonmagnetic conductors. Generally speaking, there are two kinds of inductance, namely, self-inductance and mutual inductance. The self-inductance of an electric circuit is the parameter that is associated with the flux linkage resulting from the current flowing in the circuit itself. Mutual inductance is the parameter associated with flux linkage with one electric circuit produced by the current flowing in another electric circuit. In other words, mutual inductance exists between two or more electric circuits when they have magnetic flux linkage in common.

4-2 SELF-INDUCTANCE

The voltage impressed on a series inductive circuit in which the capacitance effects are negligible and in which there are no thermal or electrolytic sources of emf, is expressed by

$$v = Ri + \frac{d\lambda}{dt}$$

where v = the applied voltage in volts
i = the current in amperes
R = the resistance in ohms
λ = the magnetic flux linkage in weber turns
t = time in seconds

The flux linkage λ may be due only to the current i in the circuit itself; in addition it may be due to currents in other circuits that are coupled magnetically with the circuit under discussion. For that matter a permanent magnet in the vicinity of the circuit may contribute flux linkage also.

If, however, the flux linkage λ is produced solely by the current i in the circuit itself and if the relationship between λ and i is a linear function, then the expression for the applied voltage becomes

$$v = Ri + L\frac{di}{dt} \qquad (4\text{-}1)$$

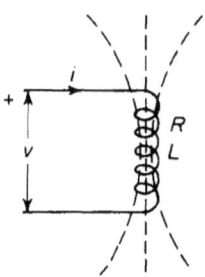

Figure 4-1. Circuit with resistance and self-inductance.

where L = self-inductance in henries. Figure 4-1 shows a schematic diagram of an inductive circuit. The dotted lines indicate the magnetic flux.

The voltage that is induced by the time variations in the current of a circuit is called the electromotive force of self-induction, or simply the emf of self-induction, and is expressed in terms of the self-inductance by

$$e = L\frac{di}{dt} \qquad (4\text{-}2)$$

The flux linkage λ can be a linear function of the current i only if the self-inductance L is constant. In circuits where the self-inductance is a variable, as for example in the case of electromagnets with variable length of air gaps as well as in some generators, the time variation of the self-inductance will affect the voltage and must be taken into account. Circuits with variable inductance will be discussed later in this chapter.

The emf induced in a circuit can always be expressed by

$$e = \frac{d\lambda}{dt} \qquad (4\text{-}3)$$

This is true whether L is a constant or a variable, or whether the flux linkage λ is produced solely by the current in that circuit or whether currents in other circuits contribute to the flux linkage as well.

Sec. 4-2   SELF-INDUCTANCE

However, if the self-inductance is constant and the only flux linkage is the one resulting from the current i in the circuit itself, then Eq. 4-2 can be equated to Eq. 4-3 with the result that

$$L\frac{di}{dt} = \frac{d\lambda}{dt} \qquad (4\text{-}4)$$

from which

$$L = \frac{d\lambda}{di} \qquad (4\text{-}5)$$

and

$$L = \frac{\lambda}{i} \qquad (4\text{-}6)$$

Equation 4-6 defines inductance as the flux linkage per ampere. Consider the toroid shown in Fig. 3-9 and assume the thickness $R_2 - R_1$ to be small compared with R_1. Then the expression for the flux produced by the current i in the winding is

$$\phi = \frac{Ni\mu w}{\pi} \frac{R_2 - R_1}{R_2 + R_1} \qquad (4\text{-}7)$$

which can be reduced to the more general form

$$\phi = \frac{\mu NiA}{l} = Ni\mathscr{P} \qquad (4\text{-}8)$$

where A is the cross-sectional area of the flux path and l is the length of the flux path. When all of the flux ϕ links all the turns N the flux linkage is

$$\lambda = N\phi$$

The flux ϕ in the toroid links all N turns, so that on the basis of Eqs. 4-6 and 4-8 the self-inductance is

$$L = \frac{\mu N^2 A}{l} = N^2 \mathscr{P} \qquad (4\text{-}9)$$

The permeability for an air core is $\mu = \mu_0 = 4\pi \times 10^{-7}$ h per m, and the expression for self-inductance if the core has a uniform cross-sectional area becomes

$$L = 4\pi \times 10^{-7} \frac{N^2 A}{l} \quad \text{h} \qquad (4\text{-}10)$$

If the core material has constant relative permeability of μ_r, then the self-inductance is given by

$$L = 4\pi \times 10^{-7} \mu_r \frac{N^2 A}{l} \tag{4-11}$$

When both sides of Eq. 4-9 are multiplied by $i^2/2$ the result expresses the energy stored in the field thus

$$W_\phi = L\frac{i^2}{2} = \mu \frac{(Ni)^2 A}{2l} \tag{4-12}$$

where

$$\frac{(Ni)^2}{l} = H^2 l \tag{4-13}$$

Substitution of Eq. 4-13 in Eq. 4-12 yields

$$W_\phi = \frac{Li^2}{2} = \frac{\mu H(HlA)}{2} = \text{Vol}\,\frac{HB}{2} \tag{4-14}$$

The term Vol $HB/2$ is the value of the energy stored magnetically in volume Vol. When Eq. 4-6 is substituted in Eq. 4-14 the energy that is stored magnetically can be expressed in terms of flux linkage and current as follows

$$W_\phi = \frac{\lambda i}{2} \tag{4-15}$$

Equations 4-14 and 4-15 apply only to the case of core materials that have constant permeability. There is the further restriction on Eq. 4-14 that the flux density B is practically uniform throughout the core.

4-3 VARIABLE SELF-INDUCTANCE

There are electric circuits in which the self-inductance is not constant but in which it varies as a result of relative linear motion or rotary motion between portions of the magnetic circuit. Relays, solenoids, and some types of electric motors and generators are examples of variable self-inductance due to one member moving with respect to another.

When the self-inductance L is a variable Eq. 4-2 does not express the induced emf because it does not take into account the time variation of L.

According to Eq. 4-6 the flux linkage is expressed in terms of self-inductance and the current by

$$\lambda = Li \qquad (4\text{-}16)$$

A time variation of flux linkage inducing the voltage

$$e = \frac{d\lambda}{dt} = L\frac{di}{dt} + i\frac{dL}{dt} \qquad (4\text{-}17)$$

Equation 4-17 shows that there are generally two components in the emf of self-induction. One of these is $L\,di/dt$ and is due to the time variation in the current, and the other, $i\,dL/dt$, is due to the time variation of the self-inductance. It is important to note that emf can be induced in an inductive circuit even when the current is constant if the inductance undergoes a variation with time.

4-4 FORCE AND TORQUE IN A CIRCUIT OF VARIABLE SELF-INDUCTANCE

The energy stored in an inductance, the magnetic circuit of which has constant permeability, is expressed by

$$W_\phi = \frac{Li^2}{2} \qquad (4\text{-}18)$$

which is true regardless of whether the inductance is constant or varies with time.

The power required to change the energy stored in the field is related to the inductance and current as follows

$$p_\phi = \frac{dW_\phi}{dt} = \frac{d\,(Li^2)}{dt\;\;2} = Li\frac{di}{dt} + \frac{i^2}{2}\frac{dL}{dt} \qquad (4\text{-}19)$$

It follows from Eq. 4-19 that the differential gain in reversible energy is

$$dW_\phi = p_\phi\,dt = Li\,di + \frac{i^2}{2}\,dL \qquad (4\text{-}20)$$

The expression for the electrical power input to the circuit during this change in the stored energy is

$$p_e = vi = \left(Ri + L\frac{di}{dt} + i\frac{dL}{dt}\right)i = Ri^2 + Li\frac{di}{dt} + i^2\frac{dL}{dt} \qquad (4\text{-}21)$$

and the differential electrical energy input therefore must be

$$dW_e = p_e \, dt = Ri^2 \, dt + Li \, di + i^2 \, dL \tag{4-22}$$

Equation 4-20 expresses the part of the electrical energy input that is stored in the magnetic field. In addition there is the energy that is converted directly into heat, namely $Ri^2 \, dt$, and that which is converted into mechanical energy. The mechanical energy output then is, as in the electromagnet of Fig. 3-25, $f \, dx$. The electrical differential energy input must therefore satisfy the following relationship

$$dW_e = Ri^2 \, dt + dW_\phi + f \, dx = Ri^2 \, dt + Li \, di + \frac{i^2}{2} dL + f \, dx \tag{4-23}$$

Equations 4-22 and 4-23 express the same amount of differential energy, and when equated to each other yield the following results

$$f \, dx = \frac{i^2}{2} dL \tag{4-24}$$

from which the expression for the force becomes

$$f = \frac{i^2}{2} \frac{dL}{dx} \tag{4-25}$$

In Eq. 4-25, f is the force *produced* by the system. This means that when dL/dx is positive, i.e., if the inductance increases with displacement, motor action results. On the other hand, a decrease in the inductance with displacement makes dL/dx negative and an external force is required resulting in generator action. If the length of the air gap in the electromagnet shown in Fig. 3-25 is increased, it is necessary to apply a positive external force. An increase in the length of air gap corresponds to a decrease in the inductance, and dL/dx is then negative.

Torque

In the case of rotary motion the mechanical energy is given by the product of torque and angular displacement as compared with that of force and linear displacement for linear motion. Figure 4-2 shows a schematic diagram of a magnetic circuit with a winding on the stator, and with the magnetic axis of the rotor displaced from that of the stator by an angle θ. The developed torque can therefore be related to the current and self-inductance by replacing the quantity $f \, dx$ in Eq. 4-24 with $T \, d\theta$,

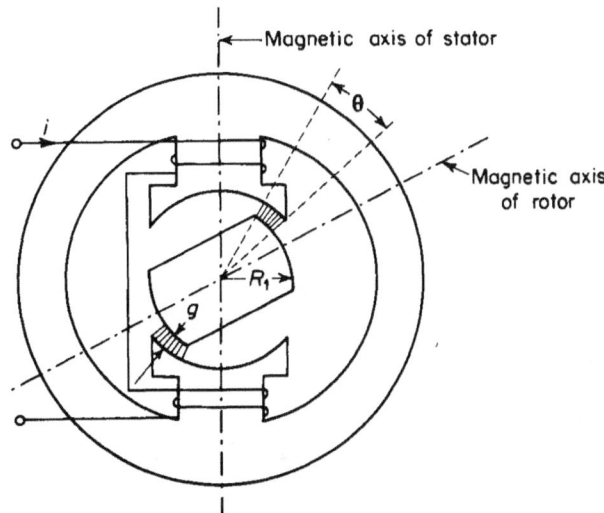

Figure 4-2. Rotary electromagnet.

where T = torque in newton meters
θ = angular displacement in radians

This results in

$$T = \frac{i^2}{2}\frac{dL}{d\theta} \qquad (4\text{-}26)$$

4-5 INDUCTANCE IN TERMS OF MAGNETIC RELUCTANCE AND MAGNETIC PERMEANCE

As mentioned previously, the magnetic circuit is similar in some respects to the electric circuit. In the case of the simple d-c circuit current, voltage and resistance are related to each other, under steady-state conditions, in accordance with Ohm's Law. The same kind of relationship is sometimes applied to magnetic circuits as follows

$$F = \phi \mathcal{R} \qquad (4\text{-}27)$$

where

F = magnetomotive force (mmf) in ampere turns (Ni)
$\phi = \lambda/N$ the equivalent flux linking all N turns of the exciting winding (4-28)
$\mathcal{R}$ = magnetic reluctance in ampere turns per weber

The relationship expressed by Eq. 4-27 can be stated in terms of the magnetic permeance as follows

$$\phi = F\mathscr{P} \tag{4-29}$$

where

$$\mathscr{P} = \frac{1}{\mathscr{R}} \tag{4-30}$$

In the case of magnetic circuits in which there is no leakage, all the flux serves to link each turn of the winding and ϕ represents the total flux. The magnetic circuit shown in Fig. 3-9 practically satisfies that condition. However, in magnetic circuits where there is magnetic leakage and the flux follows paths such that different amounts of flux link different numbers of turns in the exciting winding, the value of ϕ in Eqs. 4-27, 4-28, and 4-29 is less than the total flux. Such situations exist in air-core solenoids as well as in many circuits having ferromagnetic cores. Figure 4-3 shows a coil in which the current produces complete and partial flux linkages.

The symbol $\mathscr{P}$ stands for permeance, which is the reciprocal of reluctance. Magnetic reluctance in these relationships corresponds to resistance in the electric circuit; permeance corresponds to conductance. Both these quantities, reluctance and permeance, are functions of the magnetic permeabilities of the different parts of the magnetic circuit as well as the configurations of these parts.

Substitution of Eq. 4-28 in Eq. 4-27 yields

$$F = \frac{\lambda}{N}\mathscr{R} \tag{4-31}$$

but $F = Ni$ and $\lambda/i = L$, hence

$$Ni = \frac{\lambda \mathscr{R}}{N}$$

from which

$$L = \frac{\lambda}{i} = \frac{N^2}{\mathscr{R}} = N^2\mathscr{P} \tag{4-32}$$

For the special case in which all the flux is confined to a path of constant cross-sectional area A and the mean length of flux path is l the expression for the reluctance is reduced to

$$\mathscr{R} = \frac{l}{\mu A} \tag{4-33}$$

Magnetic pull and magnetic torques can also be expressed as functions of permeance and reluctance from Eq. 4-32

$$\frac{Li^2}{2} = \frac{\mathcal{P}(Ni)^2}{2} \qquad (4\text{-}34)$$

and the developed force can be written as

$$f = +\frac{i^2}{2}\frac{dL}{dx} = +\frac{(Ni)^2}{2}\frac{d\mathcal{P}}{dx} = \frac{F^2}{2}\frac{d\mathcal{P}}{dx} \qquad (4\text{-}35)$$

and the developed torque as

$$T = \frac{i^2}{2}\frac{dL}{d\theta} = \frac{(Ni)^2}{2}\frac{d\mathcal{P}}{d\theta} = \frac{F^2}{2}\frac{d\mathcal{P}}{d\theta} \qquad (4\text{-}36)$$

The force and torque can be expressed in terms of reluctance by replacing $\mathcal{P}$ with $1/\mathcal{R}$ in Eq. 4-36, thus

$$f = \frac{F^2}{2}\frac{d\mathcal{P}}{dx} = \frac{F^2}{2}\frac{d}{dx}\left(\frac{1}{\mathcal{R}}\right) = -\frac{F^2}{2\mathcal{R}^2}\frac{d\mathcal{R}}{dx} = -\frac{\phi^2}{2}\frac{d\mathcal{R}}{dx} \qquad (4\text{-}37)$$

and

$$T = -\frac{\phi^2}{2}\frac{d\mathcal{R}}{d\theta} \qquad (4\text{-}38)$$

In Eqs. 4-35 to 4-38 the expressions for force and torque are those developed by the electromagnetic structure, in other words, motor action is involved when there is motion of translation or rotation in the direction of the developed forces.

EXAMPLE 4-1: The rotary electromagnetic device shown in Fig. 4-3 has 1,560 turns on each stator pole. The magnetic structure is of cast steel. The dimensions are as follows

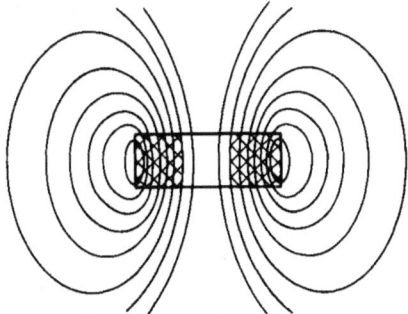

Figure 4-3. Complete and partial flux linkages with a coil carrying current.

$R_1 = 0.75$ in. radius of rounded part of rotor
$W = 1.00$ in. width of rotor- and stator-pole face at right angles to page
$g = 0.05$ in. length of single air gap
θ = angle between stator-pole tip and adjacent rotor-pole tip

(a) Neglect the effect of the iron and of flux fringing and determine the self inductance in terms of θ assuming a linear relationship between the area of the air gap and θ.*
(b) Determine the maximum torque for a current of 1.2 amp.
(c) Express the permeance as a function of the angle θ.
(d) Express the reluctance as a function of the angle θ.
(e) Express the energy stored in the air gap as a function of the angle θ for a current of 1.2 amp.

Solution:

(a) $$L = \frac{\mu N^2 A}{l}$$

$A = w\left(R_1 + \dfrac{g}{2}\right)\theta$ area of air gap in square meters

$l = 2g$ length of double air gap in meters

$\mu = 4\pi \times 10^{-7}$

$N = 1{,}560 \times 2 = 3{,}120$

$$L = \frac{4\pi \times 10^{-7}\,(3{,}120)^2\,1.00\,(0.75 + 0.025) \times 2.54 \times 10^{-2}\theta}{0.10}$$

$= 2.405\theta$ h

(b) $$T = \frac{i^2}{2}\frac{dL}{d\theta} = \frac{(1.2)^2}{2} \times 2.405$$

$= 1.73$ newton m

(c) $$\mathscr{P} = \frac{\mu A}{l} = \frac{4\pi \times 10^{-7} \times 1.00 \times 0.775 \times 2.54 \times 10^{-2}\theta}{0.10}$$

$= 2.475 \times 10^{-7}\theta$ webers per amp turn

(d) $$\mathscr{R} = \frac{1}{\mathscr{P}} = \frac{4.04 \times 10^6}{\theta}$$ amp turns per weber

(e) $$W_\phi = \frac{Li^2}{2} = 1.73\theta \ \ j$$

4-6 MUTUAL INDUCTANCE

Section 4-1 described mutual inductance as a parameter that is associated with the flux linkage produced in one circuit by the current in another circuit.

---

* Actually this is a rather rough approximation. It is evident that the circuit has some inductance even for rotor positions in which the rotor iron does not overlap the stator iron, for example, when the magnetic axes of the rotor and stator are 90° apart. The effect of fringing, which is neglected here, also changes with the angle θ.

Sec. 4-6 MUTUAL INCTANCE 173

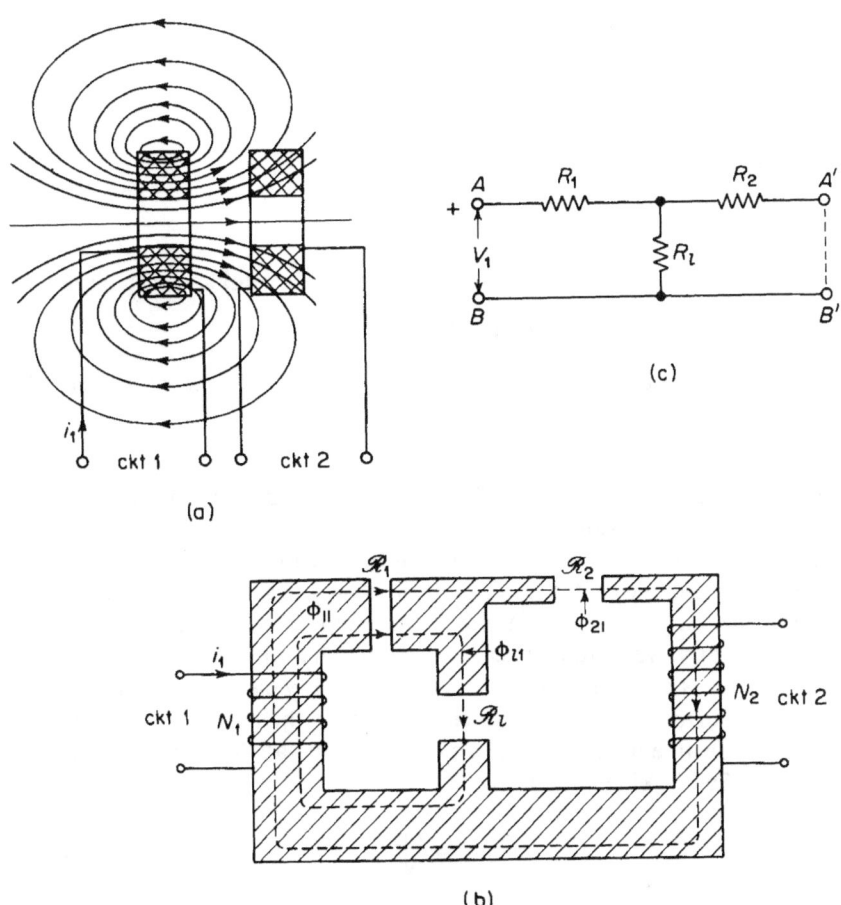

Figure 4-4. (a) Two circuits with mutual inductance; (b) equivalent magnetic circuit; (c) analogous electric circuit.

Quantitatively, the mutual inductance between two circuits may be defined as the flux linkage produced in one circuit by a current of one ampere in the other circuit. In this discussion a circuit is considered as having one or more turns. Figure 4-4 shows a schematic diagram of two circuits that are coupled inductively, i.e., magnetic flux from one circuit links the other circuit. A current i_1 in circuit 1 produces magnetic flux, some of which links some of the turns of circuit 2 and thus induces a voltage in circuit 2 when i_1 changes.

Let λ_{11} = flux linkage of circuit 1 produced by its own current i_1
λ_{21} = flux linkage of circuit 2 produced by the current i_1 in circuit 1
λ_{22} = flux linkage of circuit 2 produced by its current i_2
λ_{12} = flux linkage of circuit 1 produced by the current i_2 in circuit 2

The unit of flux linkage is the weber turn and that of inductance is the henry, which can be expressed as a weber turn per ampere. Hence

$L_{11} = \dfrac{\lambda_{11}}{i_1}$, the self-inductance of circuit 1 alone

$L_{22} = \dfrac{\lambda_{22}}{i_2}$, the self-inductance of circuit 2 alone

$L_{12} = \dfrac{\lambda_{12}}{i_2}$, the mutual inductance, based on the flux linkage with circuit 1 for unit current in circuit 2

$L_{21} = \dfrac{\lambda_{21}}{i_1}$, the mutual inductance based on the flux linkage with circuit 2 for unit current in circuit 1

The two circuits and the effect of magnetic coupling between them can be represented by the equivalent magnetic circuit and the windings N_1 and N_2 in Fig. 4-4(b), where N_1 and N_2 are the numbers of turns in circuits 1 and 2. Three air gaps having reluctances $\mathscr{R}_1$, $\mathscr{R}_2$, and $\mathscr{R}_1$ are shown in the magnetic circuit, the shaded portion of which is assumed to have infinite permeability and, consequently, zero reluctance. The resistivity of the material in the shaded portions is considered infinite so that no eddy currents can flow.

Assume a current of i_1 to flow in circuit 1 while the current in circuit 2 is zero. Let ϕ_{11} be an equivalent flux, which, in linking all N_1 turns of circuit 1, produces the flux linkage λ_{11}. Then

$$\phi_{11} = \frac{\lambda_{11}}{N_1} \qquad (4\text{-}39)$$

In Fig. 4-4(b) the equivalent flux ϕ_{11} is shown as having two components, ϕ_{21} and ϕ_{l1}, in which ϕ_{21} is the equivalent flux that, in linking all N_2 turns of circuit 2, produces the flux linkage λ_{21}. Therefore

$$\phi_{21} = \frac{\lambda_{21}}{N_2} \qquad (4\text{-}40)$$

The remaining component ϕ_{l1} is the equivalent leakage flux of circuit 1 linking all N_1 turns without linking any of the N_2 turns of circuit 2. This means that

$$\phi_{l1} = \phi_{11} - \phi_{21} \qquad (4\text{-}41)$$

The magnetic circuit of Fig. 4-4(b) is comparable to the electric circuit of Fig. 4-4(c), which is the equivalent T-circuit for any 3-terminal arrangement of pure resistances. The resistances R_1, R_2, and R_l correspond to the reluctances $\mathscr{R}_1$, $\mathscr{R}_2$, and $\mathscr{R}_l$ of the magnetic circuit, whereas the applied voltage V_1, is analogous to the mmf $N_1 i_1$ produced by the N_1-turn winding. Therefore, it is evident, on the basis of elementary electric circuit theory, that

$$\phi_{21} = \frac{\mathscr{R}_l}{\mathscr{R}_1 \mathscr{R}_2 + \mathscr{R}_1 \mathscr{R}_l + \mathscr{R}_2 \mathscr{R}_l} N_1 i_1 \qquad (4\text{-}42)$$

and that resulting flux linkage produced in circuit 2 by i_1 is

$$\lambda_{21} = \frac{\mathscr{R}_l N_1 N_2 i_1}{\mathscr{R}_1 \mathscr{R}_2 + \mathscr{R}_1 \mathscr{R}_l + \mathscr{R}_2 \mathscr{R}_l}$$

The mutual inductance based on the flux linkage with circuit 2 due the current in circuit 1 is therefore

$$L_{21} = \frac{\lambda_{21}}{i_1} = \frac{\mathscr{R}_l N_1 N_2}{\mathscr{R}_1 \mathscr{R}_2 + \mathscr{R}_1 \mathscr{R}_l + \mathscr{R}_2 \mathscr{R}_l} \qquad (4\text{-}43)$$

The mutual inductance L_{12}, based on the flux linkage with circuit 1 due the current in circuit 2, can be found by considering a current i_2 in circuit 2 while the current i_1 in circuit 1 is zero, and then going through the process used previously, or simply by interchanging the subscripts 1 and 2 in Eq. 4-43. This results in

$$L_{12} = \frac{\lambda_{12}}{i_2} = \frac{\mathscr{R}_l N_2 N_1}{\mathscr{R}_2 \mathscr{R}_1 + \mathscr{R}_2 \mathscr{R} + \mathscr{R}_1 \mathscr{R}_l} \qquad (4\text{-}44)$$

Equations 4-43 and 4-44 show the mutual inductance between two electric circuits to be reciprocal when the circuits are coupled by a homogeneous magnetic medium of constant permeability, i.e.

$$L_{12} = L_{21} \qquad (4\text{-}45)$$

In cases where there are only two magnetically coupled circuits, the letter M is used to represent mutual inductance, i.e.

$$M = L_{12} = L_{21}$$

Coefficient of coupling

The mutual inductance between two circuits can be expressed in terms of their self-inductances L_{11}, L_{22} and their coefficient of coupling k, which is a function of the reluctance of the leakage flux path. It can be seen from

Eq. 4-44 that for given values of $\mathcal{R}_1$ and $\mathcal{R}_2$ the mutual inductance increases with the reluctance $\mathcal{R}_l$ of the leakage path, becoming a maximum when $\mathcal{R}_l$ approaches infinity, the condition for perfect magnetic coupling between the two circuits. Let

$$k_1 = \frac{\phi_{21}}{\phi_{11}} = \frac{\mathcal{R}_l}{\mathcal{R}_2 + \mathcal{R}_l} \tag{4-46}$$

and

$$k_2 = \frac{\phi_{12}}{\phi_{22}} = \frac{\mathcal{R}_l}{\mathcal{R}_1 + \mathcal{R}_l} \tag{4-47}$$

from which the coefficient of coupling is, by definition

$$k \equiv \sqrt{k_1 k_2} = \frac{\mathcal{R}_l}{\sqrt{(\mathcal{R}_1 + \mathcal{R}_l)(\mathcal{R}_2 + \mathcal{R}_l)}} \tag{4-48}$$

The self-inductance of circuit 1 is

$$L_{11} = \frac{\lambda_{11}}{i_1} = \frac{N_1 \phi_1}{i_1} = \frac{(\mathcal{R}_2 + \mathcal{R}_l) N_1^2}{\mathcal{R}_1 \mathcal{R}_2 + \mathcal{R}_1 \mathcal{R}_l + \mathcal{R}_2 \mathcal{R}_l} \tag{4-49}$$

and that of circuit 2 is

$$L_{22} = \frac{\lambda_{22}}{i_2} = \frac{N_2 \phi_2}{i_2} = \frac{(\mathcal{R}_1 + \mathcal{R}_l) N_2^2}{\mathcal{R}_1 \mathcal{R}_2 + \mathcal{R}_1 \mathcal{R}_l + \mathcal{R}_2 \mathcal{R}_l} \tag{4-50}$$

The relationship between the mutual inductance and the self-inductances is given by the ratio, based on Eqs. 4-44, 4-49, and 4-50.

$$\frac{L_{12}}{\sqrt{L_{11} L_{22}}} = \frac{\mathcal{R}_l}{\sqrt{(\mathcal{R}_1 + \mathcal{R}_l)(\mathcal{R}_2 + \mathcal{R}_l)}} \tag{4-51}$$

A comparison of Eqs. 4-48 and 4-51 shows the mutual inductance to be

$$M = L_{12} = L_{21} = k \sqrt{L_{11} L_{22}} \tag{4-52}$$

The coefficient of coupling k cannot exceed unity, although values as high as 0.998 are not unusual in iron-core transformers, whereas k in air-core transformers is generally smaller than 0.5.*

---

* MIT Staff, *Magnetic Circuits and Transformers* (New York: John Wiley & Sons, Inc., 1943), 438.

Energy in the field of coupled circuits

The electromagnetic differential energy supplied to an electric circuit carrying a current i is given by

$$dW_{em} = i \, d\lambda \tag{4-53}$$

It should be remembered that electric energy over and above that required to supply the irreversible energy dissipated in the form of heat is required to maintain current in a circuit that is subjected to a variation of flux linkage. This is true whether the flux linkage λ is produced only by the current i in the circuit itself, or whether it is produced only by magnetic fields from other sources, or by fields from its own current in combination with those from other sources. The electromagnetic differential energy input to two magnetically coupled circuits is

$$dW_{em} = dW_{em_1} + dW_{em_2}$$
$$= i_1 \, d\lambda_1 + i_2 \, d\lambda_2 \tag{4-54}$$

where

$$\lambda_1 = L_{11}i_1 + Mi_2 \qquad \lambda_2 = Mi_1 + L_{22}i_2 \tag{4-55}$$

from which

$$dW_{em} = i_1 \, d(L_{11}i_1 + Mi_2) + i_2 \, d(Mi_1 + L_{22}i_2) \tag{4-56}$$

If there is no motion of one circuit relative to the other, and if there is no change in the configuration of the magnetic circuit, the self-inductances L_{11} and L_{22} as well as the mutual inductance M are constant, and no mechanical energy is supplied to or abstracted from the field. It is assumed, of course, that the inductances are independent of the currents i_1 and i_2, i.e., that the magnetic circuit is linear. Equation 4-56 can, therefore, be reduced to the form

$$dW_{em} = L_{11}i_1 \, di_1 + M(i_1 \, di_2 + i_2 \, di_1) + L_{22}i_2 \, di_2$$
$$= L_{11}i_1 \, di_1 + M \, d(i_1 i_2) + L_{22}i_2 \, di_2 \tag{4-57}$$

Since this is also the differential energy stored in the field

$$dW_{em} = dW_\phi$$

and the energy present in the field due to the currents i_1 and i_2 is

$$W_\phi = L_{11} \int_0^{i_1} i_1 \, di_1 + M \int_{0,0}^{i_1 i_2} d(i_1 i_2) + L_{22} \int_0^{i_2} i_2 \, di_2$$
$$= \tfrac{1}{2} L_{11} i_1^2 + M i_1 i_2 + \tfrac{1}{2} L_{22} i_2^2 \tag{4-58}$$

Although the inductances were assumed constant while the currents were increased from zero to i_1 and i_2 in circuits 1 and 2, Eq. 4-58 expresses the energy stored in the field for the particular values of L_{11}, L_{22}, M, i_1, and i_2 whether the inductances are constant or variable. This means that the energy stored in the field, for given values of L_{11}, L_{22}, M, i_1, and i_2, is unaffected by previous values these inductances might have had.

The development that led to Eq. 4-58 can be generalized to include n instead of two magnetically coupled circuits* expressing the energy stored in the field by

$$W_\phi = \sum_{j=1}^{n} \sum_{k=1}^{n} \tfrac{1}{2} L_{jk} i_j i_k \qquad (4\text{-}59)$$

4-7 TORQUE AND FORCE IN INDUCTIVELY COUPLED CIRCUITS†

It is evident from Eqs. 4-58 and 4-59 that the energy stored in the magnetic field associated with two or more magnetically coupled circuits can be varied by changing the current in one or more of the circuits, or by changing one or more of the self-inductances or mutual inductances. If all currents are held constant, mechanical work is involved in changing the stored energy; this necessitates a change in one or more of the inductances.

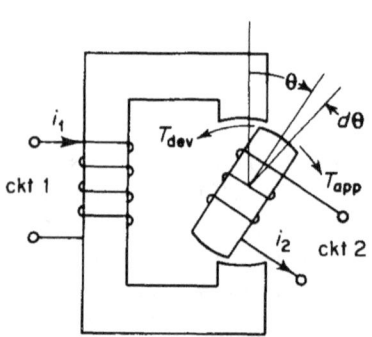

Figure 4-5. Coupled circuits with variable self- and mutual inductance.

The rotary electromagnet in Fig. 4-5 illustrates two inductively or magnetically coupled circuits and in which the reluctance of the iron is considered negligible in comparison with that of the air gaps between the stator and the rotor. When the currents i_1 and i_2 are flowing in circuits 1 and 2 in the directions shown, and with the angle θ smaller than 90°, the magnetic forces develop torque in the counterclockwise direction. Mechanical energy input is required to increase θ. This means that torque must be *applied* counterclockwise. If the angle θ is increased by the differential amount $d\theta$, and the applied torque is T, then the mechanical differential energy input is

$$dW_{\text{mech}} = T\, d\theta \qquad (4\text{-}60)$$

---

* D. C. White and H. H. Woodson, *Electromechanical Energy Conversion* (New York: John Wiley & Sons, Inc., 1959), Chap. 1.

† A. E. Fitzgerald and C. Kingsley, Jr., *Electric Machinery*, 2nd ed. (New York: McGraw-Hill Book Company, 1961), Chap. 2.

Sec. 4-7   TORQUE AND FORCE   179

If, at the same time, there is an electromagnetic differential energy input dW_{em}, the energy stored in the field will be increased by the differential amount

$$dW_\phi = dW_{em} + dW_{mech} \qquad (4\text{-}61)$$

When the differentiation in Eq. 4-56 is performed, the expression for the electromechanical differential energy becomes

$$dW_{em} = L_{11} i_1 \, di_1 + M(i_1 \, di_2 + i_2 \, di_1) + L_{22} i_2 \, di_2$$
$$+ i_1^2 \, dL_{11} + 2 i_1 i_2 \, dM + i_2^2 \, dL_{22} \qquad (4\text{-}62)$$

It should be noted that the inductances L_{11}, L_{22}, and M are not constant but are functions of the angle θ.

The energy stored in the field is given by Eq. 4-58 and it will be remembered that the energy stored in the field can be changed by a change in one or both of the currents as well as by a change in one or all of the inductances, so that the differential energy *input* to the magnetic field can be written as

$$dW_\phi = L_{11} i_1 \, di_2 + M \, d(i_1 i_2) + L_{22} i_2 \, di_2$$
$$+ \tfrac{1}{2} i_1^2 \, dL_{11} + i_1 i_2 \, dM + \tfrac{1}{2} i_2^2 + dL_{22} \qquad (4\text{-}63)$$

When Eq. 4-63 is subtracted from 4-62, then, on the basis of Eq. 4-61, the mechanical differential energy *input* to the energy stored in the field is found to be

$$dW_{mech} = -(\tfrac{1}{2} i_1^2 \, dL_{11} + i_1 i_2 \, dM + \tfrac{1}{2} i_2^2 \, dL_{22}) \qquad (4\text{-}64)$$

and, according to Eq. 4-60, the *applied* torque must be

$$T_{app} = -\left(\frac{1}{2} i_1^2 \frac{dL_{11}}{d\theta} + i_1 i_2 \frac{dM}{d\theta} + \frac{1}{2} i_2^2 \frac{dL_{22}}{d\theta} \right) \qquad (4\text{-}65)$$

The magnet develops a torque equal and opposite to the applied torque, that is

$$T_{dev} = -T_{app} = \frac{1}{2} i_1^2 \frac{dL_{11}}{d\theta} + i_1 i_2 \frac{dM}{d\theta} + \frac{1}{2} i_2^2 \frac{dL_{22}}{d\theta}$$

Similarly the developed force for a linear displacement is given by

$$f_{dev} = \frac{1}{2} i_1^2 \frac{dL_{11}}{dx} + i_1 i_2 \frac{dM}{dx} + \frac{1}{2} i_2^2 \frac{dL_{22}}{dx} \qquad (4\text{-}66)$$

When there are n-coupled circuits instead of 2-coupled circuits, and there is only one movable member in the electromagnet, as in the case of conventional motors and generators, the torque is found on the basis of Eq. 4-59 to be

$$T_{\text{dev}} = \sum_{j=1}^{n} \sum_{k=1}^{n} \frac{1}{2} i_j i_k \frac{dL_{jk}}{d\theta} \qquad (4\text{-}67)$$

If the change in inductance takes place while the currents are held constant, one-half of the electromagnetic energy input is stored in the magnetic field, whereas the other half is converted into mechanical energy. This is evident from Eqs. 4-63 and 4-64 because for constant currents, di_1 and di_2 must be zero. This equal division of stored energy and mechanical energy with constant currents is true only for linear circuits and does not hold for nonlinear magnetic circuits, i.e., where there is appreciable saturation.

On the other hand, if the flux linkages are held constant while there is a change of inductance, the electromagnetic energy input must equal zero as is shown by Eq. 4-54 for $d\lambda_1$ and $d\lambda_2 = 0$. In that case energy stored in the field is given up to be converted into mechanical energy, i.e., the mechanical energy is abstracted from the energy absorbed in the field.

4-8 FORCES IN NONLINEAR MAGNETIC CIRCUITS

The equations for force and torque which have been derived in the foregoing pages are based largely on the concept of inductance, and therefore apply strictly only to magnetic circuits that are linear. However, they may yield values that are good approximations for nonlinear magnetic circuits that are not highly saturated. Graphical analysis may be used to yield more accurate results and more valid relationships than those that make use of inductance. To keep the discussion simple, we shall consider only magnetic circuits that are excited from one winding, i.e., magnetic circuits that are singly excited. Certain conclusions, however, can be drawn from the study of singly excited circuits that apply also to multiply excited circuits.

Inductance of magnetic circuits containing iron

The concept of inductance when applied to nonlinear magnetic circuits leads to inconsistencies, depending upon how inductance is defined. The self-inductance of a circuit is defined in several ways, all of which are consistent in the case of air-core circuits. These are first

$$L = \frac{\lambda}{i} \qquad (4\text{-}68)$$

Sec. 4-8 FORCES IN NONLINEAR MAGNETIC CIRCUITS

second

$$L = e \bigg/ \frac{di}{dt} \qquad (4\text{-}69)$$

and third

$$L = \frac{2W_\phi}{i^2} \qquad (4\text{-}70)$$

In Equation 4-70, W_ϕ is the energy stored in the magnetic field.

The emf of self-induction is also expressed by

$$e = \frac{d\lambda}{dt} \qquad (4\text{-}71)$$

Another expression for self-inductance is obtained when Eq. 4-71 is substituted in Eq. 4-69, namely

$$L = \frac{d\lambda}{di} \qquad (4\text{-}72)$$

Equations 4-68, 4-70, and 4-72 do not yield the same values of inductance in a magnetic circuit containing iron or other ferromagnetic materials. This is

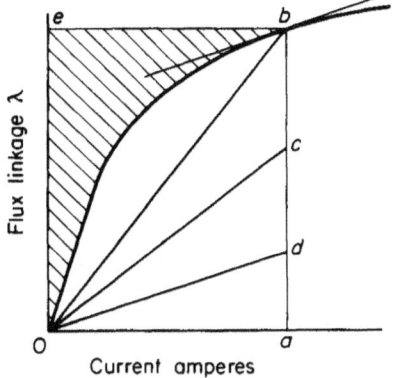

Figure 4-6. Magnetization curve.

particularly evident when there is appreciable saturation of the magnetic material.*

Figure 4-6 shows the magnetization curve of a magnetic circuit containing iron. Such a curve is typical of a-c and d-c rotating machines, chokes, transformers, and other electromagnetic devices. Let the self-inductance be evaluated on the basis of Eqs. 4-68, 4-70, and 4-72 for this magnetic circuit when the flux linkage λ has the value ab in Fig. 4-6. The current required for this value of flux linkage is oa amp.

Then, on the basis of Eq. 4-68, the apparent self-inductance is

$$L_a = \frac{\lambda}{i} = \frac{ab}{oa} \qquad (4\text{-}73)$$

and, on the basis of Eq. 4-70, the effective self-inductance is

$$L_e = \frac{2W_\phi}{i^2}$$

---

* L. T. Rader and E. C. Litscher, "Some Aspects of Inductance When Iron is Present," *Trans AIEE*, 63 (1944), 133–139.

The shaded area in Fig. 4-6 corresponds to W_ϕ in Eq. 4-70. The line oc is constructed so that the area of the triangle oac equals this shaded area. Hence

$$L_e = \frac{2W_\phi}{i^2} = \frac{ac}{oa} \qquad (4\text{-}74)$$

Finally, on the basis of Eq. 4-72, the differential self-inductance is

$$L_d = \frac{d\lambda}{di} = \frac{ad}{oa} \qquad (4\text{-}75)$$

The line od is drawn parallel to the tangent at the point b and has a slope of $\frac{d\lambda}{di}$.

It is apparent, from Eqs. 4-73, 4-74, and 4-75, that three different values of inductance are obtained in the same nonlinear circuit for a given current when three basic definitions of self-inductance are applied, although all three definitions lead to the same value for linear magnetic circuits. In spite of these inconsistencies, Eqs. 4-73, 4-74, and 4-75 lead to adequate approximations in many cases. Certain analyses of the steady-state performance of synchronous machines make use of the relationship expressed by Eq. 4-73, whereas in the cases of chokes and transformers, in which small amounts of a-c flux are superimposed on a relatively large value of d-c flux, the differential inductance in Eq. 4-75 is useful. Equation 4-74 leads to a value of inductance that is convenient, for example, as a guide for evaluating the duty imposed upon contacts in switches or circuit breakers that interrupt the current in inductive circuits. Generally, however, the problem of the nonlinear magnetic circuit is approached from standpoints that avoid the use of inductance in the analysis and instead make use of the magnetization curve. The analysis of forces and torques in nonlinear magnetic circuits is a case in point that makes use of the magnetization curve.

Force and energy relationships based on the magnetization curve

The energy stored in the magnetic field of the circuit, which has the magnetization curve shown in Fig. 4-6, is represented by the shaded area, i.e.

$$W_\phi = \int_0^\lambda i \, d\lambda = \text{area } oeb \qquad (4\text{-}76)$$

The area oba under the magnetization curve is also a product of flux linkage and current and represents the coenergy, i.e.

$$W'_\phi = \int_0^i \lambda \, di = \text{area } oba \qquad (4\text{-}77)$$

Sec. 4-8 FORCES IN NONLINEAR MAGNETIC CIRCUITS 183

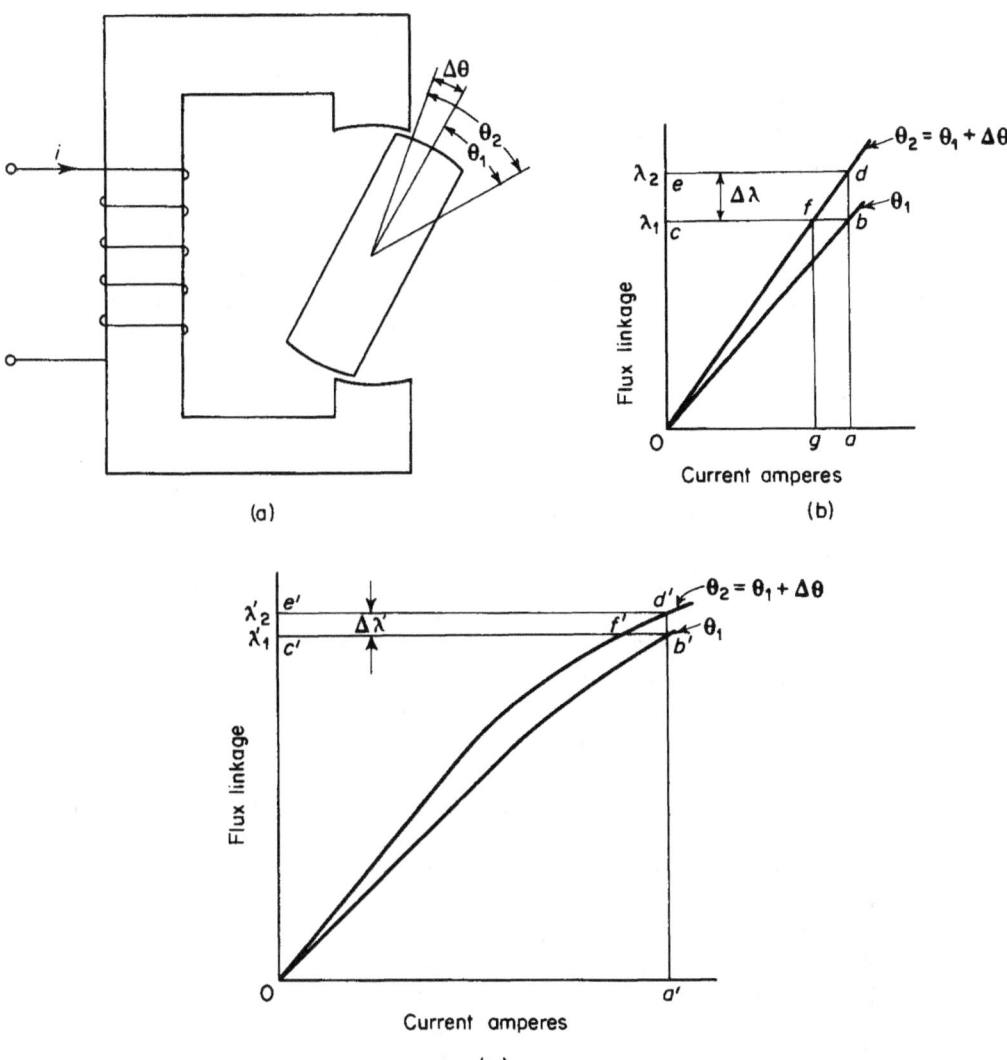

Figure 4-7. (a) Singly-excited electromagnet; (b) magnetization curves iron unsaturated; (c) magnetization curves iron saturated.

In linear magnetic circuits the energy stored in the field and the coenergy are equal, whereas in nonlinear magnetic circuits the coenergy is greater than the stored energy, as can be seen from Fig. 4-6.

Consider the rotary electromagnet shown in Fig. 4-7(a). Since there is only one winding (that on the stator), the flux linkage can be plotted against the current in that winding as in Figs. 4-7(b) and 4-7(c).

Linear Operation (Unsaturated Region). Suppose that the excitation of the electromagnet is held to values low enough so that the iron remains unsaturated so that the reluctance of the iron is negligible in comparison with that of the air gap. Then, for the angle θ_1 between the magnetic axis of the rotor and an arbitrary reference line through the center of the rotor, the magnetization curve is represented to a good degree of approximation by the straight line *ob* in Fig. 4-7(b).

If the angle is increased from θ_1 to $\theta_2 = \theta_1 + \Delta\theta$, the reluctance of the air gap is decreased and the magnetization curve is then represented by the straight line *od*. Now if the current i is held constant at the value *oa* in Fig. 4-7(b) while the displacement angle is increased from θ_1 to θ_2, the flux linkage increases from λ_1 to $\lambda_2 = \lambda_1 + \Delta\lambda$ and the electromagnetic energy input is

$$W_{\text{em}} = \int_{\lambda_1}^{\lambda_2} i \, d\lambda \qquad (4\text{-}78)$$

and is represented by the area of the rectangle *bced*. The energy originally stored in the field, i.e., for the angle θ_1, is represented by the area of the triangle *obc*, and when the area of the rectangle *bced* is added, the area *obdeo* is obtained. The final energy stored in the field, i.e., for the angle θ_2, is represented by the area of triangle *ode*, and when this area is subtracted from the area *obdeo*, the result is the area of triangle *obd*, which must represent the mechanical energy. The area of triangle *obd* is one-half that of rectangle *cbde*, which shows that, for constant current in a linear magnetic circuit, the mechanical energy output equals one-half the electromagnetic energy input, as was shown in the derivations making use of variation in inductance with displacement.

If the flux linkage is held constant while the rotor is advanced through the angle $\Delta\theta$, the current i decreases from *oa* to *og* in Fig. 4-7(b). Furthermore, the the electromagnetic energy input is zero when the flux linkage is constant. The energy stored in the field, when the angle is θ_2, is represented by the area of the triangle *ofc* and the mechanical energy output by the triangle *obf*. The mechanical energy when the flux linkage is constant is therefore abstracted from the energy stored in the field.

The average torque that is developed when the current is constant while the displacement $\Delta\theta$ occurs is

$$T_{\text{av}} = \frac{\text{Area of triangle } obd}{\Delta\theta} \qquad (4\text{-}79)$$

and for the case of constant flux linkage

$$T_{\text{av}} = \frac{\text{Area of triangle } obf}{\Delta\theta} \qquad (4\text{-}80)$$

Sec. 4-8   FORCES IN NONLINEAR MAGNETIC CIRCUITS

The same considerations, as far as force is concerned, apply to electromagnets in which the displacement is linear, say Δx. In that case, Δx is substituted for $\Delta \theta$ in Eqs. 4-79 and 4-80.

The areas of the triangles must, of course, be in terms of flux linkage and current. Thus, the area of triangle *obd* would be one half the product of $\Delta \lambda$ or $\lambda_2 - \lambda_1$ and *oa* in amperes.

Nonlinear Operation (Saturated Region). If the current in the exciting winding is increased to a value represented by oa' so that the iron becomes saturated as shown in Fig. 4-7(c), the magnetization curves for the rotor positions at θ_1 and θ_2 are no longer linear. The graphical analysis, nevertheless, may be applied just as in the case of linear operation.

If the current is held constant while the rotor undergoes a change in angular displacement of $\Delta \theta = \theta_2 - \theta_1$, the flux linkage increases from λ_1' to λ_2' or by an amount $\Delta \lambda'$ and the electromagnetic input is

$$\Delta W_{\text{em}} = i \, \Delta \lambda' \tag{4-81}$$

which is represented by the area of the rectangle $c'b'd'e'$. The energy stored in the field changes by an amount corresponding to the difference between the areas $od'e'o$ for θ_2 and $ob'c'o$ for θ_1. The mechanical energy output is then represented by the area $ab'd'o$. Similarly, if the flux linkage is held constant during the rotor displacement $\Delta \theta$, the mechanical energy output is represented by the area $ob'f'o$ since there is no electromagnetic energy input, and we have

$$\Delta W_{\text{mech}} = W_{\phi_1} - W_{\phi_2} = -\Delta W_\phi \tag{4-82}$$

where $W_{\phi 1}$ and $W_{\phi 2}$ is the energy stored in the field at θ_1 and θ_2 for λ' constant at λ_1'. The average developed torque when the flux linkage is held constant must therefore be

$$T_{\text{av}} = \frac{\Delta W_{\text{mech}}}{\Delta \theta} = -\frac{\Delta W_\phi}{\Delta \theta} \bigg|_{\lambda = \text{constant}}$$

On the other hand, when the current is held constant, the mechanical energy output equals the increase in the coenergy (area $ob'd'o$), and for constant current the average torque is

$$T_{\text{av}} = +\frac{\Delta W_\phi'}{\Delta \theta} \bigg|_{i = \text{constant}} \tag{4-83}$$

The difference between the areas $ob'd'o$ for constant current and $ob'f'o$ is the area $f'b'd'$, which becomes smaller and smaller as $\Delta \theta$ is made smaller, vanishing as $\Delta \theta \to 0$, so that torque is expressed by

$$T = -\frac{\partial W_\phi}{\partial \theta}(\lambda, \theta) = +\frac{\partial W_\phi}{\partial \theta}(i, \theta) \tag{4-84}$$

It is important to note that when the energy W_ϕ can be expressed analytically it must be explicitly in terms of the flux linkage λ and the angular position θ, whereas the coenergy W'_ϕ must be expressed explicitly in terms of the current i and the angular position θ.* This follows from the fact that the energy stored in the field is given by

$$W_\phi = \int_0^\lambda i \, d\lambda$$

and the current to produce a given value of flux linkage λ is determined not only by the value of λ, but also by that of the parameter θ, so that the current is expressed implicitly by

$$i = f(\lambda, \theta)$$

Also, the coenergy is obtained as follows

$$W'_\phi = \int_0^i \lambda \, di$$

in which the flux linkage is a function of the current and θ, or

$$\lambda = \phi(i, \theta)$$

It should be carefully noted that Eq. 4-84 expresses the *developed* torque, the direction being such as to decrease the stored energy for a given value of λ, as indicated by the minus sign, or such as to increase the coenergy for a given value of i.

In systems where translational rather than rotary displacement occurs, the developed force is

$$f = -\frac{\partial W_\phi}{\partial x}(\lambda, x) = +\frac{\partial W'_\phi}{\partial x}(i, x) \tag{4-85}$$

As mentioned previously, the stored energy and the coenergy are equal in linear systems. Then, on basis of Eq. 4-79, the torque, as $\Delta\theta \to 0$, is expressed by

$$T = +\frac{\partial W_\phi}{\partial \theta}(i, \theta) = +\frac{\partial W'_\phi}{\partial \theta}(i, \theta) \tag{4-86}$$

---

* These relationships were introduced for linear and rotational displacement of one member by A. E. Fitzgerald and C. Kingsley, Jr., *Electric Machinery*, 2nd ed. (New York: McGraw-Hill Book Company, 1961), Chap. 2.

Sec. 4-8 FORCES IN NONLINEAR MAGNETIC CIRCUITS 187

or, in translational systems

$$f = +\frac{\partial W_\phi(i, x)}{\partial x} = +\frac{\partial W'_\phi(i, x)}{\partial x} \tag{4-87}$$

Various torque and force relationships can be derived for linear systems on the basis of Eqs. 4-86 and 4-87 since the energy stored in the field, whether the system is linear or not, is determined from the integral

$$W_\phi = \int_0^\lambda i\, d\lambda = \int_0^\phi i\, d(N\phi) = \int_0^\phi Ni\, d\phi = \int_0^\phi F\, d\phi$$

The following relationships obtain for linear systems

$$\lambda = Li, \qquad F = \phi R, \qquad \phi = F\mathscr{P}$$

so that the energy stored in the field can be written as

$$W_\phi = \frac{1}{2} Li^2 = \frac{1}{2} F^2 \mathscr{P} = \frac{1}{2}\frac{F^2}{\mathscr{R}}$$

and the torque can be expressed as

$$T = \frac{\partial W_\phi(i, \theta)}{\partial \theta} = \frac{1}{2} i^2 \frac{dL}{d\theta} = \frac{1}{2} F^2 \frac{d\mathscr{P}}{d\theta} = -\frac{1}{2}\frac{F^2}{\mathscr{R}^2}\frac{d\mathscr{R}}{d\theta} = -\frac{\phi^2}{2}\frac{d\mathscr{R}}{d\theta} \tag{4-88}$$

and the force can be written as

$$f = \frac{1}{2} i^2 \frac{dL}{d\theta} = \frac{1}{2} F^2 \frac{d\mathscr{P}}{d\theta} = -\frac{1}{2}\frac{F^2}{\mathscr{R}^2}\frac{d\mathscr{R}}{d\theta} = -\frac{\phi^2}{2}\frac{d\mathscr{R}}{d\theta} \tag{4-89}$$

Various relationships for torque and force can be derived for linear systems on the basis of Eqs. 4-86 and 4-87 since the energy stored in the field (whether the system is linear or nonlinear) is given by

$$W_\phi = \int_0^\lambda i\, d\lambda = \int_0^\phi F\, d\phi \tag{4-90}$$

where ϕ is the equivalent flux linking all N turns and $F = Ni$. However, for linear systems

$$\lambda = Li, \qquad \phi = F\mathscr{P}, = \frac{F}{\mathscr{R}} \tag{4-91}$$

and when Eq. 4-91 is substituted in 4-90, the result is

$$W_\phi = \frac{1}{2} Li^2 = \frac{1}{2} F^2 \mathcal{P} = \frac{1}{2} \frac{F^2}{\mathcal{R}} \qquad (4\text{-}92)$$

Then, from Eqs. 4-86, 4-87, and 4-92, the torque and force are

$$T = \frac{\partial W_\phi}{\partial \theta}(i, \theta) = \frac{1}{2} i^2 \frac{dL}{d\theta} = \frac{1}{2} F^2 \frac{d\mathcal{P}}{d\theta} = -\frac{1}{2} \frac{F^2}{\mathcal{R}^2} = -\frac{1}{2} \phi^2 \frac{d\mathcal{R}}{d\theta} \qquad (4\text{-}93)$$

and

$$f = \frac{\partial W_\phi}{\partial x}(i, x) = \frac{1}{2} i^2 \frac{dL}{dx} = \frac{1}{2} F^2 \frac{d\mathcal{P}}{dx} = -\frac{1}{2} \frac{F^2}{\mathcal{R}^2} \frac{d\mathcal{R}}{d\theta} = -\frac{1}{2} \phi^2 \frac{d\mathcal{R}}{dx} \qquad (4\text{-}94)$$

Equations 4-93 and 4-94 show that the developed torque and developed force are in a direction such as to increase the energy stored in the field if the current is held constant. In other words, the forces on the members of the system are such as to increase the flux linkage if the current remains constant. In the case of a solenoid, the force is such as to shorten the length of the air gap, i.e., $d\mathcal{P}/dx$ is positive and $d\mathcal{R}/dx$ is negative since the permeance is inversely proportional to the length of the air gap and the reluctance is directly proportional to the gap length.

It is necessary when performing the partial differentiation of Eqs. 4-93 and 4-94 to express the stored energy W_ϕ in terms of the current i or the mmf F rather than the flux ϕ. For example, since $F = \phi\mathcal{R}$, the energy stored in the field is

$$W_\phi(\phi, \theta) = \frac{1}{2} \phi^2 \mathcal{R}$$

and if the partial derivative is taken of W_ϕ, the result is

$$\frac{\partial W_\phi(\phi, \theta)}{\partial \theta} = \frac{1}{2} \phi^2 \frac{d\mathcal{R}}{d\theta} = \frac{1}{2} \frac{F^2}{\mathcal{R}} \frac{d\mathcal{R}}{d\theta} \qquad (4\text{-}95)$$

When this result is compared with that in Eq. 4-94, it will be found that the signs are opposite. This is to be expected since, in taking the partial derivative of the stored energy with respect to θ, the flux ϕ is a parameter assumed constant. Also, a decrease in the reluctance is accompanied by a decrease in the energy stored on the field, whereas, for constant current, a decrease in the reluctance is accompanied by an increase in the stored energy.

EXAMPLE 4-2: Derive Eq. 3-71 for the force between parallel plane faces of an air gap. The derivation of Section 3-18 is based on the mechanical differential energy input required to increase the length of the air gap by

dx, so that the positive direction, as shown in Fig. 3-25, is for an *applied* rather than *developed* force. Equations 4-93 and 4-94 are for developed force.

Solution: For the present derivation, use will be made of the following relationship from Eq. 4-92

$$f = -\frac{1}{2}\phi^2 \frac{d\mathscr{R}}{dx}$$

where ϕ is the flux in the air gap and $\mathscr{R}$ is the reluctance of the air gap of the electromagnet in Fig. 3-25. Then

$$\mathscr{R} = \frac{x}{\mu_0 A} \quad \text{and} \quad \frac{d\mathscr{R}}{dx} = \frac{1}{\mu_0 A}$$

where A is the area and x is the length of the air gap. Then we have

$$f = -\frac{\phi^2}{2\mu_0 A} = -\frac{B^2 A}{2\mu_0}$$

The developed force is opposite the applied force, thus accounting for the minus sign.

Since the units for flux linkage and mmf are the weber turn and the ampere turn, the energy stored in the field can also be represented by

$$W_\phi = \int_0^\theta F\, d\theta$$

so that Eq. 4-85 can be put into the form

$$f = -\frac{\partial W_\phi}{\partial x}(\phi, x) = +\frac{\partial W'_\phi}{\partial x}(F, x) \tag{4-96}$$

Although the relationships of force and torque to the variation of the energy stored in the magnetic field with variation in displacement were derived for singly excited systems, these relationships can be extended to multiply excited systems (in which more than one winding is excited) as shown by White and Woodson.* The force is then expressed by

$$f = +\frac{\partial W'_\phi}{\partial x}(F_1, F_2, \ldots, F_n, x)$$

* D. C. White and H. H. Woodson, *Electromechanical Energy Conversion* (New York: John Wiley & Sons, Inc., 1959), 12–24.

or

$$f = +\frac{\partial W'_\phi}{\partial x}(i_1, i_2, \ldots, i_n, x) \qquad (4\text{-}97)$$

4-9 INDUCTIVE REACTANCE

When an alternating current of constant amplitude is passed through the winding of an air-core inductance, energy is alternately stored and given up every quarter cycle. If the resistance of the circuit is negligible, all of the energy that is stored in the magnetic field during one quarter cycle is reversible and is returned to the source during the following quarter cycle. The average power consumed by such a circuit is zero during a number of complete cycles.

Let L = self-inductance in henrys
$\quad i = \sqrt{2}\, I \sin \omega t$
where I = rms or effective value of the current
$\quad \omega = 2\pi f$
where f = frequency in cycles per second

If the resistance is negligible, the applied voltage is

$$\begin{aligned}
v \approx e &= L\frac{di}{dt} \\
&= L\frac{d}{dt}(\sqrt{2}\, I \sin \omega t) \qquad (4\text{-}99)\\
&= \omega L \sqrt{2}\, I \cos \omega t \\
&= \sqrt{2}\, IX \cos \omega t \\
&\approx \sqrt{2}\, V \cos \omega t
\end{aligned}$$

In Eq. 4-99, V is the rms value of the voltage applied to the self-inductance. Also

$$X = \omega L \qquad (4\text{-}100)$$

where X is the inductive reactance.

From Eqs. 4-98 and 4-99, it is evident that the current lags the voltage by 90°. The relationship between the voltage and current for a circuit of constant self-inductance and negligible resistance is shown in Fig. 4-8(a) and

Sec. 4-10 REACTIVE POWER

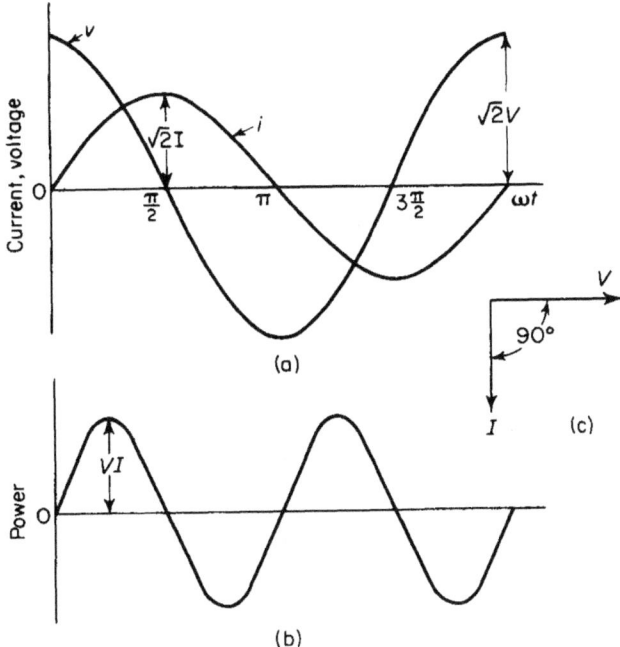

Figure 4-8. (a) Voltage and current waves, (b) power wave, (c) phasor diagram for pure self-inductance.

4-8(c). The instantaneous power is shown graphically as a function of time in Fig. 4-8(b).

4-10 REACTIVE POWER

The real power in a-c circuits under steady conditions is the average of the instantaneous power taken over an integral number of half cycles and is expressed by

$$P = VI \cos \theta \tag{4-101}$$

where θ is the angle by which the current lags or leads the voltage. It is evident from Eq. 4-101 and Figs. 4-8(b) and 4-8(c) that the real power in a purely inductive circuit is zero under steady-state conditions. However, the term *reactive power* is used to express the product of current and voltage in a circuit in which the real power is zero. In circuits where the current lags the voltage by an angle smaller than 90°, the reactive power consumed by the circuit is

$$Q = VI \sin \theta \tag{4-102}$$

In capacitive circuits the current leads the voltage, and the reactive power consumed by such a circuit is negative. Hence, capacitive circuits generate reactive power and are therefore sources of reactive power. Capacitors are used in industrial power systems to furnish reactive power.

Thus, a circuit comprised of self-inductance, L, in series with a resistance R_{eff} has an impedance of

$$Z = \sqrt{R_{\text{eff}}^2 + (\omega L)^2} \tag{4-103}$$

and

$$\cos \theta = \frac{R_{\text{eff}}}{Z} \qquad \sin \theta = \frac{\omega L}{Z} \tag{4-104}$$

furthermore

$$I = \frac{V}{Z} \tag{4-105}$$

From Eqs. 4-104 and 4-105 it follows that

$$V \cos \theta = I R_{\text{eff}} \qquad V \sin \theta = I \omega L \tag{4-106}$$

from which

$$P = I^2 R_{\text{eff}} \tag{4-107}$$

$$Q = I^2 \omega L \tag{4-108}$$

4-11 EFFECTIVE RESISTANCE AND Q-FACTOR

R_{eff} expresses the a-c resistance, which is generally greater than the d-c resistance. This is particularly true when the cross section of the wire in the winding of the self-inductance is large, or when, as in the case of some devices, such as motors and generators, the wire is embedded in slots. The leakage fluxes in transformers and stray fluxes in reactors and choke coils produce losses in the metal tanks or cases that enclose the core and windings. Such losses are reflected into the effective resistance values.

The quality factor, QF, or simply Q*, of a coil is the ratio of reactive power to real power. Hence

$$Q = QF \equiv \frac{I^2 \omega L}{I^2 R_{\text{eff}}} = \frac{\omega L}{R_{\text{eff}}} = \tan \theta \tag{4-109}$$

---

* Unfortunately the symbol for quality factor is the same as that for reactive power and care should be exercised not to confuse the one with the other.

Other definitions are

$$Q = QF \equiv \frac{2\pi(\text{energy stored in the circuit})}{\text{energy lost per cycle}} \quad (4\text{-}110)$$

$$Q = QF \equiv \frac{\text{energy stored in the circuit}}{\text{energy lost per radian}} \quad (4\text{-}111)$$

Equations 4-109, 4-110, and 4-111 are equivalent. The term, *energy lost per cycle*, represents the irreversible energy, i.e., the energy that is converted into heat and the energy stored in the circuit represents the reversible energy.

PROBLEMS

4-1 A toroid consists of a wooden core with a winding having two layers with 500 uniformly distributed turns each. The inside and outside radii of the core are 5 and 6 in. respectively, and the axial depth of the core is 4 in. Assume all the magnetic flux to be confined to the wooden core, and the mean length of the flux path to be the average of the inner and outer circumferences of the core.

  (a) If the two layers of the windings are connected in series with their polarities such that the mmfs add and the current is 1 amp, determine
    (i) The magnetic flux in the core.
    (ii) The flux linkage.
    (iii) The self-inductance of the winding.
    (iv) The energy stored in the magnetic field.
  (b) Repeat part (a) above if a current of 1 amp is passed through only one layer of the winding.
  (c) Repeat part (b) above if the two layers are connected in parallel with the polarities such that the mmfs add and if the total current is 2 amp.
  (d) Repeat part (a) for a core having a uniform relative permeability $\mu_r = 400$.

4-2 Determine the magnetic reluctance and the magnetic permeance of the toroid for the conditions specified in Problem 4-1(a), (b), (c), and (d).

4-3 A long, straight, round copper conductor of radius R m and length l m carries a current of I amp distributed uniformly over its cross section. Determine in terms of I and other suitable constants on the basis that the relative permeability of copper $\mu_r = 1$

  (a) The total magnetic flux within the conductor.
  (b) The magnetic energy density within the conductor as function of the distance from the center of the conductor.

4-4 The number of turns in the windings designated as circuits 1 and 2 in Fig. 4-4(b) is $N_1 = 100$ and $N_2 = 200$. A current $i_1 = 5$ amp in circuit 1 produces an equivalent flux $\phi_{11} = 0.0032$ weber, which links all N_1 turns. Assume the magnetic circuit to be linear, i.e., the reluctance of the iron represented by the shaded area to be negligible. Calculate

(a) The self-inductance L_{11} of circuit 1.
(b) The mutual flux ϕ_{21} if the reluctances are in the ratios $\mathscr{R}_2/\mathscr{R}_1 = 2$ and $\mathscr{R}_l/\mathscr{R}_1 = \tfrac{2}{3}$.
(c) The mutual inductance L_{21} between the two circuits.
(d) The coefficients k_1, k_2 and the coefficient of coupling k.
(e) The self-inductance L_{22} of circuit 2.
(f) The energy stored in the fields associated with each of the three reluctances.
(g) The stored energy based on $\tfrac{1}{2}L_{11}i_1^2$. [How does this compare with the sum of the three values of part (f)?]

4-5 Figure 4-9 shows an electromagnet with equal variable air gaps g. When the current in the 1,000-turn winding on the stationary member is 5 amp the flux

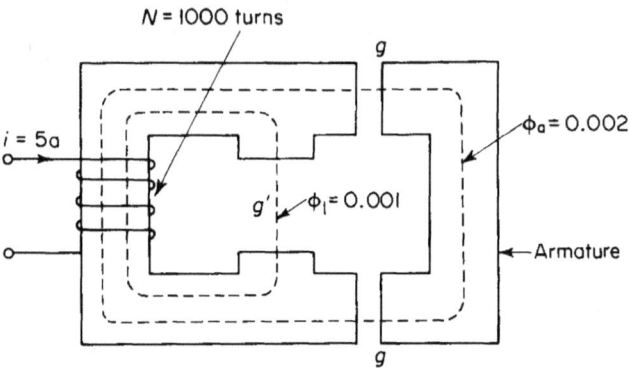

Figure 4-9. Electromagnet for Problem 4-5.

through the movable member is $\phi_a = 0.002$ weber and the leakage flux $\phi_l = 0.001$ weber. The reluctance of the iron is negligible.

The length of the variable air gaps g is increased 10 percent; the flux ϕ_a through these air gaps is held constant at 0.002 weber. Neglect the effect of fringing.

(a) Calculate for the increased length of air gaps
   (i) The current.
   (ii) The leakage flux ϕ_l.
   (iii) The total energy stored in the field of the two air gaps g.
   (iv) The stored energy associated with the leakage flux ϕ_l.
   (v) The self-inductance of the 200-turn winding.
   (vi) The energy stored in the self-inductance ($\tfrac{1}{2}Li^2$).
(b) Calculate
   (i) The mechanical energy input or output. (Which is it?)
   (ii) The electromagnetic energy input or output (Which is it?) during the increase in the length of the air gaps.

(c) If the original length of each of the gaps g is 0.125 in., calculate the force.
(d) Neglect fringing and calculate the flux density in the air gaps g.
(e) Calculate the force on the basis of area and flux density and compare with that of part (c).

HINT: Use Eq. 3-71.

(f) Calculate the average voltage induced in the winding if the time during which the air gap is increased is 0.01 sec.

4-6 When each of the variable air gaps g in Fig. 4-9 has a length of 0.125 in. and the current in the 1,000-turn winding is 5 amp, the fluxes ϕ_a and ϕ_l are 0.002 and 0.001 weber, respectively. The length of air gap g' is assumed to be fixed. Neglect fringing and the reluctance of the iron and calculate

(a) The length and the cross-sectional area of the fixed air gap g' if the cross-sectional areas of all three air gaps are equal.
(b) The permeance of the fixed air gap g'.
(c) The permeance of the two variable air gaps g in series as a function of the length of a single gap.
(d) The total permeance of the magnetic circuit as a function of the length of a single air gap g.
(e) The self-inductance of the 1,000-turn winding as a function of the length of a single air gap g.
(f) The total force on the movable member or armature based on the relationship

$$f = \frac{1}{2} i^2 \frac{dL}{dx}$$

for a current of 5 amp when the length of each air gap g is 0.125 in.
(g) The force on the faces of the fixed air gap g' for the conditions of part (f).

4-7 The self-inductance of two identical parallel straight nonmagnetic cylinders is given by

$$L = 4 \times 10^{-7}\left(\ln \frac{d}{r} + \frac{1}{4}\right) \quad \text{h per m length}$$

where d is the distance between centers (small in comparison with the length of the conductors), and where r is the radius of each conductor. During a short circuit on a power system, the current in two such conductors reaches a peak value of 100,000 amp. Calculate the force per meter length on each conductor if the radius of each conductor is 1 in. and the spacing is 6 in. between centers.

4-8 Figure 4-10 shows a rotary electromagnet.

(a) Neglect the reluctance of the iron, leakage, and fringing and determine for the rotor position in Fig. 4-10(a) when the current in the stator

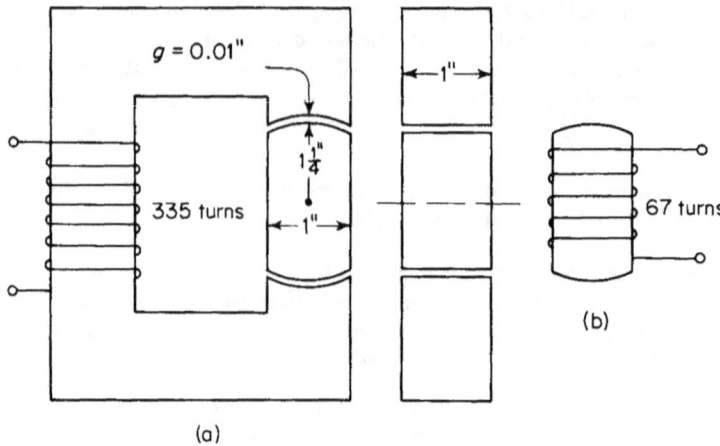

Figure 4-10. (a) Rotary electromagnet winding on stator only, for Problem 4-8, (b) rotor of rotary electromagnet with winding for Problems 4-9 and 4-10.

winding is 1.5 amp
  (i) The flux in the air gap.
  (ii) The flux in the iron.
  (iii) The self-inductance of the stator.
  (iv) The permeance of the magnetic circuit.
  (v) The reluctance of the magnetic circuit.
  (vi) The energy stored in the air gap.
  (vii) The energy stored in the iron.
(b) Repeat part (a) above, but with the rotor displaced 15°
  (i) In the clockwise direction.
  (ii) In the counterclockwise direction.
(c) What is the torque in both parts of (b) above?
(d) Repeat part (a) above when the rotor is in the position shown in Fig. 4-10(a), taking into account the reluctance of the iron (USS Electrical Annealed Sheet Steel) if the mean length of the path in the iron is 8.5 in. Use Fig. 3-12 for determining the effect of the iron.

4-9 Suppose that a winding of 67 turns were placed on the rotor [see Fig. 4-10(b)] of the electromagnet in Problem 4-8 and a current of 7.5 amp passed through this winding while the current in the stator winding was zero. Neglect leakage and fringing, but take into account the reluctance of the iron, and determine
  (a) The flux in the air gap.
  (b) The flux linkages of the rotor winding.
  (c) The self-inductance of the rotor on the basis of Eq. 4-6.
  (d) The permeance of the magnetic circuit.
  (e) The flux linking the stator winding.
  (f) The flux linkage of the stator winding due to the current in the rotor winding.

PROBLEMS   197

(g) The mutual inductance between stator and rotor on the basis of flux linkage per ampere.

4-10 In the electromagnet of Fig. 4-10(a), assume a winding of 335 turns on the stator and a winding of 67 turns on the rotor. Neglect leakage, fringing, and the reluctance of the iron. Determine

(a) The flux that links the rotor winding when the stator winding carries 1.5 amp and when there is no current in the rotor winding.

(b) On the basis of part (a) above, mutual inductance between the stator and rotor windings.

(c) The flux that links the stator winding when the rotor winding carries 1.5 amp and when there is no current in the stator winding.

(d) On the basis of part (c) above, the mutual inductance between the two windings.

(e) The self-inductance of the electromagnet when the two windings are connected series aiding and when the current is 1.25 amp. Neglect fringing, leakage, and the reluctance of the iron.

4-11 Figure 4-11 shows a cylindrical iron-clad electromagnet with a cylindrical plunger. When actuated by the current in the winding, the plunger moves in

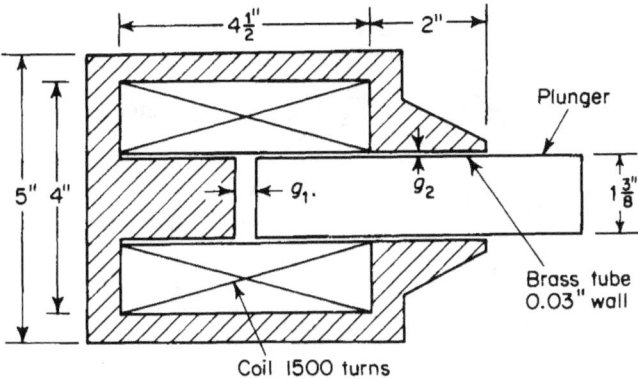

Figure 4-11. Cylindrical plunger magnet for Problem 4-11.

a brass guide tube of 0.03-in. wall thickness. The length of the air gap g_1 varies with the position of the plunger, whereas the gap g_2 is fixed by the wall thickness of the brass tube and is assumed constant of length 0.03 in. The relative permeability of brass is 1. Neglect magnetic leakage, fringing, and the reluctance of the iron.

(a) Calculate for length of $g_1 = 0.125$ and 0.500 in.
  (i) The magnetic reluctance.
  (ii) The magnetic permeance of the two gaps in series.
(b) Calculate the flux in the magnet for each air-gap length in part (a) when the winding carries a constant current of 2.2 amp.

(c) Calculate the force on the plunger for each air-gap length in part (a) when the current is 2.2 amp, based on Eq. 4-67.
(d) Express the self-inductance of the magnet as a function of g_1 when g_1 is in inches and when it is in meters.
(e) Calculate the mechanical energy associated with the movement of the plunger when the air gap g_1 is decreased from a length of 0.500 in. to 0.125 in. while the current remains constant at a value of 2.2 amp. Is motor action or generator action involved? Explain.
(f) Calculate the electrical energy required to hold the current constant at 2.2 amp while g_1 is decreased from a length of 0.500 in. to 0.125 in.
(g) Calculate the change in the energy stored in the field for parts (e) and (f) above.
(h) If the value of the current is 2.2 amp when $g_1 = 0.500$ in. and the flux remains constant, what is
(i) The value of the current when $g_1 = 0.125$ in.?
(ii) The electrical energy associated with the motion of the plunger?
(iii) The change in the stored energy from $g_1 = 0.500$ in. to $g_1 = 0.125$ in.?

4-12 Figure 4-12 shows a rotary electromagnet with a uniform air gap between the stator and the rotor. The self-inductance of the stator coils 1–1' and

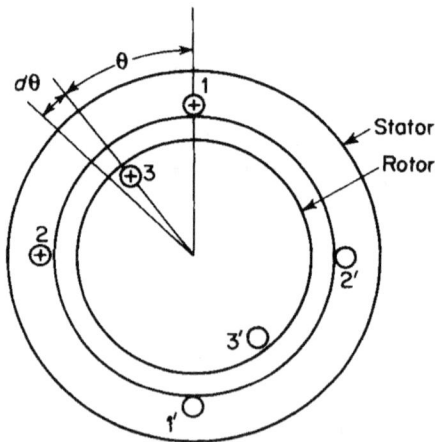

Figure 4-12. Rotary electromagnet for Problem 4-12.

2–2' is $L_{11} = L_{22} = 0.20$ h and the self-inductance of the rotor coil 3–3' is 0.10 h. The mutual inductances between the stator coils and the rotor coil are $L_{13} = 0.125[1 - (2\theta/\pi)]$ h and $L_{23} = 0.250\theta/\pi$ h when the rotor is in the position shown.
(a) Calculate the torque when the currents are in the directions indicated and have the following values

$i_1 = +10$ amp, $\quad i_2 = +5$ amp, and $i_3 = +15$ amp
for $\theta = 30°$.

(b) What is the value of θ at which the torque is zero for the values of current in part (a)?

4-13 Figure 4-13 shows the magnetization curve of an iron-core reactor. The exciting winding has 400 turns and carries a direct current of 4 amp. The

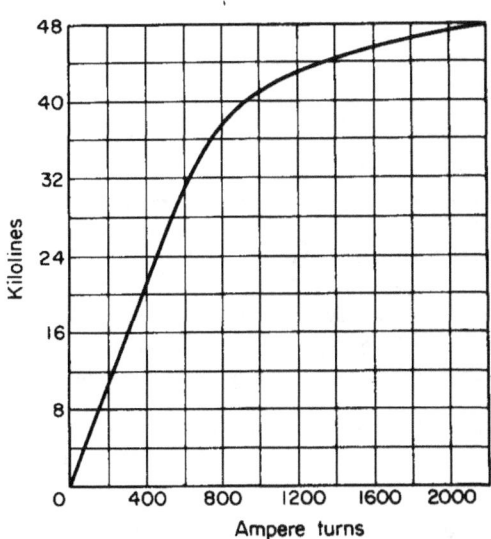

Figure 4-13. Saturation curve of magnetic circuit with air gap for Problem 4-13.

leakage flux is negligible. Determine

(a) The magnetic flux.
(b) The magnetic flux linkage λ.
(c) The apparent inductance L_a.
(d) The effective inductance L_e.
(e) The differential inductance L_d.
(f) The range of currents for which the three inductances L_a, L_e, and L_d have the same value.

4-14 The hypothetical magnetization curve shown in Fig. 4-14 approximates that of an electromagnet. Express the energy W_ϕ stored in the magnetic field and the coenergy W'_ϕ in terms of i_1, i_2, λ_1, and λ_2 for a current of i_1 amp and for a current of i_2 amp.

4-15 Find the change in the energy stored in the field, the change in the coenergy, and the electromagnetic energy input when the air gap in the electromagnet having the hypothetical magnetization curves shown in Fig. 4-15 changes from g_1 to g_2

(a) While the mmf is held constant at 6.0 amp.
(b) While the flux linkage is held constant at 0.12 weber turn.
(c) If the locus of λ vs i during the change in gap is along the line *ab*.

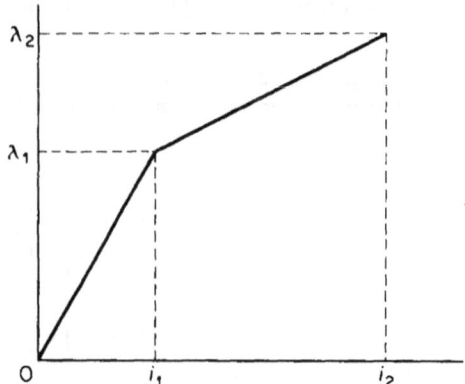

Figure 4-14. Hypothetical magnetization curve for Problem 4-14.

4-16 Calculate the average forces for parts (a), (b), and (c) in Problem 4-15 if the change in air-gap length from g_1 to g_2 is 0.10 in. Does the length of the air gap increase or decrease in going from g_1 to g_2? Explain.

4-17 The no-load current in a transformer winding is 9.1 amp when the voltage across the same winding is 6,600 v at 60 cycles. The real power is 9,300 w. Determine

(a) The reactive power.
(b) The Q of this transformer winding at no load on the basis of the real and reactive power.

4-18 A choke coil has a self-inductance of 9.0 h. The effective resistance of the winding is 250 ohms at a frequency of 1,000 cps. Determine the Q of the

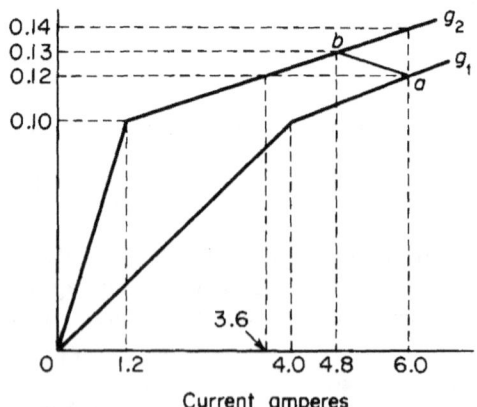

Figure 4-15. Hypothetical magnetization curves for Problems 4-15 and 4-16.

choke for a frequency of

(a) 1,000 cps.

(b) 60 cps on the basis that the self-inductance and the effective resistance have the same values as at 1,000 cps.

4-19 Prove that Eqs. 4-109, 4-110, and 4-111 are equivalent.

BIBLIOGRAPHY

Dwight, H. B., *Electrical Elements of Power Transmission Lines*. New York: The Macmillan Company, 1954.

Fano, R. M., L. J. Chu, and R. B. Adler, *Electromagnetic Fields, Energy, and Forces*. New York: John Wiley & Sons, Inc., 1960.

Fitzgerald, A. E. and C. Kingsley, *Electric Machinery*. 2nd. ed. New York: McGraw-Hill Book Company, 1959.

Karapetoff, V., *The Magnetic Circuit*. New York: McGraw-Hill Book Company, 1911.

Rosa, E. B. and F. W. Grover, "Formulas and Tables for the Calculation of Mutual and Self-inductance," NBS *Sci Paper*, 169, Washington, D.C., 1916.

Timbie, W. H. and B. Bush, *Principles of Electrical Engineering*. New York: John Wiley & Sons, Inc., 1940.

White, D. C. and H. H. Woodson, *Electrochemical Energy Conversion*. New York: John Wiley & Sons, Inc., 1959.

Winch, R. P., *Electricity and Magnetism*. Englewood Cliffs, N.J.: Prentice-Hall, Inc., 1955.

Woodruff, L. F., *Principles of Electric Power Transmission*. New York: John Wiley & Sons, Inc., 1938.

EXCITATION CHARACTERISTICS OF IRON-CORE REACTORS AND TRANSFORMERS

5

5-1 IRON-CORE REACTORS

A reactor is a coil that has substantial inductance. Reactors that operate at and below audio frequencies usually have magnetic cores of ferromagnetic material such as laminated silicon steel. Such cores may or may not contain one or more appreciable air gaps in series with the flux path. Iron-core reactors are used for a variety of purposes, among which are suppressing the a-c flux ripple in rectifier circuits, compensating long telephone circuits, limiting starting currents in motors, and stabilizing the arc in sun lamps, arc furnaces, and electric welders.

The ferromagnetic core is used because of its comparatively high relative permeability μ_r, so that even in combination with appreciably long air gaps a magnetic circuit with high permeance is obtained. This results in the desired value of inductance with a much smaller size than if an air core were used. Inductance is related to permeance as follows

$$L = N^2 \mathcal{P} \tag{5-1}$$

where L = the inductance in henrys
N = the number of turns
$\mathcal{P}$ = the permeance in webers per ampere turn

Reactors with a more-or-less fixed value of self-inductance are also known as chokes or choke coils. One of the advantages of reactors is that the energy loss is usually small compared with the energy loss of resistors that have the same ohmic value. Generally, for a given volt-ampere rating, the size of the reactor decreases with increasing frequency, whereas the size of a resistor must be the same regardless of the frequency for a given power rating. In many cases, one or more air gaps are used in series with the iron to overcome the undesirable effects of nonlinearity in the B vs H characteristic of the iron. This nonlinearity causes the inductance to vary with saturation. It also causes the wave form of the current through the reactor winding to be different from that of the flux in the core. In some cases where the current has an appreciable d-c component, an air gap is used to prevent the d-c component from saturating the core and thus reducing the reactance. In the case of an air-core reactor, the B-H characteristic is linear as $B = \mu_0 H$ and the current has the same wave form as the flux.

5-2 VOLTAGE CURRENT AND FLUX RELATIONS

Figure 5-1 is a schematic diagram of an air-core reactor. Assume the frequency of the voltage and the size of the winding to be low enough so that capacitance effects are negligible. Then if v is the instantaneous applied voltage, i the instantaneous current, R the resistance of the winding, and λ the instantaneous flux linkage, we have

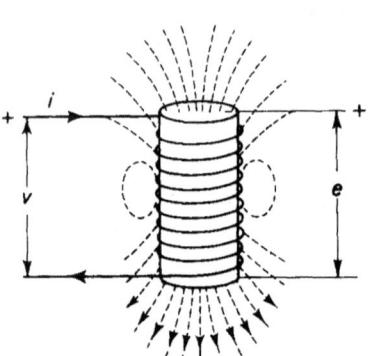

Figure 5-1. Air-core reactor.

$$v = Ri + \frac{d\lambda}{dt} \qquad (5\text{-}2)$$

or

$$v = Ri + e \qquad (5\text{-}3)$$

where e is the induced emf and equals the instantaneous change in the flux linkage λ. In Fig. 5-1, when the current is increasing while flowing in the indicated direction, the flux linkage λ is increasing and the voltage e is positive with polarities as shown. Hence

$$e = +\frac{d\lambda}{dt} \qquad (5\text{-}4)$$

If the flux linkage varies sinusoidally with time as expressed by

$$\lambda' = \lambda_m \sin \omega t \qquad (5\text{-}5)$$

where λ_m is the maximum instantaneous flux linkage, N is the number of turns in the winding of the reactor and ϕ is the equivalent flux linking all

N turns, then

$$\lambda = N\phi \tag{5-6}$$

from which it follows that

$$\phi = \phi_m \sin \omega t \tag{5-7}$$

where ϕ_m is the maximum value of the equivalent flux. Substitution of Eqs. 5-6 and 5-7 in 5-2 gives

$$v = Ri + \omega N \phi_m \cos \omega t \tag{5-8}$$

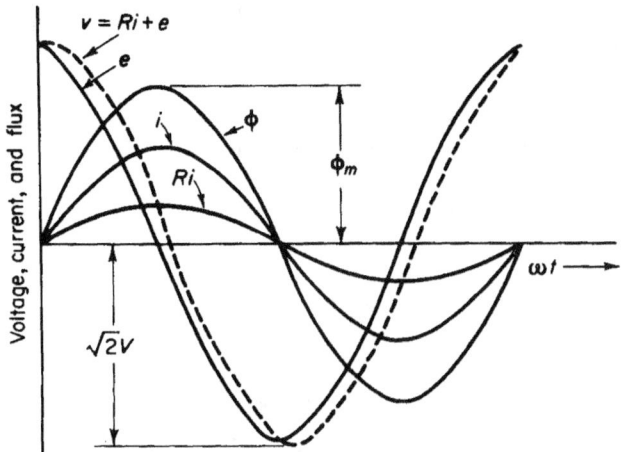

Figure 5-2. Waves of voltage, current, and flux in air-core reactor.

and the instantaneous emf is

$$e = \omega N \phi_m \cos \omega t \tag{5-9}$$

Since the magnetization curve of air is linear, the current in the case of an air-core has exactly the same wave form as the flux when expressed as a function of time. In addition, the current wave is in time phase with the flux wave. This follows from Ohm's Law of the magnetic circuit, i.e.

$$F = \phi \mathscr{R} \qquad Ni = \phi \mathscr{R} \tag{5-10}$$

Where $\mathscr{R}$ is the reluctance of the magnetic path and also constant for air. Hence, from Eqs. 5-7 and 5-10

$$i = \frac{\phi \mathscr{R}}{N} = \frac{\mathscr{R}}{N} \phi_m \sin \omega t \tag{5-11}$$

which, when expressed in terms of the effective value of the current I, becomes

$$i = \sqrt{2}\, I \sin \omega t \tag{5-12}$$

where $\sqrt{2}\, I$ is the amplitude of the current wave. Wave forms of current and voltage for an air-core reactor are shown in Fig. 5-2.

In the case of a reactor that has a ferromagnetic core of uniform cross section and where the length of flux path is great enough in relation to the cross-sectional area of the path so that the flux density may be considered uniform, the relationship between the flux ϕ and the current in the winding is not linear, but is represented graphically by a hysteresis loop. Under these conditions we have

$$\phi = BA \tag{5-13}$$

where B is the flux density and A the cross-sectional area of the flux path. Also

$$Ni = Hl$$

and

$$i = \frac{Hl}{N} \tag{5-14}$$

In the hysteresis loop, the flux density B is normally plotted against the magnetic field intensity H. Hence, when plotting ϕ against i, the effect is merely that of multiplying the horizontal scale of the loop by l/N. Both these multipliers—A and l/N—are constant and the shape of the loop is unchanged.

Figure 5-3 shows the upper half of such a hysteresis loop. The bottom of the loop has the same shape as the upper half. This loop shows the relationship between the instantaneous flux ϕ and the instantaneous current i in the winding on the basis that the eddy currents are negligible. The effect of the eddy currents is to widen the loop somewhat. This is evident from the fact that the eddy currents are in phase with the voltage and are therefore at maximum value when the flux is zero because the current wave is 90° behind the voltage wave. A loop that includes the effect of eddy currents is known as a dynamic hysteresis loop.

Suppose that the resistance of the reactor winding is negligible. (This is true for many iron-core reactors.) Furthermore, let a sinusoidal voltage

$$v = \sqrt{2}\, V \cos \omega t \tag{5-15}$$

be applied to the reactor winding. The flux in the core will also be a sine wave as expressed by Eq. 5-4. However, the flux wave lags the voltage by 90°.

Sec. 5-2 VOLTAGE CURRENT AND FLUX RELATIONS

Generally, in iron-core reactors the leakage flux is small enough so that the flux in the core may be related to the voltage as in Eq. 5-9, and we have

$$V = \omega N \phi_m \cos \omega t \tag{5-16}$$

which is a sinusoid with an amplitude of

$$V_m = \sqrt{2}\, V = \omega N \phi_m \tag{5-17}$$

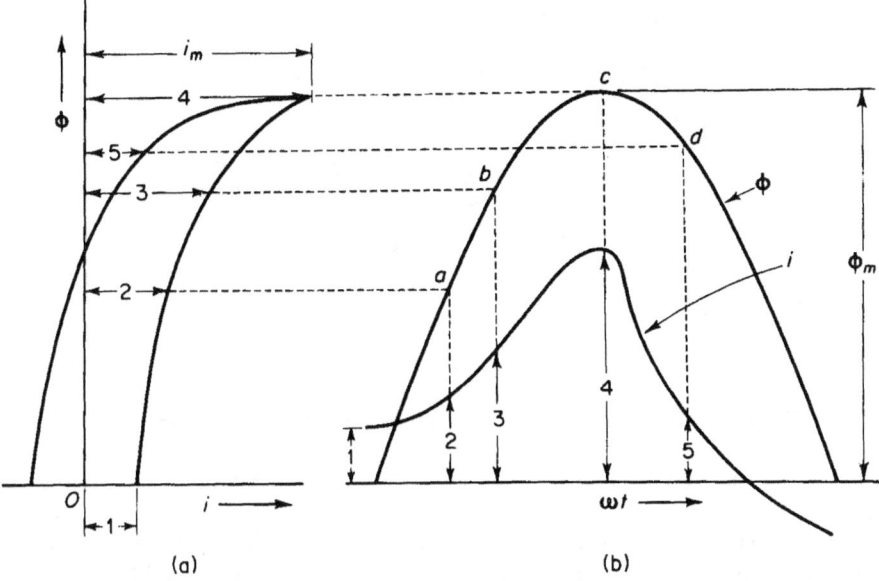

Figure 5-3. (a) Upper half of hysteresis loop; (b) sinusoidal flux and exciting current waves.

but

$$\omega = 2\pi f \text{ radians per sec} \tag{5-18}$$

where f is the frequency in cycles per second, so that

$$V \approx E = \frac{2\pi f N \phi_m}{\sqrt{2}} = 4.44\, f N \phi_m \tag{5-19}$$

from which the flux is found to be

$$\phi_m = \frac{V}{4.44\, fN}$$

The factor 4.44 in Eq. 5-19 applies only to a sinusoidal voltage wave. A more general expression is

$$\phi_m = \frac{V}{4 \times \text{form factor } fN} \tag{5-20}$$

where the form factor is the ratio of the rms value to the average value of one-half the wave. In case of a square voltage wave, the rms and average

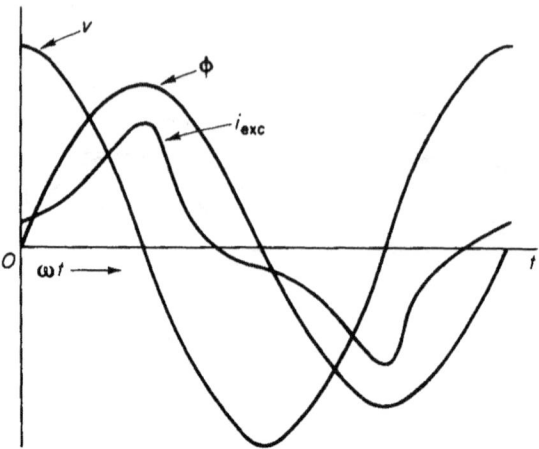

Figure 5-4. Waves of voltage, flux, and current in iron-core reactor with negligible resistance.

values are equal, and the form factor is unity. The form factor of a sine wave is 1.11.

The flux is plotted as a function of time in Fig. 5-3(b). If the hysteresis loop could be represented by a straight line, the current would have the same wave form as the flux. However, this not being the case, the current cannot have the same wave form as the flux. The wave form of the current, if eddy current is neglected, is found graphically as follows.

When the flux is zero and increasing, the instantaneous current is 1 on the hysteresis loop and is shown as 1 in Fig. 5-3(b). When the flux is a and increasing, the current is 2 on the hysteresis loop. This can be continued in the following tabulation.

Flux ϕ	a	b	c	d
Current i	2	3	4	5

Thus, by plotting the values 1, 2, 3, 4 5 for the current along the time axis in Fig. 5-3(b) on the flux values 0, a, b, c, d, the wave form of the current is obtained. In order to obtain a well-defined current wave, a greater number of points than those listed in the tabulation above are required. From Fig. 5-3 it is evident that, although the flux varies sinusoidally with time, the current is not a simple sinusoidal function of time, but is distorted. In addition to the current shown in Fig. 5-3(b), there is a small component due to the eddy currents. This component of current is practically sinusoidal and is in phase with the voltage or 90° ahead of the flux wave. Waves of voltage, flux, and current are shown for an iron-core reactor with a winding of negligible resistance in Fig. 5-4. The voltage wave is sinusoidal—so is the flux wave—but the current wave is not. The flux wave lags the voltage wave by 90° since the winding was assumed to have negligible resistance.

5-3 HARMONICS

Under steady conditions of constant voltage and constant frequency, the current in a reactor, or the exciting current (the no-load current in a transformer), is a periodic function of time. A periodic function can be represented by a Fourier series as follows

$$(t) = a_0 + a_1 \cos \omega t + a_2 \cos 2\omega t + a_3 \cos 3\omega t$$
$$+ a_4 \cos 4\omega t + \cdots + a_n \cos n\omega t + \cdots$$
$$+ b_1 \sin \omega t + b_2 \sin 2\omega t + b_3 \sin 3\omega t$$
$$+ b_4 \sin 4\omega t + \cdots + b_n \sin n\omega t \quad (5\text{-}20)$$

The constant a_0 corresponds to a d-c component. Under steady-state conditions, if the applied voltage does not contain a d-c component, the constant a_0 is equal to zero. With a-c excitation only, the exciting current has symmetrical wave form, i.e., the negative half cycle has the same shape as the positive half cycle. Figure 5-5 shows a symmetrical wave, i.e., one in which $f[t + (T/2)] = -f(t)$.

A symmetrical wave contains odd harmonics only because the presence of even harmonics leads to dissymmetry. Then, for a constant frequency, the angular velocity ω must be constant, and the wave is symmetrical when

$$f(\omega t + \omega T/2) = -f(\omega t) \quad (5\text{-}21)$$

In Eq. 5-21, T is the period or the time for the wave to undergo a complete cycle. Hence, $\omega T = 2\pi$ radians, and Eq. 5-21 can be reduced to

$$f(\omega t + \pi) = -f(\omega t) \quad (5\text{-}22)$$

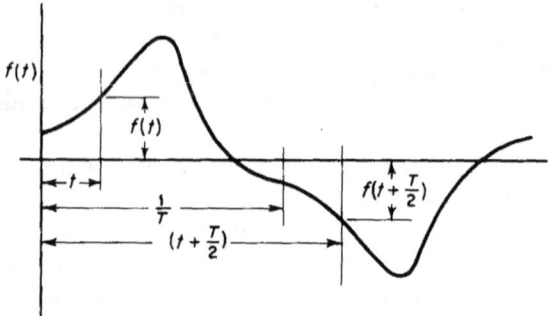

Figure 5-5. Symmetrical wave.

If the nth harmonic in the cosine series is to satisfy the conditions for symmetry

$$\cos n(\omega t + \pi) = -\cos n\omega t \tag{5-23}$$

but

$$\cos n(\omega t + \pi) = \cos n\omega t \cos n\pi \tag{5-24}$$

then, from Eqs. 5-23 and 5-24, symmetry requires that

$$\cos n\omega t \cos n\pi = -\cos n\omega t$$

and

$$\cos n\pi = -1 \tag{5-25}$$

Equation 5-24 can be satisfied only if n is odd; as for even values of n, $\cos n\pi$ is 1.
Similarly

$$\sin n(\omega t + \pi) = -\sin n\omega t \tag{5-26}$$

only if n is odd.

It follows, therefore, that only odd harmonics are present in a periodic function that is symmetrical, as is the case for an iron-core reactor or the exciting current in an iron-core transformer when the flux wave is symmetrical. Such a current can therefore be represented by omitting a_0 and all the even harmonics in Eq. 5-20 as follows

$$f(t) = a_1 \cos \omega t + a_3 \cos 3\omega t + a_5 \cos 5\omega t + \cdots$$

$$+ b_1 \sin \omega t + b_3 \sin 3\omega t + b_5 \sin 5\omega t + \cdots \tag{5-27}$$

Sec. 5-3 HARMONICS 211

Equation 5-27 can also be put into the following form

$$f(t) = c_1 \cos(\omega t + \alpha_1) + c_3 \cos(3\omega t + \alpha_3) + c_5 \cos(5\omega t + \alpha_5)$$
(5-28)
$$+ \cdots + c_n \cos(n\omega t + \alpha_n) + \cdots$$

where

$$c_n = \sqrt{a_n^2 + b_n^2} \quad (5\text{-}29)$$

$$\alpha_n = -\tan^{-1}\left(\frac{b_n}{a_n}\right) \quad (5\text{-}30)$$

Hence, if I_1 is the effective value of the fundamental component of current, and I_3, I_5, etc. are the effective values of the harmonic components, the exciting current can be expressed by

$$i_{\text{exc}} = \sqrt{2}\bigg[I_1 \cos(\omega t + \alpha_1) + I_3 \cos(3\omega t + \alpha_3)$$
(5-31)
$$+ I_5 \cos(5\omega t + \alpha_5) + \cdots + I_n \cos(n\omega t + \alpha_n) + \cdots\bigg]$$

A sinusoidal flux variation with respect to time produces a distorted current in an iron-core reactor or in iron-core transformers, which can be divided into a fundamental component and into odd harmonic components. The third harmonic is the predominating harmonic.

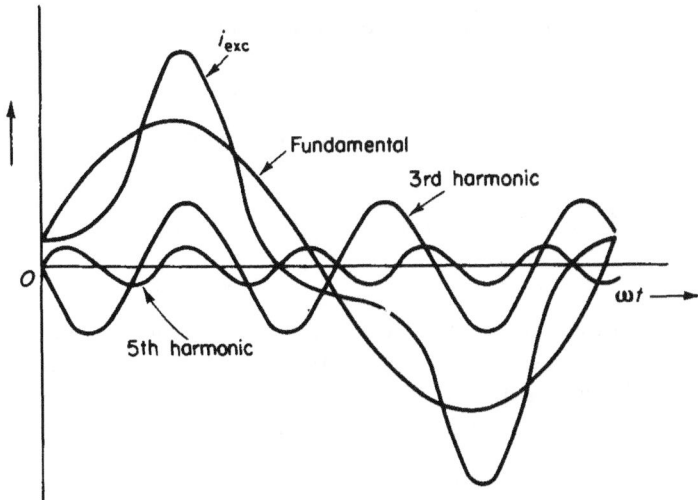

Figure 5-6. Exciting current and its fundamental, 3rd, and 5th harmonic components.

The amplitudes of the harmonics generally decrease as the order of the harmonics increases, harmonics of higher orders than the 9th usually having negligible amplitudes. Figure 5-6 shows a current wave with its fundamental, 3rd harmonic, and 5th harmonic components.

5-4 POWER

The real power consumed in a circuit is the average of the instantaneous values over a given period of time. For a-c circuits, the real power is the average taken over one or more complete cycles. Thus, if a complex voltage wave is expressed by

$$v(t) = V_0 + \sqrt{2}\, V_1 \cos(\omega t + \beta_1) + \sqrt{2}\, V_2 \cos(2\omega t + \beta_2)$$
$$+ \cdots + \sqrt{2}\, V_n \cos(n\omega t + \beta_n) + \cdots \quad (5\text{-}32)$$

and a complex current wave by

$$i(t) = I_0 + \sqrt{2}\, I_1 \cos(\omega t + \alpha_1) + \sqrt{2}\, I_2 \cos(2\omega t + \alpha_2)$$
$$+ \cdots + \sqrt{2}\, I_k \cos(k\omega t + \alpha_k) \quad (5\text{-}33)$$

the real power is then obtained as follows

$$P = \frac{1}{2\pi} \int_0^{2\pi} v(t) i(t)\, d(\omega t) \quad (5\text{-}34)$$

where the product $v(t)i(t)$ results in the following series

$$v(t)i(t) = V_0 I_0 + 2V_1 I_1 \cos(\omega t + \beta_1) \cos(\omega t + \alpha_1)$$
$$+ 2V_2 I_2 \cos(2\omega t + \beta_2) \cos(\omega t + \alpha_2) + \cdots$$
$$+ 2V_n I_n \cos(n\omega t + \beta_n) \cos(n\omega t + \alpha_n) + \cdots$$
$$+ 4V_n I_k \cos(n\omega t + \beta_n) \cos(k\omega t + \alpha_k) + \cdots$$
$$(5\text{-}35)$$

It is not necessary to integrate every term in the series represented in Eq. 5-34 in order to evaluate the series. However, by considering 9 of the several typical terms, the integral can be evaluated in a simpler manner. Let us first consider the first term in the series of Eq. 5-34 as follows

$$P_{00} = \frac{1}{2\pi} \int_0^{2\pi} V_0 I_0\, d(\omega t) = V_0 I_0 \quad (5\text{-}36)$$

Then consider the product of the nth voltage harmonic and the nth current harmonic when n may have any integral value 1, 2, 3, etc.

$$P_{nn} = \frac{1}{2\pi} \int_0^{2\pi} V_n I_n \cos(n\omega t + \beta_n) \cos(n\omega t + \alpha_n)\, d(\omega t) \qquad (5\text{-}37)$$

$$= V_n I_n \cos(\beta_n - \alpha_n)$$

There is also in Eq. 5-37 the series of cross-products represented by the term

$$4 V_n I_k \cos(n\omega t + \beta_n) \cos(k\omega t + \alpha_k)$$

which can be reduced to

$$2 V_n I_k \{\cos[(n+k)\omega t + \beta_n + \alpha_k] + \cos[(n-k)\omega t + \beta_n - \alpha_k]\} \qquad (5\text{-}38)$$

The average value of the term above, taken over a complete cycle, is zero. From this it is apparent that *the real power associated with a voltage harmonic of a given order and a current harmonic of a different order is zero.* Therefore, the expression for the power obtained by carrying out the integration of Eq. 5-33 becomes

$$P = V_0 I_0 + V_1 I_1 \cos(\beta_1 - \alpha_1) + V_2 I_2 \cos(\beta_2 - \alpha_2) + \cdots$$
$$+ V_n I_n (\cos \beta_n - \alpha_n) + \cdots \qquad (5\text{-}39)$$

In Eq. 5-38, the angle

$$(\beta_n - \alpha_n) = \theta_n \qquad (5\text{-}40)$$

or the angle by which the nth current harmonic lags or leads the nth voltage harmonic, and

$$\cos(\beta_n - \alpha_n) = \cos \theta_n = \text{PF}_n \qquad (5\text{-}41)$$

where PF_n is the power factor associated with the nth voltage harmonic and the nth current harmonic. In the case of a sinusoidal voltage and a nonsinusoidal current, the power is the product of the effective value of the voltage, the effective value of the fundamental in the current, and the cosine of the angle by which the fundamental in the current is displaced from the voltage. The same consideration applies to a sinusoidal current and nonsinusoidal voltage.

5-5 EFFECTIVE CURRENT

A nonsinusoidal current is expressed by Eq. 5-32 as follows

$$i(t) = I_0 + \sqrt{2}\, I_1 \cos(\omega t + \alpha_1) + \sqrt{2}\, I_2 \cos(2\omega t + \alpha_2) \tag{5-42}$$
$$+ \cdots + \sqrt{2}\, I_k \cos(k\omega t + \alpha_k) + \cdots$$

It should be remembered that the coefficients $\sqrt{2}\, I_1$, $\sqrt{2}\, I_2$, $\sqrt{2}\, I_k$ are the amplitudes of the various harmonic components in the current. If the current represented by Eq. 5-42 were passed through a noninductive resistance of R ohms, the voltage across that resistor would become

$$v(t) = I_0 R + \sqrt{2}\, I_1 R \cos(\omega t + \alpha_1) + \sqrt{2}\, I_2 R \cos(2\omega t + \alpha_2) \tag{5-43}$$
$$+ \cdots + \sqrt{2}\, I_k \cos(k\omega t + \alpha_k)$$

The average power, as read by an indicating wattmeter, consumed by the noninductive resistance R, can be expressed either as

$$P = \frac{1}{2\pi} \int_0^{2\pi} v(t) i(t)\, d(\omega t) \tag{5-44}$$

or as

$$P = I^2 R \tag{5-45}$$

where I in Eq. 5-45 is the effective value of the current expressed by Eq. 5-42. Hence

$$I^2 R = \frac{1}{2\pi} \int_0^{2\pi} v(t) i(t) \tag{5-46}$$

which, from Eq. 5-38, becomes

$$I^2 R = I_0 R I_0 + I_1 R I_1 + I_2 R_1 I_2 + \cdots + I_n R I_n + \cdots$$
$$= (I_0^2 + I_1^2 + I_2^2 + \cdots + I_n^2 + \cdots) R$$

or

$$I = \sqrt{I_0^2 + I_1^2 + I_2^2 + \cdots + I_n^2 + \cdots} \tag{5-47}$$

5-6 CORE-LOSS CURRENT AND MAGNETIZING CURRENT

The fundamental, 3rd, and 5th harmonic components of the current in a reactor are shown in Fig. 5-6. Such a current could also be the exciting current or no-load current in a transformer. The fundamental component i_{exc_1} is shown in Fig. 5-7 along with the flux wave ϕ and induced voltage wave e. Sinusoidal flux variation is assumed. In Section 5-4 it was shown

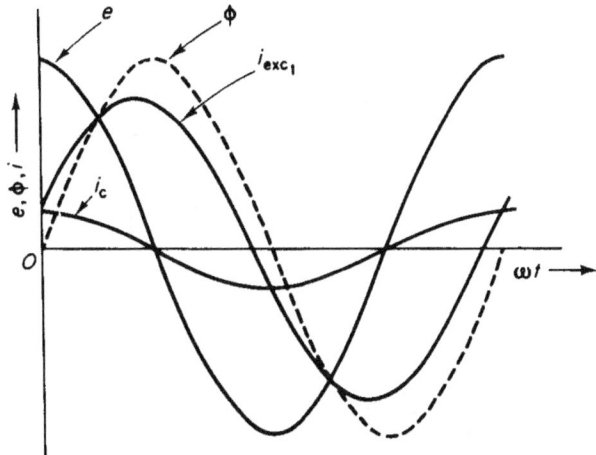

Figure 5-7. Core-loss current, fundamental component of exciting current, flux, and induced voltage waves.

that in the case of sinusoidal voltage the harmonics in the current wave do not contribute to the real power. Therefore, core loss, i.e., hysteresis and eddy-current losses, can be taken into account by taking the product of the induced emf and the component of the fundamental in the current that is in phase with the induced emf. This component of current is called the core-loss current and is represented by i_c in Fig. 5-7.

If the current wave representing the core-loss current i_c is subtracted from the total fundamental component of current i_{exc_1}, the result is a sine wave of current that is in phase with the flux wave ϕ. If there were no harmonics, this latter component of current would be the magnetizing current. Under these conditions the two currents, i.e., core-loss and magnetizing currents, as well as the flux and the induced voltage, could be represented by the phasor diagram shown in Fig. 5-8. However, the magnitude of the exciting current is actually greater than that resulting from the fundamental component, as it does include all the harmonic components. Since the harmonics in the exciting current do not contribute to the real power if the induced voltage is sinusoidal, the core-loss component I_c in Fig. 5-8 is not affected by inclusion or exclusion of the harmonics. In accordance with Eq. 5-46, the exciting current has an rms or effective value of

$$I_{\text{exc}} = \sqrt{I_c^2 + I_{\text{mag}_1}^2 + I_3^2 + I_5^2 + \cdots} \tag{5-48}$$

It must be kept in mind that under a-c excitation, in the absence of a d-c component, only odd harmonics are present in the exciting current.

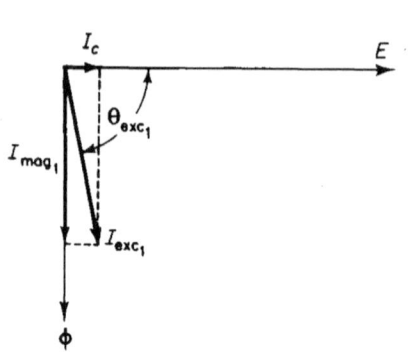

Figure 5-8. Phasor diagram of induced voltage, flux, core-loss current, and fundamental components of exciting current and magnetizing current.

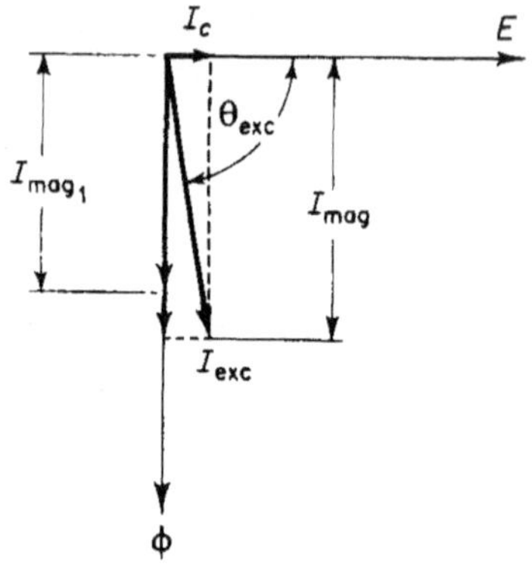

Figure 5-9. Phasor diagram including harmonics in the exciting current.

The following relationship is obtained from Eq. 5-48

$$\sqrt{I_{\text{mag}_1}^2 + I_3^2 + I_5^2 + \cdots} = \sqrt{I_{\text{exc}}^2 - I_c^2} \tag{5-49}$$

Generally, phasor diagrams apply to sinusoidal quantities, but may be used to include nonsinusoidal currents and voltages if these are replaced by equivalent sinusoidal quantities, i.e., sinusoids that have the same rms or effective values. If the exciting current is represented by a phasor I_{exc} as in Fig. 5-9 and is treated as an equivalent sinusoidal current, the phasor difference expressed by the right-hand side of Eq. 5-49 leads to a phasor that lags the core-loss current by 90°. The equivalent sinusoid represented by this phasor is called the *magnetizing current*, and from Eq. 5-49 is seen to be

$$I_{\text{mag}} = \sqrt{I_{\text{mag}_1}^2 + I_3^2 + I_5^2 + \cdots} \tag{5-50}$$

Equation 5-50 suggests that the harmonic components be included in the magnetizing component of the exciting current. This facilitates the development of a phasor diagram and a simple equivalent circuit. A phasor diagram including the harmonics in the magnetizing current is shown in Fig. 5-9.

EXAMPLE 5-1: The following quantities were measured at no load on the 240-volt winding of a 30-kva, 2400/240-volt, 60-cycle transformer at rated frequency

$$\text{Volts} = 240$$

Exciting current $= 2.53$ amp

$$\text{Power} = 164 \text{ w}$$

Determine the core-loss current and the magnetizing current. Neglect the resistance of the transformer winding.

Solution: The exciting current $I_{exc} = 2.53$ amp, and

$$I_c = \frac{P_c}{V} = \frac{164}{240} = 0.683 \text{ amp}$$

$$I_{mag} = \sqrt{I_{exc}^2 - I_c^2} = \sqrt{(2.53)^2 - (0.683)^2}$$
$$= 2.44 \text{ amp}$$

5-7 EQUIVALENT CIRCUITS

On the basis of Section 5-6, equivalent circuits are used to represent iron-core reactors in circuit analysis. A reactor may be represented by one of several different equivalent circuits, all of which are approximate. The phasor diagram of Fig. 5-9 suggests a parallel circuit as shown in Fig. 5-10(a). This equivalent circuit does not take into account the resistance of the reactor winding. However, the induced voltage E is produced by the flux linkage with the reactor winding. The instantaneous voltage applied to the winding of a reactor is given by Eq. 5-2 and 5-3 as

$$v = Ri + \frac{d\lambda}{dt} \qquad v = Ri + e$$

where e is the instantaneous induced voltage. In the case of sinusoidal voltage and current, the relationship in Eq. 5-3 can be expressed in phasor form as

$$V = RI + jE \tag{5-51}$$

where $V =$ applied voltage (rms)
$I =$ current (rms)
$E =$ induced voltage (rms)
$R =$ resistance of the reactor winding

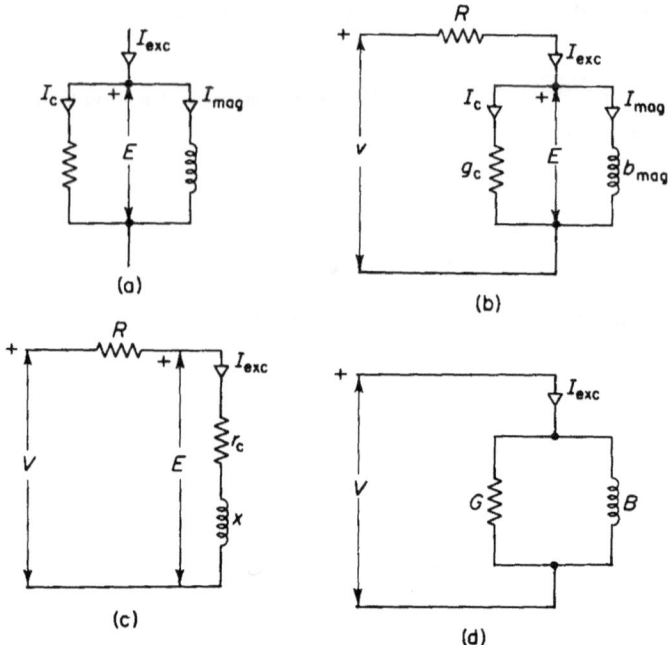

Figure 5-10. Equivalent circuits for iron-core reactors.

Although Eq. 5-3 is exact, the equivalent circuit and Eq. 5-51 are approximations since the current is not sinusoidal. Nevertheless, these approximations are very useful in many analyses, and the errors resulting therefrom are usually negligible. The equivalent circuits of Figs. 5-10(b) and 5-10(c) satisfy Eq. 5-51. In Fig. 5-10(d) the circuit parameters G and B include the effect of the winding resistance. In Fig. 5-10(c)

$$R + r_c = \frac{P}{I_{\text{exc}}^2} \tag{5-52}$$

where P is the real power in watts. In the same figure

$$x = \sqrt{(V/I_{\text{exc}})^2 - (R + r_c)^2} \tag{5-53}$$

It is customary in the equivalent series circuit to lump $R + r_c$ into a resistance R_a, the apparent resistance. In Fig. 5-10(d)

$$G = \frac{P}{V^2} \tag{5-54}$$

and

$$B = \sqrt{(I_{\text{exc}}/V)^2 - G^2} \qquad (5\text{-}55)$$

The parameters r_c, x, G, and B are not constant and a given value for any of them is valid only for the particular frequency and degree of magnetic saturation for which it was obtained. The terms r_c and G arise from core loss, which is a function of frequency and flux density. Similarly, x and G are functions of frequency and of magnetic permeability.

For example, if the voltage is increased while the frequency is held constant, the magnetic flux density increases, causing the magnetic saturation to increase, thereby producing an increase in the value of the susceptance B and a corresponding decrease in the reactance x. If the core losses varied as the square of the voltage, the conductance G would remain constant and the resistance r_c would decrease. However, the core losses vary at a rate that is somewhat less than the square of the voltage. Consequently, the conductance G decreases with increasing voltage.

5-8 EFFECT OF AIR GAPS

An iron-core reactor is much smaller than an air-core reactor of the same value of inductance and current rating. However, the inductance of an

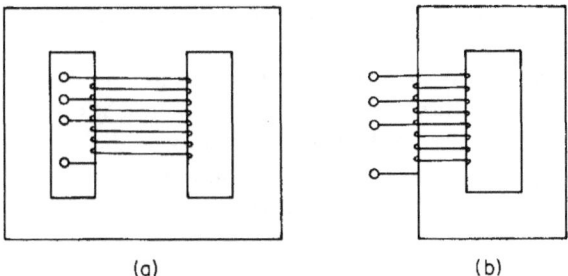

Figure 5-11. Reactors with tapped windings. (a) Shell-type; (b) core-type.

air-core reactor is constant regardless of the degree of the magnetic flux density. The inductance of an iron-core reactor varies with flux density, its value decreasing with increasing magnetic saturation.

Reactors are used for controlling current in apparatus such as arc furnaces and a-c welders. In such arrangements the winding of the reactor is generally provided with taps, so that different numbers of turns may be excited and different values of inductance or reactance obtained thereby. However, for a given tap setting and at constant frequency, the inductance should not vary excessively with variations in the voltage across the reactor. Figures 5-11(a) and 5-11(b) are schematic representations of reactors with tapped windings such that three different values of inductance may be had in each. One or more air gaps in series with the flux path will make the characteristic of the reactor more nearly linear.

In some applications the distortion in the current, caused by the nonlinear characteristic of the iron, is objectionable and the reactor is built with a small air gap in the iron circuit. The effect of the air gap is to increase the fundamental component of the magnetizing current without materially affecting the core-loss current. The air gap then has the effect of smoothing out the current waves.

Reactors that are used for decreasing the ripple in rectified current generally have air gaps in their cores to prevent the d-c component of the current from saturating the iron and making for a low value of inductance, thus imparing the effectiveness of the reactor.

Effect of air gap on core size and on stability of reactors

Reactors that, for example, are used for controlling current in arc furnaces or a-c welders carry no appreciable d-c component of current under steady-state conditions and no d-c magnetic saturation will result. Nevertheless, if such a reactor is to be stable, the inductance for a given tap setting should remain nearly constant at a given frequency for normal fluctuations in the applied voltage. This means that if there are no air gaps in the core of the reactor, the operation must be such that the flux density is confined to the linear portion below the knee of the magnetization curve.

Figure 5-12 shows the manner in which variations in the voltage applied to a given number of turns in an iron-core reactor affect the inductance. If N is the number of turns in the reactor winding, A the cross-sectional area of the core at right angles to the flux path, and l the length of the flux path, the maximum instantaneous flux linkage is

$$\lambda_m = NAB_m \tag{5-56}$$

where B_m is the maximum instantaneous flux density and the induced voltage is

$$E = 4.44 fNAB_m \tag{5-57}$$

also, the maximum value of the exciting current is

$$I_m = \frac{H_m l}{N} \tag{5-58}$$

where H_m is the magnetic field intensity required to produce the maximum flux density B_m in the core.

When hysteresis and eddy-current effects are neglected, the vertical axis of the BH curve of the iron can be provided with two scales in addition to the B scale. These are for the maximum flux linkage λ_m and the rms voltage E, and are obtained simply by multiplying the B-scale by the area turns to give λ, and by $4.44f$ times the area turns to give E, as shown in Eqs. 5-56 and 5-57. The horizontal axis can likewise be provided with a scale to give the maximum

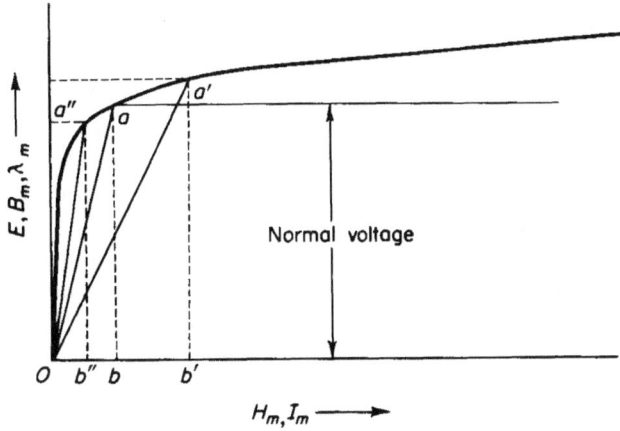

Figure 5-12. Magnetization curve of iron-core reactor.

value I_m of the exciting current by multiplying the H_m scale by the ratio of length to turns as indicated in Eq. 5-58.

In Fig. 5-12, ab represents the normal voltage when projected on the voltage scale of the vertical axis. When ab is projected on the flux linkage scale of the vertical axis, it represents the normal value of the maximum instantaneous flux linkage. The current required to produce this value of flux linkage is represented by ob projected on the current scale of Fig. 5-12. Although we are dealing with a nonlinear magnetic circuit, for the present purpose let the inductance be defined as the flux linkage per ampere. Then the normal inductance is given by

$$L = \frac{ab}{ob} \tag{5-59}$$

Suppose, because of line fluctuations or some other reason, the voltage rises from its normal value ab to that of $a'b'$. The inductance for this increased voltage is now less than normal, as shown by

$$L' = \frac{a'b'}{o'b'} \tag{5-60}$$

If, on the other hand, the voltage drops to a value of $a''b''$, the inductance will be greater than normal, rising to a value of

$$L'' = \frac{a''b''}{o''b''} \tag{5-61}$$

This shows that, when there are no air gaps and the iron is operated in the saturated region of the magnetization curve, relatively small variations in the voltage can lead to sizeable changes in the inductance of the reactor. This seems to suggest that the operation should be confined to the region below the knee of the magnetization curve where the characteristic is nearly linear. An alternative to this might be operation in region of high magnetic saturation to keep the inductance fairly constant. This, however, means high flux densities and corresponding large core losses, which are objectionable because of heat dissipation and poor efficiency. However, operation in the unsaturated region requires a core of excessive mass, as is shown in the following. If the resistance of the winding is neglected, the applied voltage and the induced emf are equal. Then, if E is the rated rms value of the voltage and I the rated rms value of the current, we have

$$\frac{E}{I} = \omega L \tag{5-62}$$

where L is the inductance of the reactor and ω is $2\pi f$ radians per sec. Equation 5-62 neglects the effects of the nonlinearity of the iron. These effects, however, are small enough in the unsaturated region so that they may be neglected in this discussion.

The value of inductance alone does not determine the size of the reactor. The size is also a function of current and frequency. In order to take these into account, multiply both sides of Eq. 5-62 by I^2 to give

$$EI = \omega L I^2 \tag{5-63}$$

which expresses the volt ampere rating of the reactor for a given frequency.

The magnetic energy stored in the iron at any instant is proportional to the square of the instantaneous current and is a maximum when the current reaches its maximum. Hence

$$W_{\phi_{max}} = \frac{L(\sqrt{2}\,I)^2}{2} = LI^2 \tag{5-64}$$

However, the maximum energy stored in the field on the basis of Eq. 3-72 is expressed by

$$W_{\phi_{max}} = \text{Vol}\,\frac{B_m H_m}{2} \tag{5-65}$$

where Vol is the volume of the core.

A comparison of Eqs. 5-63, 5-64, and 5-65 shows that

$$\text{Vol} = \frac{2EI}{\omega B_m H_m} \tag{5-66}$$

If B_m and H_m are in *webers* per square inch and ampere turns per inch respectively, the volume in cubic inches per volt ampere is given by

$$\frac{\text{Vol}}{EI} = \frac{2}{2\pi f B_m H_m} = \frac{1}{\pi f B_m H_m} \tag{5-67}$$

When B_m is expressed in lines per sq in. and H_m in ampere turns per in. according to the Mixed English System of Units, the volume in cubic inches per volt-ampere is expressed by

$$\frac{\text{Vol}}{EI} = \frac{10^8}{\pi f B_m H_m} = \frac{3.18 \times 10^7}{f B_m H_m} \tag{5-68}$$

EXAMPLE 5-2:
(a) Determine the volume of a core comprised of U.S.S. Annealed Electrical Sheet Steel (see Fig. 3-12) for a reactor rated at 90 v, 200 amp, and and 60 cps. The maximum flux density B_m is to be 80,000 lines per sq in. Neglect the effects of air gaps introduced by the joints in the core.
(b) What is the weight of the core if its density is 0.276 lb per cu in.?

Solution:
A flux density of $B_m = 80$ kilolines per sq in. requires a magnetic field intensity H_m of about 10 amp turns per in. for annealed sheet steel according to Fig. 3-13.

(a) $$\text{Vol} = \frac{3.18 \times 10^7\, EI}{f B_m H_m}$$

$$= \frac{3.18 \times 10^4 \times 90 \times 200}{60 \times 80 \times 10} = 11{,}930 \text{ cu in.}$$

(b) Weight $= 11{,}930 \times 0.276 = 3{,}285$ lb

Pounds per volt-ampere $= 1.825$

The values computed in Example 5-2 show the size of the core to be excessive when the operation is restricted to the region below the knee of the magnetization curve. Actually, a value of 80 kilolines per sq in. is well in the knee of the curve and a lower value of flux density, perhaps 60 kilolines or less, is required to insure stability.

The excessive amount of iron is due to the high value of the relative permeability in the unsaturated region. Since $B_m = \mu_r \mu_0 H_m$, Eq. 5-68 can be expressed in terms of B_m^2 as

$$\frac{\text{Vol}}{EI} = \frac{3.18 \mu_r \mu_0 \times 10^7}{f B_m^2} \qquad (5\text{-}69)$$

where μ_r is the relative permeability of the core material and μ_0 is the permeability of free space having the value of 3.19 in the Mixed English System of Units. Hence, Eq. 5-69 can be reduced to

$$\frac{\text{Vol}}{EI} = \frac{1.015 \mu_r \times 10^8}{f B_m^2} \qquad (5\text{-}70)$$

where B_m is expressed in maxwells or lines per square inch.

Equation 5-70 shows that, for a given value of maximum flux density B_m, the volume of the core is directly proportional to the relative permeability μ_r. This relationship suggests the use of air, or some material with magnetic characteristics similar to those of air, instead of iron for the core. Air alone or other nonmagnetic material will not do, as it is extremely difficult to arrange the exciting winding in a manner such as to confine the flux path to a definite volume of air. This can be achieved by means of a toroid, an arrangement that is impractical for most reactors. For that reason, as well as for structural considerations, both iron and air are used, the main function of the iron being to impart the proper size and configuration to the one or more air gaps that may be required in the core and to accommodate the exciting winding. The air gap gives the magnetic circuit a more linear characteristic, thus reducing the variation of inductance with voltage fluctuation.

The value of μ_r is very nearly unity for air. If the air gap or air gaps are so proportioned in relation to the iron that the reluctance of the iron is negligible compared with that of the air, and if the dimensions of the one or more gaps are such that the effect of fringing can be neglected, then expression for the volume of air gap per volt ampere is obtained by substituting $\mu_r = 1.00$ in Eq. 5-70. This yields

$$\frac{\text{Vol}_{\text{air}}}{EI} = \frac{1.015 \times 10^8}{f B_m^2} \qquad (5\text{-}71)$$

EXAMPLE 5-3: Determine the volume of air to be used in the core of a reactor of the same rating as the one in Example 5-1 if the maximum flux density is to be 80 kilolines per sq in. Neglect the reluctance of the iron, leakage, and fringing.

Sec. 5-8   EFFECT OF AIR GAPS   225

Solution:
The reactor has the following rating

$$E = 90 \text{ v}, \quad I = 200 \text{ amp}, \quad f = 60 \text{ cps}$$

When these values are substituted in Eq. 5-71, the volume of the air is found to be

$$\text{Vol}_{\text{air}} = \frac{1.015 \times 10^8 \times 90 \times 200}{60 \times (80{,}000)^2} = 4.76 \text{ cu in.}$$

Although Eq. 5-71 is useful for determining the volume of the air in the core, it gives no indication as to the configuration, i.e., length and area, of the one or more air gaps. The cross sections of reactor cores are generally rectangular and the air gaps are usually between plane parallel surfaces of iron. So once the volume is known, the length and cross-sectional area can be determined from the number of turns in the exciting winding, the current, the voltage, and the maximum flux density as follows. On the basis of sinusoidal voltage and current we have

$$4.44 f N A_a B_m = E \tag{5-72}$$

and

$$A_a = \frac{E}{4.44 f N B_m} \tag{5-73}$$

also

$$H_m = \frac{N\sqrt{2}\,I}{l_a} = \frac{B_m}{\mu_0} \tag{5-74}$$

from which

$$l_a = \frac{\mu_0 \sqrt{2}\,NI}{B_m} \tag{5-75}$$

In the Mixed English System, the area in square inches is

$$A_a = \frac{E \times 10^8}{4.44 f N B_m} \tag{5-76}$$

and the length in inches is

$$l_a = \frac{3.19\sqrt{2}\,NI}{B_m} = \frac{4.51\,NI}{B_m} \tag{5-77}$$

Equations 5-76 and 5-77 show that in a reactor, of a given frequency rating, volt ampere rating, and flux density B_m, the cross-sectional area of the core is proportional to the volts per turn and the length of air gap is proportional to the number of turns. The volume of the winding is a function of the number of turns; the volume of the core is a function of the volts per turn. The coordination of these two volumes is usually dictated by considerations of economical size and construction of the reactor.

Effect of air gap on wave form

The introduction of an air gap in a given iron circuit increases the fundamental component I_{mag_1} in Fig. 5-9 without appreciably affecting the core-loss

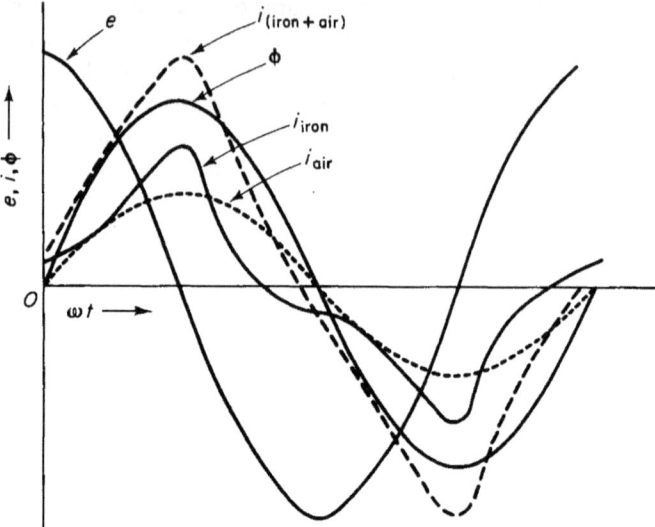

Figure 5-13. Effect of air gap on wave form of exciting current.

current I_c or the component $(I_{mag} - I_{mag_1})$ in the magnetizing current resulting from the harmonics. Thus, the fundamental component is increased without increasing the harmonics, a condition that leads to a more nearly sinusoidal wave form for the exciting current. If the increase in the fundamental component of the magnetizing current produced by the air gap is I_{air}, Eq. 5-47 becomes

$$I_{exc} = \sqrt{I_c^2 + (I_{mag_1} + I_{air})^2 + I_3^2 + I_5^2 + \cdots} \qquad (5\text{-}78)$$

Figure 5-13 shows the components of exciting current I_{iron} for the iron, i_{air} for the air gap and the total exciting current $i_{(iron + air)}$ in a reactor or transformer with an iron core containing an air gap. The induced emf e

Sec. 5-8   EFFECT OF AIR GAPS   227

and the flux ϕ are both sinusoidal. The total exciting current $i_{(\text{iron} + \text{air})}$ has a better wave form than the exciting current for the iron alone. It is important to note that the air-gap component current is in phase with the flux wave and is therefore also in phase with the fundamental harmonic of the magnetizing component in the exciting current required by the iron alone. Therefore, the magnetizing current I_{air} for the air gap adds directly to the fundamental harmonic I_{mag_1} as indicated in Eq. 5-78.

Use of air gap to prevent d-c saturation

There are circuits in which reactors or chokes and transformers carry components of alternating current and direct current simultaneously. This is particularly true of many circuits associated with electronic devices. A typical circuit that carries alternating current and direct current simultaneously is a single-phase, dry-type rectifier in which alternating current is converted into unidirectional current, which is generally a pulsating direct current unless provision is made for smoothing out the ripples. In such circuits iron-core reactors, known as chokes, are used to suppress the ripples so that a smooth direct current is produced. The characteristics of a choke should be such that the choke offers a relatively low amount of resistance to the direct current, while at the same time possessing high reactance. If the direct current in the winding of such a choke develops a large enough mmf to saturate the iron appreciably, the inductance will be correspondingly low. As a result, a choke of excessive size is required for obtaining the desired value of inductance for the alternating component of current. Direct-current saturation can be made negligible by the use of an air gap in series with the flux path. However, if the length of the air gap is too great, the magnetic reluctance is excessive and the inductance is correspondingly low for a given size of core and winding.

The effect of an air gap on current ripple is shown in Figs. 5-14(a) and 5-14(b). For a given d-c component i_{dc} in the current and a given a-c component ϕ_{ac} in the flux or flux ripple, the a-c component in the current i_{ac} is smaller [as shown in Fig. 5-12(b)] when the core has an air gap than when it has none [as indicated in Fig. 5-14(b)]. The presence of the air gap therefore makes the reactor or choke more effective in suppressing the a-c component or ripple in the current by increasing the value of L_d, the differential self-inductance.

In applications where the a-c flux is small in relation to the d-c flux, the length of the air gap should be such that the air-gap line intersects the magnetization curve of the iron near the bottom of the knee, as shown in Fig. 5-15.

The calculation of the a-c inductance of chokes in which a-c excitation is superimposed on d-c excitation is outside the scope of this text. Such

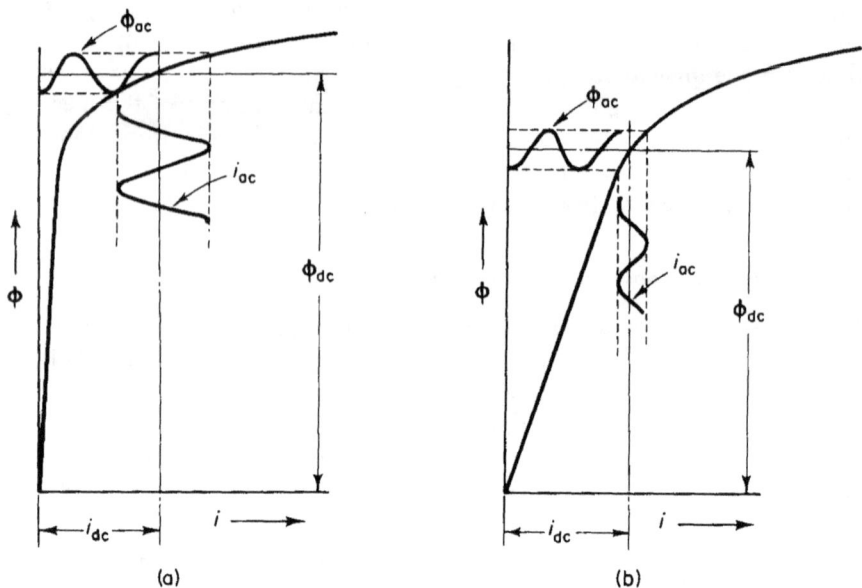

Figure 5-14. Effect of air gap in reactor core on current ripple (a) without air gap, (b) with short air gap.

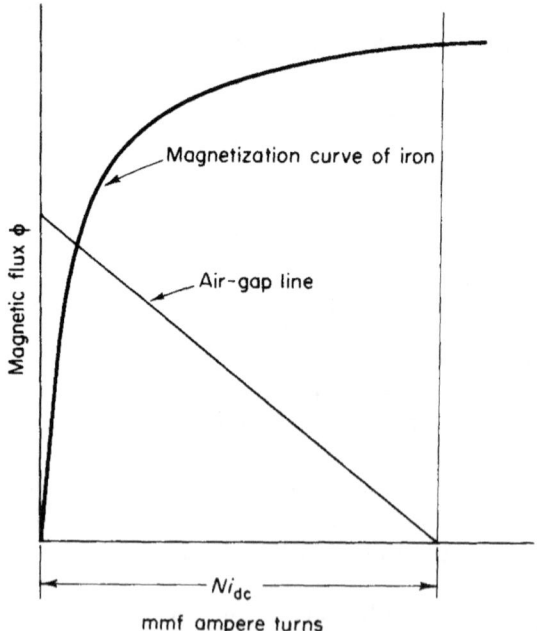

Figure 5-15. Approximate location of air-gap line for preventing d-c saturation in a reactor.

Sec. 5-9   TIME CONSTANT AND RATING OF REACTORS   229

calculations are not analytically straightforward, but are based upon empirical data.*

5-9 TIME CONSTANT AND RATING OF REACTORS AS FUNCTIONS OF VOLUME

Time constant

The time constant of an inductive circuit is defined as the ratio L/R where L is the inductance and R the resistance of the circuit and both L and R are assumed to be constant. If a reactor has an air gap in its core and the

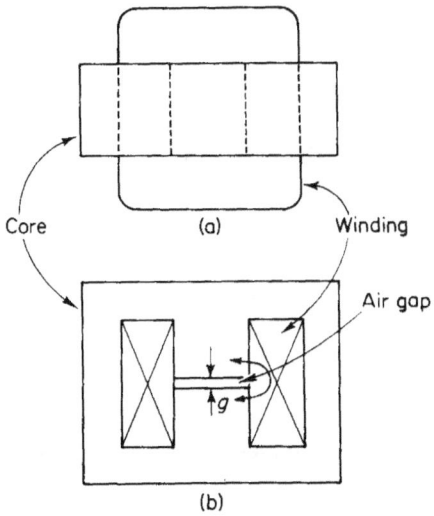

Figure 5-16. Three-legged reactor with air gap. (a) Top view; (b) side view.

reluctance of the iron, hysteresis, and eddy currents are neglected, then the inductance is constant; and if the winding is at a constant temperature, its resistance is constant also. Consider, on that basis, an iron-core reactor with an air gap of length g as shown in Fig. 5-16. Let the effective cross-sectional area of the air gap be A_g and the net area occupied by the conductors in the winding be $k_w A_w$, where k_w is a space factor and A_w is the area of the window in the core, i.e., the area of the space between the center leg and

---

* For treatments of this problem, see C. R. Hanna, "Design of Reactances and Transformers Which Carry Direct Current," AIEE *Trans.*, 46 (1927) 155–158; V. E. Legg, "Optimum Air Gaps For Various Magnetic Materials in Cores of Coils Subject to Superposed Direct Current," AIEE *Trans.*, 64 (1945) 709–712; and Landee, R. W., *et al.*, *Electronic Designers' Handbook.* New York: McGraw-Hill Book Company, 1957.

one outer leg of the core in Fig. 5-16. The space factor k_w is less than unity because not all of the window area A_w is occupied by conductor material of the winding since some space is occupied by electrical insulation and voids between turns, winding, and core.

When the reluctance of the iron and magnetic leakage are neglected, the self-inductance is, according to Eq. 4-11

$$L = \frac{\mu_0 N^2 A_g}{g}$$

where N is the number of turns in the reactor winding.

The resistance of the winding, which is a conductor of uniform cross section, is given by

$$R = \frac{\rho l}{A} \tag{5-79}$$

Where ρ is the resistivity of the conductor material, usually copper, l the length of the winding, and A the cross-sectional area of the conductor material. The length is

$$l = N l_t \tag{5-80}$$

Where l_t is the mean length of the turns and the area

$$A = \frac{k_w A_w}{N} \tag{5-81}$$

From Eqs. 5-79, 5-80, and 5-81 it follows that the resistance of the winding is

$$R = \frac{\rho N^2 l_t}{k_w A_w} \tag{5-82}$$

so that the time constant

$$\frac{L}{R} = \frac{\mu_0 k_w A_w A_g}{\rho l_t g} \tag{5-83}$$

If all linear dimensions of the reactor core and air gap are increased by a factor k, the areas A_w and A_g are each increased by k^2, whereas the lengths l_t and g are increased directly as k, so that the time constant L/R for a given configuration of core and winding increases as k^2 or as the $\frac{2}{3}$ power of the volume, i.e.

$$\frac{L}{R} \propto \text{Vol}^{2/3} \tag{5-84}$$

Sec. 5-9   TIME CONSTANT AND RATING OF REACTORS   231

It should be noted that the time constant is independent of the number of turns because the inductance and resistance of the winding both vary as the square of the turns for a given space factor k_w.

Rating

For given values of frequency, flux density, and current density, the volts per turn are proportional to the cross-sectional area of the core and the ampere turns are proportional to the area occupied by the winding, so that

$$\frac{V}{N} = 4.44 f B_m A_{\text{core}} \tag{5-85}$$

where A_{core} is the cross-sectional area of the core or of the air gap corrected for fringing, and

$$NI = J k_w A_w \tag{5-86}$$

where J is the rated current density in the winding. The volt-ampere rating of the reactor is the product of Eqs. 5-85 and 5-86

$$VI = 4.44 f B_m J k_w A_w A_{\text{core}} \tag{5-87}$$

For a given configuration, an increase in all linear dimensions by a factor k increases the product $A_w A_{\text{core}}$ by k^4 and the volt-ampere rating of the reactor increases as the $\frac{4}{3}$ power of the volume or of the weight. Therefore

$$VI \propto \text{Vol}^{4/3} \quad \text{or} \quad VI \propto \text{Weight}^{4/3}$$

from which it follows that the volt-amperes per unit weight or volt-amperes per pound is related to weight as

$$\frac{VI}{\text{Weight}} \propto \text{Weight}^{1/3}$$

or   Volt-ampere per pound $\propto k$ \hfill (5-88)

The rated volt-amperes per pound then varies directly as the linear dimension. With fixed current density the I^2R losses in the winding vary at a given temperature directly as the volume of the winding, and when the flux density and frequency are fixed, the core loss varies directly as the volume of the core. It follows, therefore, that under rated conditions, the heat generated in the reactor is proportional to the volume of the reactor, or k^3. However, the radiating surface varies as k^2 so that the temperature rise for a given rating varies directly as the linear dimension, or as k. This indicates that if frequency, flux density, current density, and configuration remain fixed, greater provision in proportion to k must be made to maintain the same temperature rise in the core and windings as the size and rating of the reactor are increased. For small ratings, normal radiation and convection dissipate the heat generated in the core and windings without excessive temperature rise, but beyond a certain rating it becomes necessary to facilitate cooling by means of forced air, circulation of oil, or some other method, when operating at the same current density and flux density.

PROBLEMS

5-1 A 400-cycle emf of 100 v is applied to the winding of the electromagnet in Fig. 3-17. Assume a stacking factor 0.95 for the laminated core and neglect the resistance of the winding.

(a) Determine the maximum flux density B_m in the core if the wave form of the voltage is sinusoidal.
(b) Neglect the effect of air gaps in the core and determine the max value of the exciting current if the core is composed of U.S.S. Annealed Electrical Sheet Steel (see Fig. 3-11).
(c) Repeat parts (a) and (b) for a 400-cycle emf of 130 v.
(d) Repeat parts (a) and (b) for a core thickness of 1.25 in. instead of 1.50 in.
(e) Repeat parts (a) and (b) for a core thickness of 1.75 in. instead of 1.50 in.
(f) Using the dimensions of the core as shown in Fig. 3-17 determine the 400-cycle voltage that will give twice the max value of current as in part (b).
(g) Using the dimensions of the core shown in Fig. 3-17, determine the frequency that will give twice the max value in part (a) at 100 v.

5-2 A reactor has a laminated core of U.S.S. Annealed Electrical Sheet Steel of the dimensions shown in Fig. 3-17. Assume a stacking factor of 0.95 for the core and neglect the resistance of the winding.

(a) Determine the number of turns in the exciting winding such that the max flux density B_m is 90,000 lines per sq in. for a 60-cycle emf of 240 v.
(b) Neglect air gaps in the core and determine the max value of the current for the condition of part (a) above.
(c) If the winding in parts (a) and (b) is composed of two identical parts, i.e., equal number of turns and equal amounts of copper, what value of 60-cycle voltage should be applied to the two halves of the winding connected in parallel such that the max flux density B_m is 90,000 lines?
(d) What is the max value of the current supplied by the source for the condition of part (c) above?
(e) What is the max value of the current in each half of the winding for the condition of part (c) above?

5-3 Repeat Problem 5-1(a) and (b) if the voltage is a square wave.

5-4 A certain reactor has a core of net cross section A_0. The length of the flux path in the core is l_0. The exciting winding has N_0 turns and is rated at V_0 v at a frequency of f_0 cps. The peak value of the current is I_{m_0}. The core losses are P_{c_0} and the flux density is B_{m_0}.

Another reactor of the same construction, except that all its linear dimensions are 1.5 times those of the reactor in Problem 5-4, operates at the same value of flux density B_{m_0} and at the same voltage V_0 and the same frequency f_0. Determine

(a) The number of turns in the winding of the larger reactor in terms of the turns N_0 of the first reactor.
(b) The max value of the current in the larger reactor in terms of I_{m_0}.
(c) The core loss in the larger reactor in terms of P_{c_0}.

5-5 A sinusoidal voltage impressed on one winding of a transformer has a value such that the maximum flux density B_m in the core is 10,000 gausses. Data for the upper half of the hysteresis loop for a maximum flux density of 10,000 gausses are tabulated below.

B-Gausses	H-Oerstedts	B-Gausses	H-Oerstedts
0	0.55	9,000	1.17
2,000	0.65	8,500	0.83
4,000	0.78	8,000	0.57
6,000	1.03	7,000	0.18
7,000	1.23	6,000	−0.07
8,000	1.50	4,000	−0.33
9,000	1.89	2,000	−0.46
10,000	2.39	0	−0.55
9,500	1.67		

Plot the complete hysteresis loop and the wave form of the exciting current as shown in Fig. 5-3, neglecting the effect of eddy current and the effects of joints in the magnetic circuit.

NOTE: Joints increase the magnetic reluctance of the core because in those regions the flux enters small air gaps.

5-6 Repeat Problem 5-5, but for sinusoidal exciting current.

5-7 A reactor with a winding of negligible resistance has an exciting current that is expressed approximately by

$$i_{\text{exc}} = \sqrt{2}[10.4 \cos(377t - 79°) + 4.50 \cos(1,131t + 90°) + 1.10 \cos(1,885t - 90°)]$$

when the applied voltage is

$$v = \sqrt{2}[120 \cos 377t]$$

Determine

(a) The real power in watts.
(b) The effective value of the exciting current.

5-8 Plot waves of the voltage and the fundamental, 3rd harmonic, and 5th harmonic components of the exciting current along with the total exciting current in Problem 5-7.

5-9 Using the data of Problem 5-7

(a) Determine
  (i) The effective value of the core-loss current.
  (ii) The effective value of the fundamental component in the magnetizing current.
  (iii) The effective value of the total magnetizing current.
(b) Draw to scale a phasor diagram, as shown in Fig. 5-9, showing the voltage, exciting current, core-loss current, fundamental component of magnetizing current, the total magnetizing current, and the phase angle θ_{exc}.
(c) Write the equation for the instantaneous magnetizing current.

5-10 A Transformer with 450 primary turns, when energized from a 120-v, 60-cycle line, draws a no-load current of 1.5 amp. The no-load power is 82 w. Neglect the resistance of the winding and leakage flux and determine

(a) The core-loss current.
(b) The magnetizing current.
(c) The maximum flux in the core.
(d) The exciting admittance Y.
(e) The exciting conductance G.
(f) The exciting susceptance B.

5-11 Draw an equivalent series circuit for the transformer of Problem 5-10 based on the circuit of Fig. 5-10(c). Evaluate r_c and x.

5-12 The primary winding of the transformer in Problem 5-10 has a resistance of 0.035 ohm.

(a) Find the value of the induced emf E.
(b) Show the equivalent circuit of the transformer and its primary, based on Fig. 5-10(b), indicating the numerical values of all the quantities shown in Fig. 5-10(b).
(c) Repeat part (b) above, but for Fig. 5-10(c).

5-13 How are the magnitudes of the quantities in parts (b) and (c) of Problem 5-12 affected by an increase in the frequency if the voltage is held constant?

5-14 Determine the volume of air gap in a 60-cycle, 75-v, 20-amp reactor that is designed to operate at a maximum flux density 85 kilolines per sq in. Neglect the reluctance of the iron, fringing and leakage.

5-15 The reactor of Problem 5-14 has 130 turns. Determine

(a) The length of air gap.
(b) The cross-sectional area of the core.

5-16 Neglect iron losses and determine

(a) The current rating.
(b) The voltage rating of the reactor in Problem 5-14 for operation at 400 cps at $B_m = 85$ kilolines per sq in.

5-17 The magnetic flux density in the reactor of Problem 5-7 is 80,000 lines per sq in. and the winding has 40 turns. An air gap of 0.040 in. is put into the iron. Determine

(a) The effective value of the component of magnetizing current produced by the air gap.
(b) The effective value of the total magnetizing current.
(c) The effective value of the exciting current.
(d) The real power expressed in watts.
(e) The reactive power in vars.

5-18 A choke that carries a direct-current component of 0.50 amp has a core comprised of E and I laminations assembled to form three butt joints as

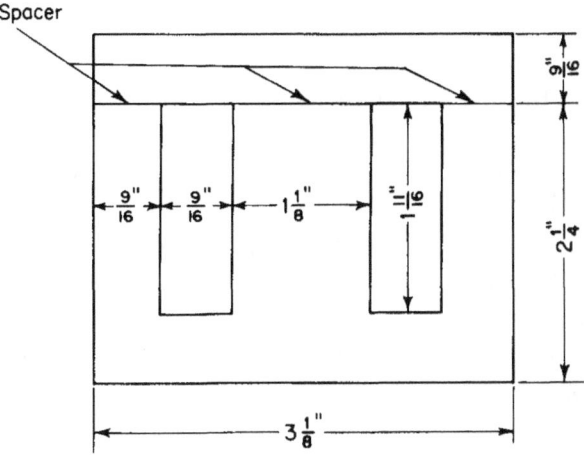

Figure 5-17. Laminations in reactor core for Problem 5-18.

shown in Fig. 5-17. The winding consists of 1,100 turns of No. 25 AWG enamel wire. The core is stacked to a depth of 3.60 in. with a stacking factor of 0.94. The magnetic characteristics of the core material, i.e., U.S.S. Annealed Electrical Sheet Steel is shown graphically in Fig. 3-11. Estimate the smallest size of nonmetallic spacer to be placed in each of the three butt joints of the core such that saturation due to the direct current in the winding is negligible.

5-19 A certain 60-cycle, iron-core reactor has a hysteresis loss of 4 percent, eddy-current loss of 1 percent, and a copper (I^2R) loss of 5 percent in terms of its volt-ampere rating when operating under rated conditions of current, voltage, and frequency. If this reactor is operated at a frequency of 50 cps instead of 60 cps, but at the same current density and flux density as for normal 60-cycle operation, what is

(a) The 50-cycle volt-ampere rating in terms of the 60-cycle volt-ampere rating?

(b) The percent core loss and the percent copper loss in terms of the 50-cycle volt-ampere rating?

5-20 A reactor has the same configuration and lamination thickness as that of Problem 5-19 with twice the volt-ampere rating at 60-cycles when operating at the same current density and flux density as that of Problem 5-19 under rated conditions. Calculate the following ratios for the larger reactor in terms of the smaller

(a) The linear dimension k.
(b) The number of turns N.
(c) The size of the conductor in the winding.
(d) The temperature rise of the winding, neglecting the change in winding resistance with copper temperature.

BIBLIOGRAPHY

Ball, J. D., "The Unsymmetrical Hysteresis Loop," AIEE *Trans.*, 34, Part 2 (1915) 2693–2715.

Hanna, C. R., "Design of Reactances and Transformers Which Carry Direct Current," AIEE *Trans.*, 46 (1927) 155–158.

Landee, R. W., et al., *Electronic Designers' Handbook*. New York: McGraw-Hill Book Company, 1957.

Legg, V. E., "Optimum Air Gaps for Various Magnetic Materials in Cores of Coils Subject to Superposed Direct Current," AIEE *Trans.*, 64 (1945) 709–712.

M.I.T. Staff, *Magnetic Circuits and Transformers*. New York: John Wiley & Sons, Inc., 1943.

THE TRANSFORMER

6

The transformer is largely responsible for the prevalence of the a-c power system. In the earliest systems electric power was generated, transmitted, distributed, and utilized in the form of direct current. The limitations of the d-c generator restricted operation to low values of generating voltages as well as transmission voltages, which not only severely restricted the amounts of power that could be transmitted efficiently to small values by present-day standards but also limited the transmission of even these small amounts of power to relatively short distances. Because of its high efficiency, exceeding 99 percent in ratings of 10,000 kva and above, and also because of its almost unlimited ratio of voltage transformation, the transformer makes it possible to generate power at voltages consistent with the most economical design of generators, 22,000 v being a common generating voltage in the larger systems. The transformer also makes it possible to step these voltages up to values approaching a million v for efficient transmission over distances as great as several hundred miles, and eventually to step these voltages down to values at which the power can be utilized most effectively. Most household appliances operate at about 120 v.

The transformer is also a vital component in many low-voltage, low-power applications—as in electronic circuits, where it may be used to provide the optimum ratio of load impedance to source impedance so that signals may be amplified with small distortion, or in some cases to effect maximum power transfer from a source to a load. It is not unusual to have a source of several thousand ohms impedance supplying a load that has an impedance as low as several ohms and still get maximum power transfer by means of an impedance-matching transformer.

The transformer may be defined as a device in which two or

6-1 INDUCED EMFs

For its operation, the transformer depends on the emfs that are induced in its various circuits by a common magnetic flux. According to Faraday's Law, the emf induced in a fixed circuit is proportional to the time rate of change of the magnetic flux. If there are N turns of wire in series, through all of which the flux undergoes the same rate of change, the induced emf is

$$e = -N\frac{d\phi}{dt} \quad (6\text{-}1)$$

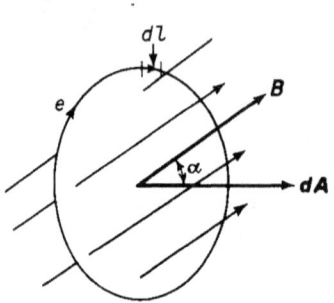

Figure 6-1. Direction of induced emf around a fixed closed path when the flux through the surface bounded by the path is increasing.

The minus sign is in accordance with Lenz's Law and shows that the induced emf e would produce a current in such a direction as to oppose any *change* in the flux ϕ through the circuit. It is important to note that the opposition is to the *change* in the flux and not to the flux itself. The sense of the induced emf e in relation to an *increasing* flux through a closed path is shown in Fig. 6-1. The vector $d\mathbf{A}$ in Fig. 6-1 represents an elemental area through which the flux density is represented by the vector $\mathbf{B}$ at the angle α to the normal of the surface associated with the area vector $d\mathbf{A}$. If E is the magnitude of the electric field intensity due to the varying magnetic flux, then the emf around the closed path is

$$e = \oint E \cos\theta\, dl \quad (6\text{-}2)$$

where $E \cos\theta$ is the component of the electric field intensity parallel to the path at a given point. This component of $\mathbf{E}$ has the same sense as e. If the integration in Eq. 6-2 is taken around the complete path once, the result is the emf per turn. This emf, however, exists whether the path is occupied by a conducting material or by free space or some other nonconducting material. Hence, we have from Eq. 6-1

$$e = \oint E \cos\theta\, dl = -\frac{d\phi}{dt} \quad (6\text{-}3)$$

as the volts per turn.

The flux ϕ, which links the area A enclosed by the path in Fig. 6-1, is obtained by integrating the normal component of the flux density **B** over the area A so that

$$\phi = \int_A B \cos \alpha \, dA \qquad (6\text{-}4)$$

and

$$\frac{d\phi}{dt} = \frac{d}{dt} \int_A B \cos \alpha \, dA \qquad (6\text{-}5)$$

If the configuration of the path is fixed, the area A is constant, and only the quantity $B \cos \alpha$ is free to change with time, thus permitting differentiation with respect to time under the integral, and

$$\frac{d\phi}{dt} = \int_A \frac{d(B \cos \alpha)}{dt} \, dA \qquad (6\text{-}6)$$

From Eqs. 6-1, 6-2, and 6-6 it follows that

$$e = \oint E \cos \theta \, dl = -\int_A \frac{d(B \cos \alpha)}{dt} \, dA \qquad (6\text{-}7)$$

which can be expressed in vector notation as

$$e = \oint \mathbf{E} \cdot \mathbf{ds} = -\int \frac{d\mathbf{B}}{dt} \cdot \mathbf{dA} \qquad (6\text{-}8)$$

6-2 THE TWO-WINDING TRANSFORMER

The two-winding transformer is one in which two windings are linked by a common time-varying magnetic flux. One of these windings, known as the primary, receives power at a given voltage from a source; the other winding, known as the secondary, delivers power, usually at a value of voltage different from that of the source, to the load. The roles of the primary and secondary windings can be interchanged. However, in iron-core transformers a given winding must operate at a voltage that does not exceed its rated value at rated frequency—otherwise the exciting current becomes excessive.

Transformers that operate in the audio-frequency range generally have iron cores of a basic construction similar to those of the reactors and electromagnets discussed in Chapters 3 and 4. Figures 6-2(a) and 6-2(b) show the arrangement of cores and coils in transformers with wound cores. These cores are each wound of a long continuous strip of sheet steel in the direction in which the metal was rolled during manufacturing. In these transformers, the core is magnetized in the direction of rolling, thus making for lower core loss and lower exciting current than when the magnetization is in a direction across that of rolling. The wound core feature does not lend itself to the construction of large power transformers; also, in the case of small audio transformers, it is more economical to build the core out of laminations in

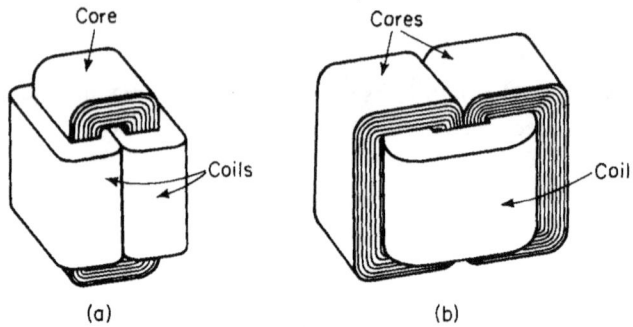

Figure 6-2. Transformers with wound cores. (a) Core type; (b) shell type.

the form of punchings. A conventional type of transformer is shown in Fig. 6-3.

The windings of transformers are generally of copper, which, in small transformers, is round wire. In larger transformers, the copper is a ribbon, making for better utilization of space. However, aluminum is coming into use for transformer windings also, particularly in the low-voltage windings of distribution transformers. The conductor is in the form of a sheet of foil in which the ratio of width to thickness is usually 100:1 or larger.*

In most conventional iron-core transformers the core provides the main path for the magnetic flux, although some flux, known as leakage flux, leaks through the space between the windings, between the core and windings, and through some of the turns in one winding without linking the other winding. At no load, and under steady conditions, practically all of the flux is confined to the core of iron-core transformers and the leakage flux is usually negligible. Hence, the only appreciable flux that can link each of the turns in the windings at no load must be the flux in the core even if the turns are not wound tightly around the core, and even if the planes of the turns are not exactly at right angles to the flux. Therefore, at no load, the integral on the right-hand side of Eq. 6-4, which expresses the flux linking each turn, can yield only the value of the flux ϕ in the core itself, except for

---

* E. A. Goodman, "Sheet Windings," *Allis-Chalmers Electrical Review*, 2nd Quarter 1963.

the negligible no-load leakage flux. Therefore, the no-load voltage induced in each turn of the windings, according to Eq. 6-7, is

$$e_{\text{turn}} = -\int_A \frac{d(B \cos \alpha)}{dt} dA = -\frac{d\phi}{dt}$$

(6-9)

where ϕ is the flux in the core.

The rms value of the voltage induced in an N-turn winding by a sinusoidally time-varying flux of amplitude ϕ_m is given by Eq. 5-19 as

$$E = 4.44 f N \phi_m$$

This is sometimes called the transformer voltage formula. It can be extended, with suitable modifications, to the analysis of devices other than reactors and transformers, such as motors and generators.

6-3 VOLTAGE RATIO, CURRENT RATIO, AND IMPEDANCE RATIO IN THE IDEAL TRANSFORMER

Figure 6-3. Cut-away view of pole-type distribution transformer. (Courtesy of Westinghouse Electric Corporation.)

The ideal transformer is an imaginary one that has no core losses and no leakage fluxes. The windings have no resistance and the core has infinite magnetic permeability. Such a transformer would therefore have no losses and no exciting current.

Figure 6-4 is a schematic diagram of a two-winding, shell-type transformer. Assume this transformer to be ideal. Then the mean path of the flux would be approximately as indicated by the broken lines whether the transformer is operating under load or at no load. It must be kept in mind that Fig. 6-4 is schematic, and that if the transformer were ideal the magnetic coupling between the windings would need to be perfect. The coupling between the windings shown in Fig. 6-4 is rather loose, but the windings are shown this way for reasons of simplicity. If the windings were arranged in an actual shell-type transformer as shown in Fig. 6-4, the mmfs due to the load current would produce sizeable leakage fluxes along the paths indicated in Fig. 6-7(a), although at no load the leakage flux would be small. All the flux in the transformer would be confined almost entirely to the core, as indicated by the broken lines in Fig. 6-4. An arrangement that is generally used to promote tight magnetic coupling between windings is shown for a shell-type transformer in Fig. 6-8, and for a core-type transformer in Fig. 6-9.

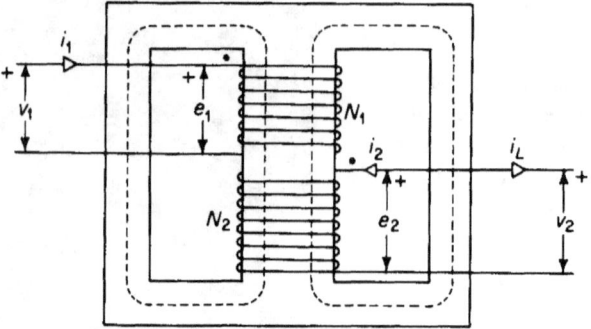

Figure 6-4. Two-winding, shell-type transformer.

Leakage fluxes and their effects on the characteristics of transformers are taken up later on in this chapter.

Voltage ratio and transformer polarity

Returning to the discussion of the ideal transformer with N_1 primary turns and N_2 secondary turns, let the instantaneous voltage applied to the primary winding be v_1 v, and let the polarity be as shown in Fig. 6-4. The time-varying flux in the core of the transformer induces an instantaneous emf e_1 in the primary winding, and, since the windings of an ideal transformer are assumed to be without resistance, e_1 must equal v_1. Then

$$v_1 = e_1 = N_1 \frac{d\phi}{dt} \qquad (6\text{-}10)$$

The sense of the induced emf e_1 in Eq. 6-10 and in Fig. 6-4 is opposite that in Eq. 6-9, accounting for the omission of the minus sign in Eq. 6-10.

In the ideal transformer there is no leakage flux. The flux everywhere in the core has the same value and is changing at the same rate at a given instant. Thus, all of the flux links both windings of the ideal transformer and is considered as the mutual flux ϕ. Hence, the voltage induced in the secondary winding is

$$e_2 = N_2 \frac{d\phi}{dt} \qquad (6\text{-}11)$$

and has the polarity shown in Fig. 6-4. Since the secondary winding of the ideal transformer is assumed to have no resistance, the secondary terminal voltage v_2 must equal the induced secondary voltage e_2 even when the transformer carries load. Hence

$$v_2 = e_2 = N_2 \frac{d\phi}{dt} \qquad (6\text{-}12)$$

The ratio of primary to secondary turns N_1/N_2 of a transformer is called the turns ratio. Division of Eq. 6-10 by Eq. 6-12 shows that, in an ideal transformer, the voltage ratio equals the turns ratio, thus

$$\frac{v_1}{v_2} = \frac{e_1}{e_2} = \frac{N_1}{N_2} \qquad (6\text{-}13)$$

Also, if V_1, E_1, V_2, and E_2 are the corresponding primary and secondary voltage phasors, we have

$$\frac{V_1}{V_2} = \frac{E_1}{E_2} = \frac{N_1}{N_2} \qquad (6\text{-}14)$$

It is important to note that the induced emfs e_1 and e_2 as shown in Fig. 6-4 have like polarities. Therefore, the induced voltage phasors E_1 and E_2 are in phase with each other and not 180° out of phase. In the ideal transformer, the terminal voltage phasors V_1 and V_2 are also in phase with each other for the convention adopted in Fig. 6-4. If, in Fig. 6-4, the plus sign of either e_1 or e_2—but *one of them only*, were placed at the bottom of the winding, then the induced emf phasors E_1 and E_2 would be 180° out of phase with each other. The black dots near the tops of the two windings are called polarity marks. These indicate that the upper terminals, or the marked terminals, have like polarities, just as the unmarked terminals have like polarities—but opposite those of the marked terminals. One way of establishing the significance of polarity marks is by considering the magnetic effect on the core when positive currents enter both windings either at the marked terminals or when positive currents enter both windings at the unmarked terminals. In either case, the mmfs of the two currents in the two windings add.

Current ratio

In the transformer of Fig. 6-4, the instantaneous primary current and instantaneous load current are designated as i_1 and i_L. The instantaneous primary current i_1 is shown *entering* the marked terminal of the primary winding, and the instantaneous load current i_L is shown leaving the marked terminal of the secondary winding. However, it is conventional, when writing the equations for two or more circuits that are coupled magnetically, to assume the instantaneous currents are flowing simultaneously either all into or all out of the marked terminals of the various coupled circuits. Thus, in a two-winding transformer, if the instantaneous primary and secondary currents flowing into the marked terminals are both positive or both negative, each of the two currents acting by itself would magnetize the core in the same direction, i.e., the mmfs of the two windings would be additive. Therefore, in accordance with this convention, the instantaneous secondary current i_2 is shown flowing into the marked terminal of the secondary of the transformer in Fig. 6-4, whereas the instantaneous load current i_L is shown flowing out of the marked secondary terminal. This means that i_2 and i_L are equal and opposite, and we have

$$i_L = -i_2 \qquad (6\text{-}15)$$

The relationship between these currents can be readily established on the basis that the exciting current in the ideal transformer is zero. Further, the capacitances between windings, between the windings and the core and tank, and between the turns of a winding can usually be neglected in the case of audio-frequency range transformers because the capacitive currents at these frequencies are generally small in relation to the rated current. Hence, capacitance effects in the windings of the ideal transformer are neglected.

Since the exciting current in an ideal transformer is zero, the permeability of the core in an ideal transformer must be infinite. This means that the net mmf of the windings must be zero under load as well as at no load. Therefore, the mmfs of the two windings resulting from the currents as shown in Fig. 6-4 must be equal, and we have for the ideal transformer

$$N_1 i_1 = N_2 i_L$$

giving a current ratio of

$$\frac{i_1}{i_L} = \frac{N_2}{N_1} \qquad (6\text{-}16)$$

The current ratio of the ideal transformer is therefore N_2/N_1, and, if the current phasors I_1 and I_L are used, we have

$$\frac{I_1}{I_L} = \frac{N_2}{N_1} \qquad (6\text{-}17)$$

The current phasors I_1 and I_L in the ideal transformer are in phase with each other in accordance with the directions assumed in Fig. 6-4. Generally, when writing coupled-circuit equations, the instantaneous secondary current

Sec. 6-3 VOLTAGE, CURRENT, AND IMPEDANCE RATIO

i_2 is shown flowing into the marked side of the secondary winding at the same instant that the instantaneous primary current i_1 flows into the marked side of the primary winding. This means that the current $i_L = -i_2$, and that the phasor I_2 is 180° out of phase from the phasors I_L and I_1.

Impedance ratio

In many problems that involve transformers it is convenient to refer the impedance in one side of the transformer to the other side of the transformer.

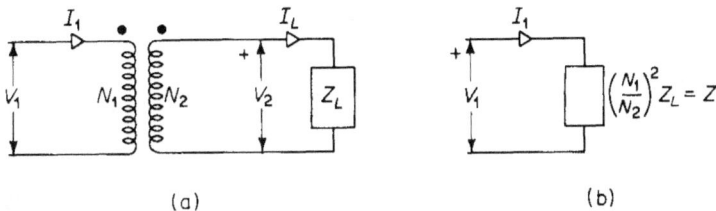

Figure 6-5. Equivalent circuits of (a) ideal transformer and connected load impedance; (b) load impedance referred to primary side of transformer.

For example, Fig. 6-5(a) shows an ideal transformer with a load impedance of Z_L ohms connected across its secondary terminals. The load impedance referred to the primary is the value of impedance that, if connected directly across the source of voltage V_1, would draw the same value of current I_1 as the transformer with its connected load impedance Z_L. This is shown in Fig. 6-5(b).

When Eq. 6-14 is divided by Eq. 6-17 the result is

$$\frac{V_1 I_L}{V_2 I_1} = \left(\frac{N_1}{N_2}\right)^2$$

or

$$\frac{V_1}{I_1} = \left(\frac{N_1}{N_2}\right)^2 \frac{V_2}{I_L} \qquad (6\text{-}18)$$

But the load impedance is

$$Z_L = \frac{V_2}{I_L} \qquad (6\text{-}19)$$

and an impedance that would draw a current of I_1 amp when connected directly across the source of voltage of V_1 v must have a value of

$$Z_1 = \frac{V_1}{I_1} \qquad (6\text{-}20)$$

A comparison of Eqs. 6-18, 6-19, and 6-20 shows that

$$Z_1 = \left(\frac{N_1}{N_2}\right)^2 Z_L \qquad (6\text{-}21)$$

where Z_1 is the value of the load impedance referred to the primary of the transformer. The impedance ratio is, therefore

$$\frac{Z_1}{Z_L} = \left(\frac{N_1}{N_2}\right)^2 \qquad (6\text{-}22)$$

or the square of the turns ratio of the transformer.

The turns ratio of a transformer is sometimes represented by the letter a, which means that

$$a = \frac{N_1}{N_2} \qquad (6\text{-}23)$$

and the impedance ratio is

$$\frac{Z_1}{Z_L} = a^2 \qquad (6\text{-}24)$$

EXAMPLE 6-1: An electronic amplifier supplies the voice coil of a loudspeaker through a 20-to-1-ratio output transformer. The impedance of the voice coil is 9.6 ohms. Assume the transformer to be ideal and determine the value of the impedance that the transformer and loud-speaker place across the output of the amplifier.

Solution:

Turns ratio $a = N_1/N_2 = 20$. Impedance ratio $= a^2 = 400$ $Z_1 = a^2 Z_L = 400 \times 9.6 = 3,840$ ohms.

6-4 EQUIVALENT CIRCUIT OF THE TRANSFORMER

The equivalent circuit shown in Fig. 6-5(a) is for an ideal transformer, and is therefore limited to problems that require only a rather approximate evaluation of transformer performance in general. The efficiency of the ideal transformer is unity, and, for a given primary voltage, the secondary voltage of the ideal transformer is constant regardless of the variations in the load. A realistic transformer, however, has real power losses and reactive power losses. The real power losses include not only the core loss but also the copper losses due to the resistance of the windings; the reactive power

Sec. 6-4   EQUIVALENT CIRCUIT OF THE TRANSFORMER

losses include the volt-amperes associated with the magnetizing current that is required by the flux in the core as well as the volt-amperes required to produce the leakage flux whose path is largely in air. As a result of the real power losses, the efficiency of a transformer is less than unity, and, because of the real power losses in combination with the reactive power losses, the secondary voltage will undergo changes in value with variations in the load even though the primary voltage is constant. It is therefore necessary to modify the equivalent circuit of the ideal transformer to take into account the imperfections of the realistic transformer.

Exciting current, conductance, and susceptance

The general subject of exciting current, core-loss current and magnetizing current was taken up in Chapter 5 in connection with reactors. Figure 5-10 showed equivalent circuits for a reactor. When capacitance effects are neglected, as is done in this discussion, a transformer operating at no load performs like a reactor and the only current that flows is in the primary. This no-load primary current is the exciting current. However, if the secondary is closed to a load, such that a current I_L flows in the secondary winding, then the primary must carry a component $N_2 I_L / N_1 = I_L'$ in addition to the exciting current I_{exc}. The component $N_2 I_L / N_1$ in the primary current is required to overcome the mmf $N_2 I_L$ of the current in the secondary winding. It is a simple matter to modify the equivalent circuit of the ideal transformer to take into account the exciting current. This is done by connecting the parallel circuit, which consists of the exciting admittance $y_{exc_1} = g_1 - jb_1$, across the primary of the ideal transformer, as in Fig. 6-6(a). Then

$$I_{exc_1} = y_{exc_1} E_1 \tag{6-25}$$

and

$$I_{mag_1} = b_1 E_1 \tag{6-26}$$

also

$$I_{c_1} = g_1 E_1 \tag{6-27}$$

where I_{exc_1}, I_{mag_1} and I_{c_1} are the exciting current, the magnetizing current and core-loss current, respectively, in the primary winding. Under load, the primary current is the phasor sum expressed by

$$\mathbf{I}_1 = \mathbf{I}_{L'} + \mathbf{I}_{exc_1} \tag{6-28}$$

This addition is shown graphically in the phasor diagram of Fig. 6-6(b).

248 THE TRANSFORMER Chap. 6

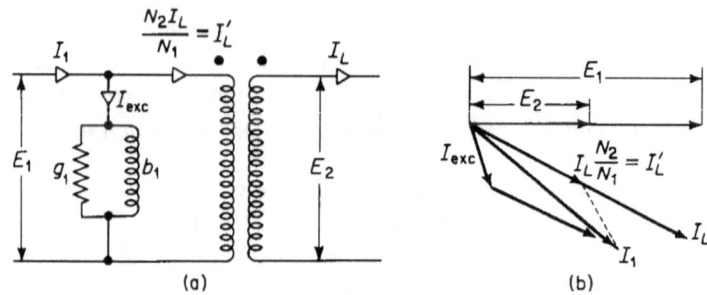

Figure 6-6. (a) Equivalent circuit of a transformer taking into account exciting current; (b) phasor diagram for induced emfs and currents in equivalent circuit.

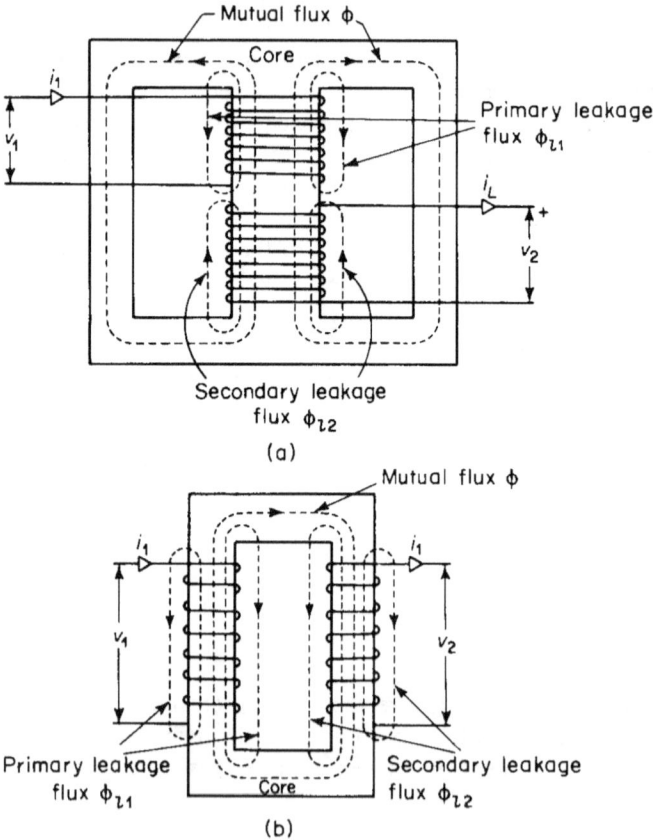

Figure 6-7. Mutual flux path and leakage flux paths in (a) shell-type transformer, (b) core-type transformers. The magnetic coupling of the windings in both transformers is loose, resulting in high leakage.

Sec. 6-4 EQUIVALENT CIRCUIT OF THE TRANSFORMER   249

Leakage flux and leakage reactance

It was pointed out in Section 6-3 that in an actual iron-core transformer all the magnetic flux is not confined completely to the core, especially under load conditions, i.e., when the secondary is delivering appreciable current. Although Eqs. 6-7 and 6-8 could theoretically be made to take into account the mutual flux and the leakage fluxes to give the total induced emf in each

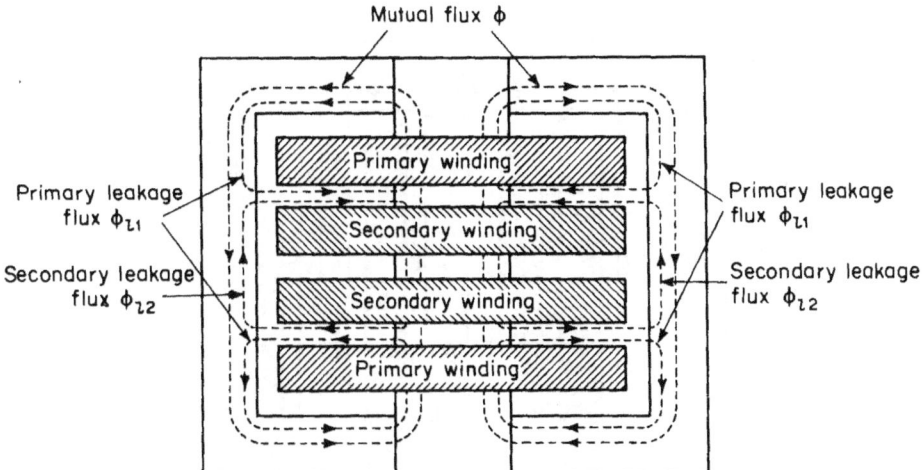

Figure 6-8. Winding arrangement in shell-type transformer showing approximate paths taken by mutual flux and primary and secondary leakage fluxes.

of the two windings, it is more convenient to divide the magnetic flux into three components. These are

1. The mutual flux ϕ, which is confined almost completely to the core and which is common to both the primary and the secondary windings. This flux transforms the power from primary to secondary and is due to the resultant mmf of the primary and secondary windings.
2. The primary leakage flux ϕ_{l1}, due to the mmf of the primary winding and which links the primary without linking the secondary.
3. The secondary leakage flux ϕ_{l2}, due to the mmf of the secondary winding and which links the secondary without linking the primary winding.

Although Figs. 6-4, 6-7, 6-8, and 6-9 represent iron-core transformers, the magnetic flux in air-core transformers can also be divided into the three components, namely, mutual flux and primary and secondary leakage fluxes, even though the air-core transformer does not have a discretely defined magnetic core.

The arrangement of the windings shown schematically in Figs. 6-7(a) and 6-7(b) make for high magnetic leakage, since the leakage flux can spread out over a relatively large area. When the windings are arranged as shown in Figs. 6-8 and 6-9, the magnetic leakage is relatively small, since the leakage flux is restricted to paths of relatively small cross sections, and hence, of correspondingly high magnetic reluctance. Although Figs. 6-7, 6-8, and 6-9

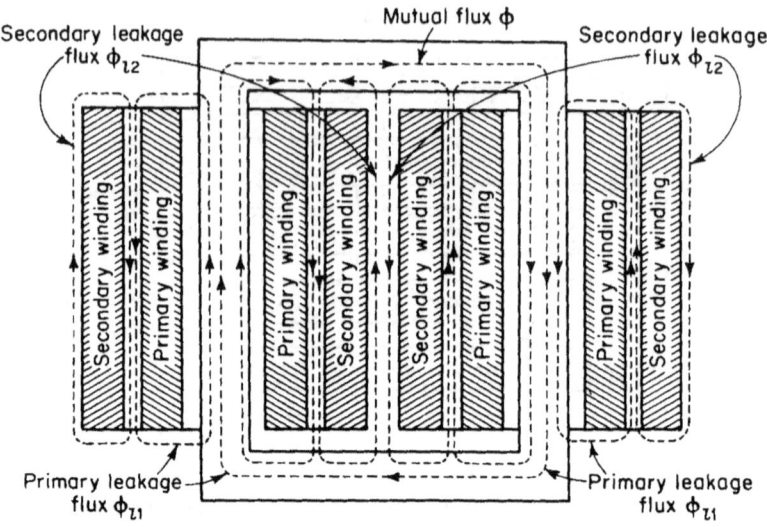

Figure 6-9. Winding arrangement in core-type transformer showing approximate paths taken by mutual flux and primary and secondary leakage fluxes.

show separately defined paths for the three component fluxes, these fluxes do not exist as independent entities, but merge into a common flux pattern. Nevertheless, the concept of these three component fluxes facilitates the analysis of transformer performance.

The actual flux paths are such that the leakage fluxes are distributed throughout the windings rather than concentrated into spaces between the windings. As a result, each winding is subjected to partial flux linkages. This means that differing amounts of leakage flux link the different turns of a given winding. Therefore, the leakage fluxes ϕ_{l1} and ϕ_{l2} are equivalent leakage fluxes; the former links all N_1 turns of the primary and the latter links all N_2 turns of the secondary. Thus, if λ_{l1} = the primary leakage flux linkage, and λ_{l2} = the secondary leakage flux linkage, then

$$\phi_{l1} = \frac{\lambda_{l1}}{N_1} \qquad (6\text{-}29)$$

Sec. 6-4 EQUIVALENT CIRCUIT OF THE TRANSFORMER

and

$$\phi_{l2} = \frac{\lambda_{l2}}{N_2} \tag{6-30}$$

Furthermore, if ϕ is the equivalent mutual flux linking all the turns of both windings, then the total equivalent fluxes linking the primary and the secondary windings, respectively, are

$$\phi_1 = \phi_{l1} + \phi \tag{6-31}$$

$$\phi_2 = \phi_{l2} + \phi \tag{6-32}$$

and the primary and secondary terminals voltages are

$$v_1 = R_1 i_1 + N_1 \frac{d\phi_1}{dt} = R_1 i_1 + N_1 \frac{d\phi_{l1}}{dt} + N_1 \frac{d\phi}{dt} \tag{6-33}$$

$$v_2 = R_2 i_2 + N_2 \frac{d\phi_2}{dt} = R_2 i_2 + N_2 \frac{d\phi_{l2}}{dt} + N_2 \frac{d\phi}{dt} \tag{6-34}$$

The primary leakage flux linkage and the secondary leakage flux linkage are

$$\lambda_{l1} = N_1 \phi_{l1} \tag{6-35}$$

$$\lambda_{l2} = N_2 \phi_{l2} \tag{6-36}$$

and from these the primary leakage inductance and secondary leakage inductance are, respectively

$$L_{l1} = \frac{\lambda_{l1}}{i_1} \tag{6-37}$$

$$L_{l2} = \frac{\lambda_{l2}}{i_2} \tag{6-38}$$

so that

$$L_{l1} i_1 = \lambda_{l1} = N_1 \phi_{l1}$$

$$L_{l2} i_2 = \lambda_{l2} = N_2 \phi_{l2}$$

and

$$L_{l1} \frac{di_1}{dt} = N_1 \frac{d\phi_{l1}}{dt} \tag{6-39}$$

$$L_{l2} \frac{di_2}{dt} = N_2 \frac{d\phi_{l2}}{dt} \tag{6-40}$$

When Eqs. 6-39 and 6-40 are substituted in Eqs. 6-33 and 6-34, the primary and secondary voltages are expressed by

$$v_1 = R_1 i_1 + L_{l1}\frac{di_1}{dt} + N_1\frac{d\phi}{dt} \qquad (6\text{-}41)$$

$$v_2 = R_2 i_2 + L_{l2}\frac{di_2}{dt} + N_2\frac{d\phi}{dt} \qquad (6\text{-}42)$$

Equation 6-42 can be rewritten in terms of i_L on the basis of Eq. 6-15 to read

$$v_2 = -R_2 i_L - L_{l2}\frac{di_L}{dt} + N_2\frac{d\phi}{dt} \qquad (6\text{-}43)$$

Since the leakage flux paths are largely in air, the leakage inductances L_{l1} and L_{l2} are practically constant.*

If a sinusoidal voltage, having an effective or rms value of V_1 v, is applied to the primary of an air-core transformer when the secondary is connected to a linear load impedance, the primary current I_1 and the secondary load current I_L will also be sinusoidal. The quantities I_1 and I_L are the rms values of the primary current and secondary load current. It was mentioned in Chapter 5 that the exciting current in iron-core transformers contains harmonics, and is therefore not sinusoidal. The exciting current, however, is small enough in comparison with the rated primary current of the transformer so that its effect on the wave form of the primary and secondary currents is usually negligible in the normal range of load. Thus, if the impedance of the load in the secondary is linear, the primary and secondary currents in an iron-core transformer are practically sinusoidal for normal load conditions if the voltage applied to the primary is sinusoidal. Equations 6-41 and 6-43 can be rewritten in terms of phasor quantities as follows

$$\mathbf{V}_1 = R_1 \mathbf{I}_1 + jX_{l1}\mathbf{I}_1 + \mathbf{E}_1 \qquad (6\text{-}44)$$

$$\mathbf{V}_2 = -R_2 \mathbf{I}_L - jX_{l2}\mathbf{I}_L + \mathbf{E}_2 \qquad (6\text{-}45)$$

* In transformers designed for high leakage reactance, the leakage inductances are not linear. The effects of such nonlinearities are treated by H. W. Lord, "An Equivalent Circuit for Transformers in Which Nonlinear Effects are Present," *Trans AIEE*, 78 Part I (1959), 580–586.

Sec. 6-4 EQUIVALENT CIRCUIT OF THE TRANSFORMER

where $V_1 =$ the effective applied primary terminal voltage
$V_2 =$ the effective secondary terminal voltage
$E_1 =$ the effective primary voltage induced by the mutual flux
$E_2 =$ the effective secondary voltage induced by the mutual flux
$I_1 =$ the effective primary current
$I_L =$ the effective secondary load current
$R_1 =$ the resistance of the primary winding
$R_2 =$ the resistance of the secondary winding
$X_{l1} = 2\pi f L_{l1} =$ the primary leakage reactance
$X_{l2} = 2\pi f L_{l2} =$ the secondary leakage reactance
$f =$ the frequency in cycles per second

The exact equivalent circuit

Figure 6-6(a) shows the equivalent circuit of an ideal transformer modified to take into account the exciting current of an iron-core transformer. Equations 6-44 and 6-45 suggest further modifications of the ideal transformer

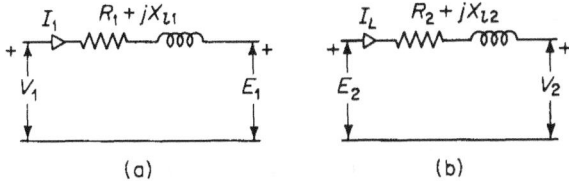

Figure 6-10. Equivalent circuits of (a) primary leakage impedance (Eq. 6-44), and (b) secondary leakage impedance (Eq. 6-45).

by taking into account the effect of the resistance and leakage reactance of the windings. Equation 6-44 and 6-45 are satisfied by the relationships shown in the equivalent circuits of Figs. 6-10(a) and 6-10(b) respectively. The equivalent circuit that completely represents an iron-core transformer in which capacitance effects are negligible and in which the leakage inductances are linear combines the circuits of Figs. 6-10(a) and 6-10(b) with that of Fig. 6-6(a) as shown in Fig. 6-11.

A phasor diagram based on the exact equivalent circuit of Fig. 6-11(b) is shown in Fig. 6-12. In this phasor diagram θ_L is the power factor angle of the load connected to the secondary terminals of the transformer. Also, the flux is shown lagging the induced emf by an angle of 90° in accordance with the phasor diagram of Fig. 5-9.

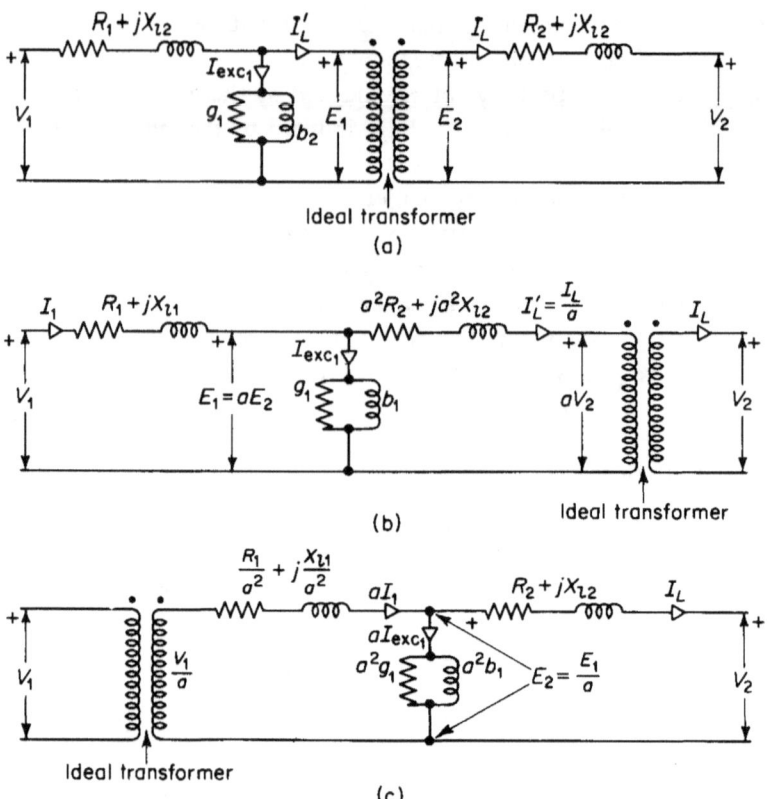

Figure 6-11. Exact equivalent circuits of a transformer.

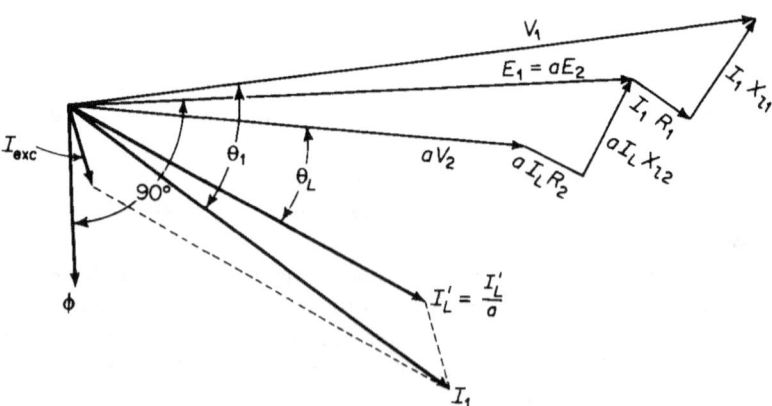

Figure 6-12. Phasor diagram for exact equivalent circuit [Figure 6-11(b)].

Sec. 6-4 EQUIVALENT CIRCUIT OF THE TRANSFORMER

EXAMPLE 6-2: The constants of a 150-kva, 2,400/240-v, 60-cps transformer are as follows

Resistance of the 2,400-v winding $R_1 = 0.216$ ohm
Resistance of the 240-v winding $R_2 = 0.00210$ ohm
Leakage reactance of the 2,400-v winding $X_{l1} = 0.463$ ohm
Leakage reactance of the 240-v winding $X_{l2} = 0.00454$ ohm
Exciting admittance on the 240-v side $y_{\text{exc}} = 0.0101 - j0.069$ mho

(a) Draw the equivalent circuit on the basis of Fig. 6-11(a) showing the numerical values of the leakage impedances and the exciting admittance expressed in complex form.

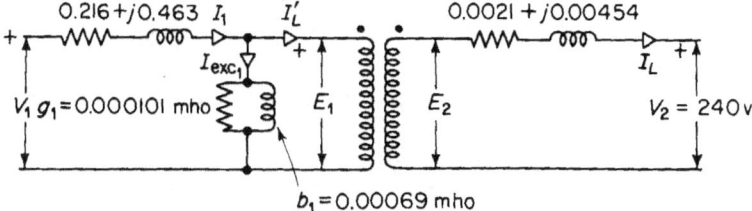

Figure 6-13. Equivalent circuit for transformer of Example 6-2.

(b) Determine the values of the emfs E_1 and E_2 induced by the equivalent mutual flux, the exciting current I_{exc}, the primary current, I_1 at 0.80 power factor, current lagging, and the applied primary voltage when the transformer delivers rated load at rated secondary voltage.

Solution:
(a) The equivalent circuit is shown in Fig. 6-13.
(b) The rated secondary terminal voltage is 240 v and the rated load (rated output) of the transformer is 150,000 va at 240 v. The rated load current is therefore

$$I_L = 150{,}000 \div 240 = 625 \text{ amp}$$

Let the phasor that represents the secondary terminal voltage V_2 lie on the axis of reals as shown in Fig. 6-12, then

$$V_2 = 240 + j0 \text{ v}$$

and the load current expressed as a phasor is

$$I_L = 625(0.80 - j0.60) = 500 - j375 \text{ amp}$$

The secondary induced voltage E_2 in Fig. 6-12 must equal the phasor sum of the secondary terminal voltage plus the voltage drop across the secondary leakage impedance, thus

$$E_2 = V_2 + I_L(R_2 + jX_{l2})$$
$$= 240 + j0 + (500 - j375)(0.00210 + j0.00454)$$
$$= 240 + j0 + 1.05 + 1.70 + j2.27 - j0.79$$
$$= 242.75 + j1.48 = 242.8 \underline{/0.35°}$$

Since the primary induced voltage E_1 and the secondary induced voltage E_2 are both produced by the mutual flux, their ratio E_1/E_2 must equal the turns ratio $a = N_1/N_2 = 10$, and we have

$$E_1 = aE_2 = 10 \times 242.8 \underline{/0.35°} = 2{,}428 \underline{/0.35°} = 2{,}428 + j14.8$$

The load component $I_{L'}$ of the primary current expressed in phasor form is

$$I_{L'} = \frac{I_L}{a} = \frac{500 - j375}{10} = 50.0 - j37.5 \text{ amp}$$

The primary current is the phasor sum of the load component $I_{L'}$ and the exciting current I_{exc}. The exciting current produces the mutual flux, so to speak, and is obtained by multiplying the primary induced voltage by the exciting admittance referred to the primary side, i.e.

$$I_{exc_1} = E_1 Y_{exc_1}$$

The value of the admittance in the given data is as measured on the secondary side. This admittance is referred to the primary side by making use of the impedance ratio a^2. It should be remembered that an impedance is transferred from the secondary side to the primary side of a transformer by multiplying its value by the impedance ratio. Further, impedance is the reciprocal of admittance, and on transferring the exciting admittance from the secondary to the primary side, we have

$$\frac{1}{Y_{exc_1}} = a^2 \left(\frac{1}{Y_{exc_2}} \right)$$

or

$$Y_{exc_1} = \frac{1}{a^2}(Y_{exc_2}) = \frac{0.0101 - j0.069}{(10)^2}$$
$$= (1.01 - j6.9)\,10^{-4} = 6.98 \times 10^{-4} \underline{/-81.67°} \text{ mho}$$

Sec. 6-4   EQUIVALENT CIRCUIT OF THE TRANSFORMER

The primary exciting current is therefore

$$\mathbf{I}_{exc} = Y_{exc_1}\mathbf{E}_1 = 6.98 \times 10^{-4} \,\underline{/-81.67°} \times 2{,}428 \,\underline{/0.35°}$$
$$= 1.70 \,\underline{/-81.32°} = 0.257 - j1.68$$

From Fig. 6-12 it is evident that the primary current is

$$\mathbf{I}_1 = \mathbf{I}_{L'} + \mathbf{I}_{exc}$$
$$= 50.0 - j37.5 + 0.257 - j1.68 = 50.257 - j39.18$$
$$= 63.6 \,\underline{/-37.9°}$$

The primary applied voltage is the sum of the primary induced voltage and the voltage drop across the primary leakage impedance. Hence,

$$\mathbf{V}_1 = \mathbf{E}_1 + \mathbf{I}_1(R_1 + jX_{l1})$$
$$= 2{,}428 + j14.8 + (50.257 - j39.18)(0.216 + j0.463)$$
$$= 2{,}428 + j14.8 + 10.86 + 18.14 + j23.27 - j8.46$$
$$= 2{,}457 + j29.5 = 2{,}457 \,\underline{/0.69°} \text{ v}$$

All three of the equivalent circuits shown in Fig. 6-11 lead to exactly the same values of primary current, $\mathbf{I}_1$, and primary applied voltage, $\mathbf{V}_1$ as were obtained in the solution of Example 6-2.

The approximate equivalent circuit

Calculations involving transformers seldom require the accuracy of the exact equivalent circuit and can be simplified by using the approximate equivalent circuits shown in Fig. 6-14. The equivalent circuits in Fig. 6-14 neglect the effect of the exciting current in the voltage drop across the primary leakage impedance. However, the error in the value computed for the primary applied voltage V_1 is quite small, as is evident from the fact that in Example 6-2 the primary applied voltage V_1 is 2,457 v, as compared with a value of 2,428 v for the primary induced emf E_1. Here we have a difference of 29 v in the magnitude of the two voltages V_1 and E_1 due to rated current. The exciting current, which has a magnitude of only about 1.7/62.5, or less than 3 percent of rated current, would account for a difference of only about 1 v or only 1 part in about 2,400. The exciting current that results from impressing the primary terminal voltage on the exciting admittance is only a few percent greater than that produced by the primary induced voltage E_1. Since the exciting current is only a few percent of rated current in iron-core transformers, a change of a few percent in the exciting current

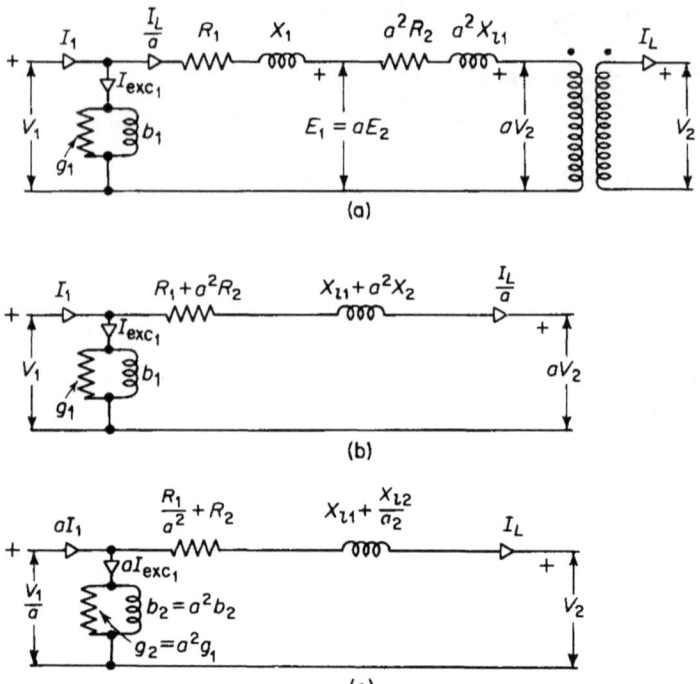

Figure 6-14. Approximate equivalent circuits.

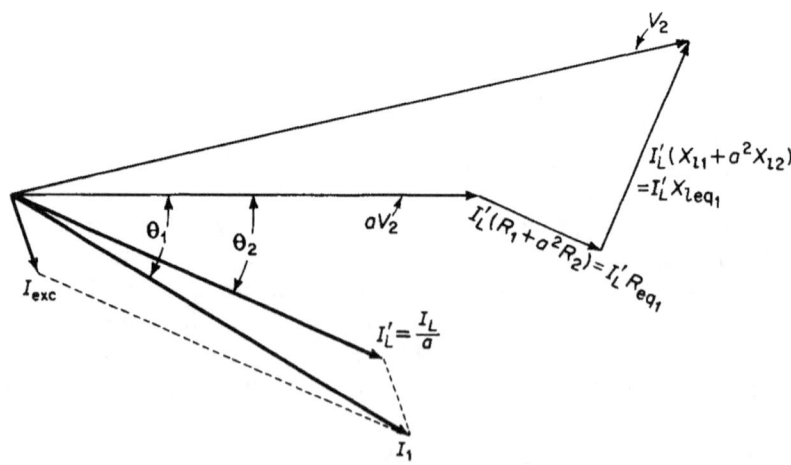

Figure 6-15. Phasor diagram for approximate equivalent circuit [Figure 6-13(a)].

Sec. 6-5 OPEN-CIRCUIT AND SHORT-CIRCUIT TESTS

itself is a very small change when compared with the value of the rated current. In view of the rather small effect on the voltage drops in the transformer windings and on the exciting current produced by moving the exciting admittance to the part of the circuit that corresponds to the primary terminals of the transformer, the approximate equivalent circuit is valid for most transformer calculations. The phasor diagram for the approximate equivalent circuit is shown in Fig. 6-15.

6-5 OPEN-CIRCUIT AND SHORT-CIRCUIT TESTS, EXCITING ADMITTANCE, AND EQUIVALENT IMPEDANCE

The constants of the transformer for the approximate equivalent circuit can be determined from the open-circuit and short-circuit tests.

The open-circuit test is made to obtain the values of the exciting-admittance, conductance, and susceptance. In the case of power transformers or constant-voltage transformers operating at one specified frequency, the open-circuit test consists in the application of rated voltage at rated frequency usually to the low-voltage winding with the high-voltage winding open circuited. Measurements are made by means of indicating instruments (voltmeter, ammeter, and wattmeter) of voltage, current, and power. Actually, the open-circuit test can be made by applying rated voltage at rated frequency to the high-voltage winding with the low-voltage winding open circuited. However, it is usually more convenient to work on the low-voltage side because the lower voltage is easier to handle.

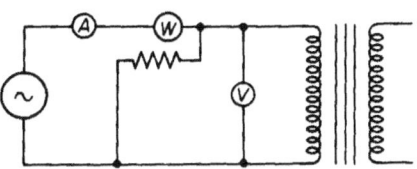

Figure 6-16. Open-circuit test with indicating instruments.

The quantities involved in most communication transformers are generally too small for wattmeter and ammeter measurements. Methods involving the use of a-c bridges or other suitable devices are employed.
In Fig. 6-16

let I_{exc} = the exciting current as read by the ammeter, A*
  V = the applied voltage as read by the voltmeter, V
  P_{oc} = power as measured with the wattmeter, W

---

* When the exciting current is so small that the current in the voltmeter and the current in the potential coil of the wattmeter are appreciable, the voltmeter and the wattmeter potential coil should be disconnected from the circuit when the exciting current is measured.

Then the exciting admittance is

$$Y_{exc} = \frac{I_{exc}}{V} \tag{6-46}$$

and the exciting conductance is

$$g = \frac{P_{oc}}{V^2} \tag{6-47}$$

from which the exciting susceptance is found to be

$$b = \sqrt{y_{exc}^2 - g^2} \tag{6-48}$$

EXAMPLE 6-3: The following data, corrected for instrument losses, were obtained in an open-circuit test made on the 240-v winding of the transformer in Example 6-2.

$$\text{Volts} = 240$$
$$\text{Current} = 16.75 \text{ amp}$$
$$\text{Power} = 580 \text{ w}$$

Determine the exciting-admittance, conductance, and susceptance.

Solution:
The exciting admittance is

$$Y_{exc} = \frac{I_{exc}}{V} = \frac{16.75}{240} = 0.0698 \text{ mho}$$

and the exciting conductance is

$$g = \frac{P_{oc}}{V^2} = \frac{580}{(240)^2} = 0.0101 \text{ mho}$$

from which the exciting susceptance is found to be

$$b = \sqrt{y_{exc}^2 - g^2}$$
$$= \sqrt{(0.0698)^2 - (0.0101)^2} = 0.069 \text{ mho}$$

The short-circuit test yields results from which the equivalent impedance, equivalent resistance, and equivalent leakage reactance of the transformer can be evaluated. In Fig. 6-14(b) the equivalent impedance of the transformer referred to the primary is

$$Z_{eq_1} = R_{eq_1} + jX_{eq_1} \tag{6-49}$$

Sec. 6-5 OPEN-CIRCUIT AND SHORT-CIRCUIT TESTS 261

in which the equivalent resistance referred to the primary of the transformer is

$$R_{eq_1} = R_1 + a^2 R_2 \tag{6-50}$$

and the equivalent leakage reactance referred to the primary is

$$X_{eq_1} = X_{l1} + a^2 X_{l2} \tag{6-51}$$

The equivalent impedance referred to the secondary is

$$Z_{eq_2} = \frac{Z_{eq_1}}{a^2} = \frac{R_{eq_1}}{a^2} + j\frac{X_{eq_1}}{a^2} \tag{6-52}$$

from which

$$R_{eq_2} = \frac{R_{eq_1}}{a^2} = \frac{R_1}{a^2} + R_2 \tag{6-53}$$

and

$$X_{eq_2} = \frac{X_{eq_1}}{a^2} = \frac{X_{l1}}{a^2} + X_{l2} \tag{6-54}$$

It is the general practice, in the short-circuit test, to short circuit the low-voltage winding and to apply voltage, at rated frequency, such that rated current flows in the transformer. Measurements are made of input current, power, and voltage, using indicating instruments, i.e., ammeter, wattmeter, and voltmeter as shown in Fig. 6-17. Again, in the case of communication transformers, an a-c bridge method or an arrangement adapted for measuring the smaller quantities associated with such transformers is used.

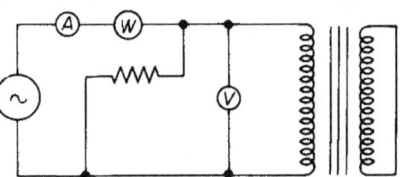

Figure 6-17. Short-circuit test with indicating instruments.

The short-circuit test could also be made by short circuiting the high-voltage side and making the measurements on the low-voltage side. This, however, is more inconvenient since the applied voltage is low and the measured current high. These quantities are handled and measured more readily on the high-voltage side.

For the short-circuit test

let V_{sc} = applied voltage as read by the voltmeter, V (in Fig. 6-17)
 I_{sc} = input short-circuit current as read by the ammeter, A (in Fig. 6-17)
 P_{sc} = input power as read by the wattmeter, W (in Fig. 6-17)

The voltage V_{sc} is low in comparison with the rated voltage. It can be shown that, in the case of the transformer of Example 6-2, the short-circuit voltage is less than 3 percent of rated value. On that basis, the exciting current can be neglected and the short-circuit impedance Z_{sc} assumed to equal the equivalent series impedance of the transformer. Hence, the equivalent series impedance is

$$Z_{eq} \approx Z_{sc} = \frac{V_{sc}}{I_{sc}} \qquad (6\text{-}55)$$

and the equivalent series resistance is practically equal to the short-circuit resistance

$$R_{eq} \approx R_{sc} = \frac{P_{sc}}{I_{sc}^2} \qquad (6\text{-}56)$$

Also, the equivalent leakage reactance of the transformer is

$$X_{eq} \approx X_{sc} = \sqrt{Z_{sc}^2 - R_{sc}^2} \qquad (6\text{-}57)$$

EXAMPLE 6-4: A short-circuit test with the low-voltage winding of the transformer in Example 6-2, short circuited, yielded the following data

$$\text{Volts} = 63.0$$
$$\text{Current} = 62.5 \text{ amp}$$
$$\text{Power} = 1,660 \text{ w}$$

Determine

(a) The equivalent primary impedance.
(b) The equivalent primary resistance.
(c) The equivalent primary leakage reactance of the transformer.

Solution:

(a) $\quad Z_{eq_1} = \dfrac{63}{62.5} = 1.008 \text{ ohm}$

(b) $\quad R_{eq_1} = \dfrac{1,660}{(62.5)^2} = 0.425 \text{ ohm}$

(c) $\quad X_{eq_1} = \sqrt{(1.008)^2 - (0.425)^2} = 0.915 \text{ ohm}$

6-6 TRANSFORMER LOSSES AND EFFICIENCY

The losses in a transformer are the core loss due to hysteresis and eddy currents in the core, as explained in Chapter 5; and the copper losses in the windings and stray losses due to eddy currents induced by the leakage fluxes in the tank and other parts of the structure. The sum of the copper losses* and the stray losses is known as *load losses* and is determined from

---

* The term *copper losses* is still sometimes used instead of *load losses*, and when so used is meant to include the stray losses.

the short-circuit test, being I^2R_{eq}. The core loss is determined from the open-circuit test.

The power efficiency of a device is the ratio of the useful power output to the power input, the power input being equal to the useful power output plus the power losses. Thus

$$\text{Efficiency} = \frac{\text{output}}{\text{input}} \quad (6\text{-}58)$$

$$= \frac{\text{input} - \text{losses}}{\text{input}} = 1 - \frac{\text{losses}}{\text{input}} \quad (6\text{-}59)$$

The rated load efficiency of transformers is generally quite high (as high as 90 percent for transformers, as small as 1 kva) and since the losses can be determined readily, Eq. 6-59 is preferred to Eq. 6-58 for calculations of efficiency for reasons of accuracy.

Methods of testing transformers and of calculating their performance have been developed to a high degree and are specified in the ASA Standards.* These include corrections for temperature and other refinements that are not within the scope of this text.

EXAMPLE 6-5: Determine the efficiency of the transformer in Examples 6-2, 6-3, and 6-4

(a) At rated load 0.80 power factor.
(b) At one-half rated load 0.60 power factor.

Solution:
It is important to note that the transformer delivers rated load when it delivers its rating, *at rated voltage*, in *volt-amperes, regardless of power factor*.

(a) $\qquad$ Output = 150,000 × 0.80 = 120,000 w

Rated current = 62.5 amp (on the high-voltage side)

The equivalent resistance of the transformer referred to the high-voltage side from Example 6-4 is

$$R_{eq_1} = 0.425 \text{ ohm}$$

and the load loss at rated current is

$$I_1^2 R_{eq_1} = (62.5)^2 \times 0.425 = 1,660 \text{ w}$$

which is also the value measured in the short-circuit test at rated current.

* ASA Standards, *Transformers, Regulators and Reactors*, Bull C57. New York: American Standards Association, 1948.

The no-load loss at rated voltage is from Example 6-3

$$P_{oc} = 580 \text{ w}$$

The total loss then is $1,660 + 580 = 2,240$ w and the power input

$$= \text{output} + \text{losses}$$
$$= 120,000 + 2,240$$
$$= 122,240 \text{ w}$$

The rated-load efficiency at 0.80 power factor is therefore

$$\text{Eff} = 1 - \frac{\text{losses}}{\text{input}} = 1 - \frac{2,240}{122,240}$$
$$= 1 - 0.0183 = 0.9817$$

(b)   One-half rated load $= 150,000 \times 1/2 = 75,000$ va

$$\text{Output} = 75,000 \times 0.60 = 45,000 \text{ w}$$

The core loss is assumed to remain unchanged as the secondary voltage is again at rated value. Hence

$$P_{oc} = 580 \text{ w}$$

The load losses, however, vary as the square of the current and, therefore, at one-half rated current, are equal to one-fourth the load losses at full-load current.
Hence

$$I^2 R_{eq} = \tfrac{1}{4} \times 1,660 = 415 \text{ w}$$

and the total losses are $580 + 415 = 995$ w. The efficiency, then, at one-half rated load 0.60 power factor, is

$$\text{Eff} = 1 - \frac{\text{losses}}{\text{input}} = 1 - \frac{995}{45,000 + 995}$$
$$= 1 - \frac{995}{45,995}$$
$$= 0.9784.$$

A power transformer may undergo a considerable variation in load during a 24-hr period. There may be intervals during which the transformer carries a substantially rated load and others during which the transformer carries only a small part of its rating. So if the efficiency of operation is to be taken over such a 24-hr period, the *energy efficiency* is the proper criterion. The energy efficiency for a given period is the ratio of the kilowatt-hour output to the kilowatt-hour input and depends not only on the characteristics of the transformer but also on the load cycle.

Sec. 6-6   TRANSFORMER LOSSES AND EFFICIENCY   265

For instance, to compute the energy efficiency for a 24-hr period, or the *all-day efficiency*, it is necessary to integrate the core loss and the load losses over a 24-hr period. The core losses are practically constant since the voltage of a power transformer is nearly constant. The energy in kilowatt hours associated with the core loss is merely the product of the core loss in kilowatts and the time in hours during which the transformer is energized. It is more difficult to compute the energy loss due to the load losses (I^2R) when the load is fluctuating. However, if the daily load schedule is known and minor fluctuations of the load can be neglected, the load curve can be approximated by a step curve, and the energy losses can be calculated for a 24-hr period as shown in Example 6-5 below.

EXAMPLE 6-5: The load schedule for a 24-hr period on a 30-kva, 2,400/240 v, 60-cycle transformer is as follows

Hours	Kw	PF
10	2.5	0.75
8	12.5	0.80
3	20.0	0.85
3	25.0	0.90

The data taken from open-circuit and short-circuit tests follow

High Side open, measurements on low-voltage side		
Volts	Amperes	Watts
240	2.45	154

Low Side short circuited, measurements on high side		
Volts	Amperes	Watts
61.1	12.5	706

Determine the energy efficiency for a daily load cycle.

Solution:
The core losses are constant for the 24-hr period, and the energy dissipated in core loss is therefore

$$\frac{154 \times 24}{1,000} = 3.70 \text{ kwhr}$$

The load losses are proportional to the square of the current, i.e., to the square of the kva load and the energy dissipated in the load loss are computed in the following table.

					Load loss	
Hours	Kw	Kwhr	PF	Kva	Kw	Kwhr
10	2.5	25	0.75	3.33	0.00871	0.087
8	12.5	100	0.80	15.63	0.192	1.536
3	20.0	60	0.85	23.50	0.433	1.299
3	25.0	75	0.90	27.80	0.606	1.818

The load loss is proportional to the square of the kva load; hence, at 3.33 kva, the load loss is

$$\left(\frac{3.33}{30.0}\right)^2 \times 706 = 8.71 \text{ w}$$

and for 15.63 kva

$$\left(\frac{15.63}{30.0}\right)^2 \times 706 = 192 \text{ w}$$

$$\begin{aligned}
\text{Total output} &= 260 \text{ kwhr} \\
\text{Total load loss} &= 4.74 \text{ kwhr} \\
\text{Total core loss} &= 3.70 \text{ kwhr} \\
\hline
\text{Input} &= 268.44 \text{ kwhr}
\end{aligned}$$

$$\text{Eff} = 1 - \frac{8.44}{268.44} = 1 - 0.0314 = 0.9686$$

6-7 VOLTAGE REGULATION

An important measure of the performance of a transformer is its voltage regulation. The voltage regulation of a transformer is the change in the secondary voltage from its rated value at full load to that at no load in terms of rated secondary voltage when the primary voltage is held constant. Voltage regulation is usually expressed in percent, and is given by the relationship

$$\text{percent regulation} = \frac{E_{oc_2} - V_2}{V_2} \times 100 \tag{6-60}$$

where V_2 is the rated secondary voltage at full load and E_{oc_2} is the no-load secondary voltage with the same value of primary voltage for both full load and no load. It is important to note that only the magnitudes rather than the phasor quantities of the voltages are involved in Eq. 6-60.

The regulation of the transformer in Example 6-2 for full-load current at 0.80 power factor lagging is, according to Eq. 6-60

$$\text{percent regulation} = \frac{245.7 - 240}{240} \times 100 = 2.38$$

It is not necessary to carry out the calculations for transformers in as much detail as was done in Example 6-2. The approximate equivalent circuit is an adequate basis for calculations of regulation and affords a great deal of simplification. If the numerator and denominator in Eq. 6-60 are both multiplied by the ratio a, the result is

$$\text{percent regulation} = \frac{aE_{oc_2} - aV_2}{aV_2} \times 100 \qquad (6\text{-}61)$$

But $aE_{oc_2} = V_1$, the primary applied voltage, and we can now apply the phasor diagram of Fig. 6-15 for determining the regulation.

EXAMPLE 6-6: Find the regulation of the transformer in Example 6-2 for rated load 0.80 PF, current lagging, on the basis of the data in Example 6-4.

Solution:
The short-circuit data in Example 6-4 are the basis for the part of the approximate equivalent circuit that is involved in computing voltages, and therefore regulation.

$$\mathbf{V}_1 = a\mathbf{V}_2 + (R_{eq_1} + jX_{eq_2})\frac{\mathbf{I}_L}{a}$$

$$a\mathbf{V}_2 = 2{,}400 + j0 \text{ v}$$

$$\frac{\mathbf{I}_L}{a} = 62.5(0.8 - j0.6) = 50.0 - j37.5 \text{ amp}$$

$$R_{eq_1} + jX_{eq_1} = 0.425 + j0.915 \text{ ohm}$$

$$\mathbf{V}_1 = 2{,}400 + j0 + (0.425 + j0.915)(50.0 - j37.5)$$
$$= 2{,}400 + j0 + 21.25 + j45.75 - j15.95 + 34.3$$
$$= 2{,}455.55 + j29.8$$
$$= 2{,}456 \;\underline{/0.70°}$$

$$\text{percent regulation} = \frac{2{,}456 - 2{,}400}{2{,}400} \times 100$$
$$= 2.33$$

This is a good check against the value of 2.38 based on the exact equivalent circuit.

Certain types of loads operate most efficiently, and with long life, at their rated value of voltage and frequency. When operating appreciably above rated value, the load's life may become shortened, as in the case of incandescent lamps; on the other hand, operation at subnormal voltages leads to reduced efficiency, and, in the case of motors, may cause overheating at rated load. Such loads should be supplied by transformers that have small values of regulation. There are other types of loads that operate at nearly constant current, such as series lighting systems and electric welding arcs. Each such constant-current load is usually supplied from its own individual transformer—one that has a high value of regulation.

6-8 AUTOTRANSFORMERS

In cases where the ratio of transformation does not differ greatly from unity, and in cases that do not require the secondary winding to be isolated from the primary winding, a great saving in size can be effected by using autotransformers in place of the two-circuit transformers discussed previously in this chapter. An ordinary two-circuit transformer can be converted to an autotransformer by connecting its windings 1 and 2 in series with each other as shown in Fig. 6-18. In the autotransformer of Fig. 6-18, consider winding 1 to be between points a and b and winding 2 to be between points b and c. The voltage $\mathbf{V}_H$ on the high side of the autotransformer is the phasor sum of the terminal voltages of windings 1 and 2, i.e.

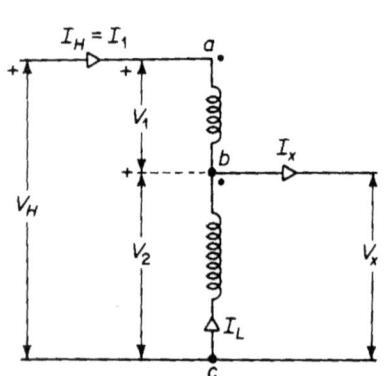

Figure 6-18. Autotransformer connections.

$$\mathbf{V}_H = \mathbf{V}_2 + \mathbf{V}_1 = \mathbf{V}_x + \mathbf{V}_1 \quad (6\text{-}62)$$

The phase angle between the primary and secondary voltage of a power transformer is small. In the case of the transformer in Example 6-2, the angle between the primary and secondary voltage phasors is smaller than 1°, although phase angles as high as 10° are possible. Nevertheless, it can be assumed for most practical purposes that the arithmetic sum of the voltages $\mathbf{V}_1$ and $\mathbf{V}_2$ equals the phasor sum, i.e.

$$V_H \approx V_2 + V_1 = V_x + V_1 \quad (6\text{-}63)$$

In addition, the ratio of the terminal volts V_2 and V_1 is within a few percent of the turns ratio, as indicated in the transformer of Example 6-2. Hence

$$\frac{V_H}{V_x} \approx \frac{V_x + V_1}{V_x} \approx \frac{N_2 + N_1}{N_2} \tag{6-64}$$

The exciting ampere turns required by the flux in the core is the phasor difference between the ampere turns of the two windings. Generally, in iron-core transformers the exciting current is small in comparison with the rated current. (In the transformer of Example 6-2 the ratio of exciting current to rated current is 1.70 ÷ 62.5 or 2.72 percent.) Then, if the exciting current is neglected, the ampere turns of the two windings must be equal, and we have

$$N_1 \mathbf{I}_1 = N_1 \mathbf{I}_H \approx N_2 \mathbf{I}_L$$

$$\mathbf{I}_H = \mathbf{I}_1 \approx \frac{N_2}{N_1} \mathbf{I}_L \tag{6-65}$$

The current $\mathbf{I}_x$, which flows in the low-voltage terminals, is the sum of the secondary current $\mathbf{I}_L$ and the primary current $\mathbf{I}_1$ as shown in Fig. 6-18, so that

$$\mathbf{I}_x = \mathbf{I}_1 + \mathbf{I}_L \tag{6-66}$$

A comparison of Eqs. 6-63 and 6-64 shows that

$$\frac{\mathbf{I}_H}{\mathbf{I}_x} \approx \frac{N_2}{N_1 + N_2} \tag{6-67}$$

When the ratio of transformation is near unity, the transformer has a much higher rating when operating as an autotransformer than when operating as a 2-circuit transformer. This is shown as follows

$$\text{autotransformer rating} = V_H I_H \approx (V_x + V_1) I_1 \tag{6-68}$$

$$\text{2-circuit transformer rating} = V_1 I_1 \tag{6-69}$$

$$\frac{\text{autotransformer rating}}{\text{2-circuit transformer rating}} \frac{V_x}{V_1} + 1 \tag{6-70}$$

Thus, if the voltage rating as a 2-circuit transformer is 2,400/240, the volt-ampere rating as a 2,640/2,400-v autotransformer is eleven times that of the volt-ampere rating as a 2-winding transformer.

6-9 INSTRUMENT TRANSFORMERS

The instruments and relays associated with protective and control devices are usually connected in the secondary circuits of instrument-voltage and instrument-current transformers of a-c power circuits that operate at voltages in excess of a few hundred volts. Consider a 3-phase generator rated at 22,000 v between terminals with an output of 100,000 kva. The line-neutral voltage of such a generator is $22,000 \div \sqrt{3} = 12,700$ v and the rated line current is $100,000,000 \div (\sqrt{3} \times 22,000) = 2,625$ amp per phase. Obviously, it would be extremely impractical to apply such values of voltage and current directly to voltmeters, ammeters, wattmeters, relays, and other control devices. For that reason, use is made of *instrument potential transformers*, which step the voltage down to about 115 secondary v and *instrument current transformers*, which step the current down to a rated value of 5 amp or less. Potential transformers are usually connected line to neutral in 3-phase installations, although line-to-line connections are not uncommon. The secondaries of instrument-potential transformers and instrument-current transformers are grounded for reasons of safety. Grounding eliminates the hazard of raising the secondary winding to a high potential through capacitance coupling with the primary winding. In addition, as a matter of safety, the secondary circuit of a current transformer should never be opened while under load, because when the secondary is opened there are no secondary ampere turns opposing the primary ampere turns, and all of the primary current becomes exciting current, and this may induce a very high voltage in the secondary winding.

6-10 VARIABLE-FREQUENCY TRANSFORMERS

Iron-core transformers are operated over a wide range of audio frequencies in communication circuits such as radio-frequency transmitters, radio receivers, and in conjunction with telephone circuits. Generally, the function of transformers in such arrangements is to couple an audio-frequency source, which may have relatively high impedance, usually in the form of resistance, to a load circuit, which may have a relatively low value of impedance, so as to obtain maximum power transfer or optimum performance. For example, the electronic circuit in a radio receiver may give its optimum performance, from the standpoint of power output, with negligible distortion when feeding into a load impedance of about 4,000 or 5,000 ohms, whereas, the apparent impedance of the load, such as the voice coil of a loud-speaker, is only about 10 ohms. A properly designed transformer, with an impedance ratio of about 400, connected between the electronic circuit and the loud-speaker would assure good performance. Such a transformer is known as an output transformer, and in this case would have a ratio of $a = \sqrt{400} = 20$. Iron-core transformers are also used to isolate one audio-frequency circuit from another so as to prevent the d-c component of current in one of these circuits from flowing into the other circuit. Although the wave forms of speech, music, and other sounds are very complicated, the audio-frequency range is considered to extend from about 16 to 20,000 cps, which are the effective limits of audibility for a steady-state sinusoidal signal.

If the ratio and the phase angle between the primary and secondary voltages were unaffected by frequency, signals would be transmitted from the primary of the transformer to the secondary without distortion. This would be an ideal situation. However, the leakage reactance ωL_{eq} increases with frequency and the exciting admittance decreases with increasing frequency. Both of these effects influence the ratio and phase angle. However, in the usual electronic circuit, the transformer leakage impedance is considerably smaller than the resistance of the source and the load, so that the effects of frequency are mitigated by the impedances of the source and the load. The core loss can be neglected in output transformers. The distributed capacitances between the turns and layers of each winding, as well as between windings (and between each winding and case and core) can also usually be neglected.

The equivalent circuit* shown in Fig. 6-19(a) is based on these considerations and includes, in addition to the transformer leakage impedance and magnetizing inductance, the resistance of the source and the load, all referred to the transformer primary. The circuit of Fig. 6-19(a) is valid throughout the normal range of audio frequencies. Further simplifications in the equivalent circuit are shown in Figs. 6-19(b), 6-19(c), and 6-19(d).

Figure 6-19(b) applies to the middle-frequency range. In this range the frequencies are low enough so that the leakage reactance ωL_{eq_1} is negligible in comparison with the resistance of the entire circuit, and at the same time the frequencies are high enough so that the current through the magnetizing reactance $\omega a M$ is negligible in relation to the load current.

Frequencies of such high values that the leakage reactance ωL_{eq_1} of the transformer becomes appreciable are said to be in the high-frequency range.† The equivalent circuit in Fig. 6-19(c) applies to the high-frequency range. Since the current through the magnetizing reactance $\omega a M$ is already negligible in the middle-frequency range, it is even smaller proportionately in the high-frequency range and can, therefore, again be neglected.

---

* F. E. Terman and R. E. Ingebretsen, "Output Transformer Response," *Electronics*, January, 1936, 30–32. See also F. E. Terman, *Radio Engineer's Handbook* (New York: McGraw-Hill Book Company, 1943) 385–393.

† This term should not be confused with *radio frequency*. *High-frequency range* means in the upper range of audio frequency.

Below the middle range of frequencies lies the low-frequency range in which the magnetizing current through the mutual reactance $\omega a M$ becomes appreciable. However, since the frequencies are now below the middle range, the leakage reactance ωL_{eq_1} is again negligible and the equivalent circuit of Fig. 6-19(d) now applies.

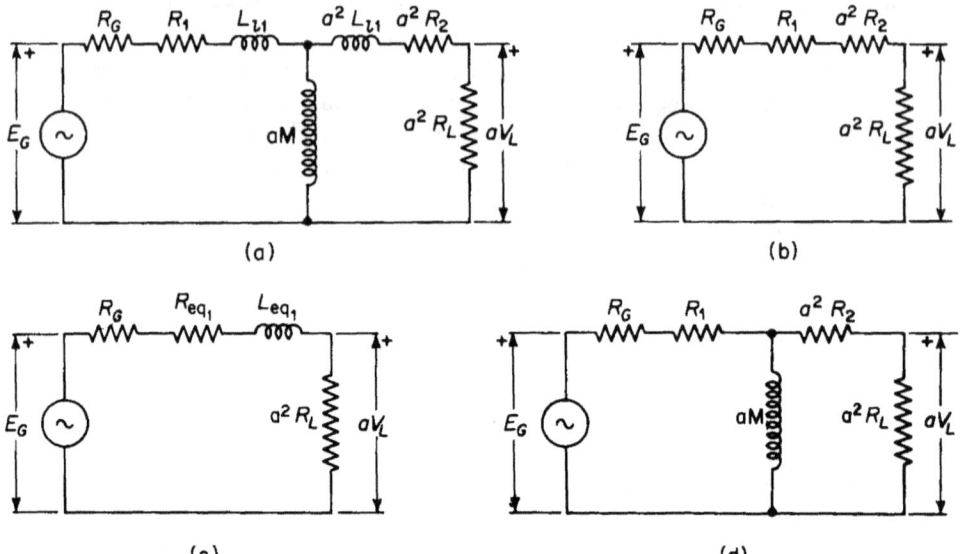

Figure 6-19. Equivalent circuits of an output transformer. (a) Complete representation; (b) approximated for middle-frequency range; (c) approximated for high-frequency range; (d) approximated for low-frequency range.

Signals at frequencies within the middle range are all transmitted at about the same voltage ratio and with negligible phase shift. This means that such signals undergo negligible distortion when transmitted through the output transformer. From Fig. 6-19(b) the load voltage is

$$V_L = \frac{1}{a} \frac{a^2 R_L}{R_G + R_1 + a^2(R_2 + R_L)} \qquad (6\text{-}71)$$

and the voltage ratio can be abbreviated to

$$\frac{V_L}{E_G} = \frac{1}{a} \frac{R'_L}{R'_{se}} \qquad (6\text{-}72)$$

in which

$$R'_L = a^2 R_L$$

Sec. 6-10   VARIABLE-FREQUENCY TRANSFORMERS   273

and
$$R'_{se} = R_G + R_1 + a^2(R_2 + R_L)$$

The frequency components of signals, which are in the high-frequency range, are not all transmitted to the load at the same voltage ratio nor at the same phase angle between the generated voltage and the load voltage. The higher the frequency of a component or harmonic, the greater is the attenuation, i.e., the smaller is the voltage ratio V_L/E_G and the greater is the phase shift between the voltages V_L and E_G. As a result, a signal, which is a composite of a number of frequencies some of which are in the high-frequency range, undergoes distortion because of the transformer leakage inductance L_{eq_1}. The voltage ratio and phase angle for the high-frequency range are expressed, on the basis of Fig. 6-19(c), by

$$\frac{V_L}{E_G} = \frac{1}{a} R'_L \frac{\underline{/\tan^{-1}(-\omega L_{eq_1}/R'_{se})}}{\sqrt{(R'_{se})^2 + (\omega L_{eq_1})^2}} \qquad (6\text{-}73)$$

The ratio of transformation in the low-frequency range decreases with decreasing frequency. The phase shift is opposite that of the high-frequency range, increasing in the opposite direction as the frequency decreases. On the basis of the equivalent circuit in Fig. 6-19(d), the voltage ratio at low frequencies is

$$\frac{V_L}{E_G} = \frac{1}{a}\frac{R'_L}{R'_{se}} \frac{1}{\sqrt{1 + (R'_{par}/\omega a M)^2}} \underline{/\tan^{-1}(R'_{par}/\omega a M)} \qquad (6\text{-}74)$$

where
$$R'_{par} = \frac{(R_G + R_1)[a^2(R_2 + R_L)]}{R_G + R_1 + a^2(R_2 + R_L)}$$

Distortion results in the low-frequency range because the signal strength decreases with decreasing frequency and the phase shift increases with decreasing frequency.

Since the signals in the middle-frequency range undergo negligible distortion, the voltage ratio in that range may be considered a norm with which the voltage ratios in the other two frequency ranges are compared. Hence, division of the amplitudes in Eq. 6-73 by that in Eq. 6-72 yields

High-frequency relative voltage ratio

$$= \frac{1}{\sqrt{1 + (\omega L_{eq_1}/R'_{se})^2}} \qquad (6\text{-}75)$$

and when Eq. 6-74 is divided by Eq. 6-72, the result is

Low-frequency relative voltage ratio

$$= \frac{1}{\sqrt{1 + (R'_{\text{par}}/\omega aM)^2}} \qquad (6\text{-}76)$$

The power output of the circuit is

$$P_0 = \frac{V_L^2}{R_L} \qquad (6\text{-}77)$$

and the frequencies at which the power output is one-half that in the middle range for a given value of E_G are called the *half-power points*. At these frequencies the square of the relative voltage ratio equals $\frac{1}{2}$.

The high half-power frequency f_h is found by equating the square of the right-hand side of Eq. 6-75 to $\frac{1}{2}$ as follows

$$\frac{1}{1 + (\omega L_{\text{eq}_1}/R'_{\text{se}})^2} = \frac{1}{2}$$

from which we get

$$\omega_h L_{\text{eq}_1} = \omega L_{\text{eq}_1} = R'_{\text{se}}$$

and

$$f_h = \frac{\omega_h}{2\pi} = \frac{R'_{\text{se}}}{2\pi L_{\text{eq}_1}} \qquad (6\text{-}78)$$

Similarly, the half-power frequency for the low-frequency range is found to be

$$f_l = \frac{R'_{\text{par}}}{2\pi aM} \qquad (6\text{-}79)$$

It is customary in practice to take the value of the short-circuit self-inductance L_{sc_1} as L_{eq_1} in Eq. 6-78 and the open-circuit inductance L_{oc_1} as aM in Eq. 6-79 since $L_{\text{oc}_1} = L_{l1} + aM$ and the primary leakage inductance L_{l1} is small in comparison with the open-circuit inductance L_{oc_1}. Equations 6-78 and 6-79 can therefore be approximated to

$$f_h = \frac{R'_{\text{se}}}{2\pi L_{\text{sc}_1}} \qquad (6\text{-}80)$$

and

$$f_l = \frac{R'_{par}}{2\pi L_{oc_1}} \qquad (6\text{-}81)$$

The following ratio is a measure of the width of the band between the half-power frequencies

$$\frac{f_h}{f_l} \approx \frac{L_{oc_1}}{L_{sc_1}} \frac{R'_{se}}{R'_{par}} \qquad (6\text{-}82)$$

The geometric mean frequency is

$$f_0 = \sqrt{f_h f_l} \qquad (6\text{-}83)$$

A graphical representation* of the frequency and phase characteristics of output transformers is shown in Fig. 6-20.

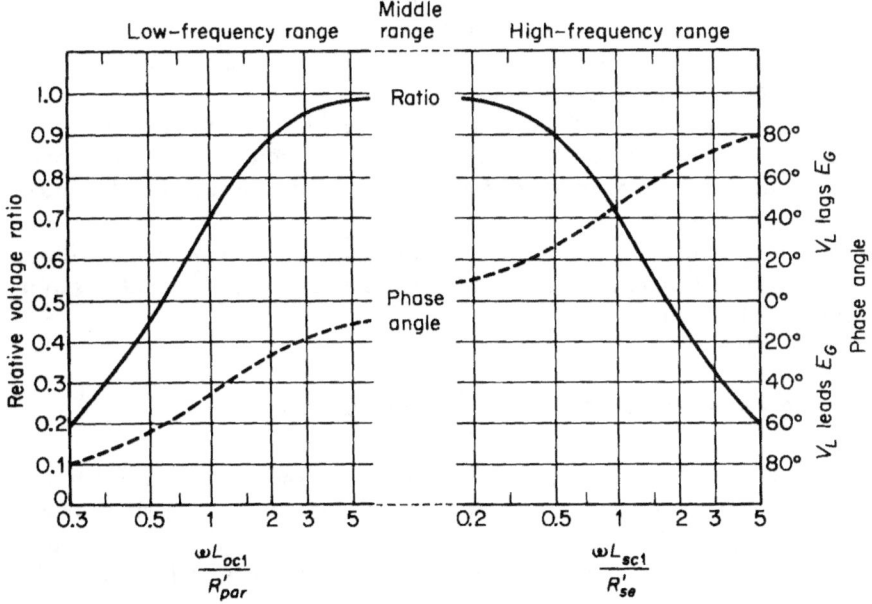

Figure 6-20. Universal frequency and phase characteristics of output transformers.

The geometric mean frequency for output transformers used in radio receivers is roughly 500 cps. The greater the band width for the proper

* Reprinted from MIT Staff, *Magnetic Circuits and Transformers.* New York: John Wiley & Sons, Inc., 1943.

value of geometric mean frequency, the smaller is the distortion of a signal that contains harmonics throughout the audio-frequency spectrum. This means that a high value of open-circuit inductance L_{oc_1} and low value of short-circuit inductance L_{sc_1}, i.e., tight magnetic coupling between windings, is desirable. It is difficult to increase the band width of an audio transformer with a core of a given size and configuration.

The open-circuit inductance L_{oc_1} can be increased by increasing the number of turns in the windings, which means going to a smaller wire size and, therefore, an increase in the resistance. In addition, increasing the turns also increases the short-circuit inductance L_{sc_1} correspondingly. Hence, the band width remains about the same, as is evident from Eq. 6-82. However, the geometric mean frequency would increase, which means a shift in the band between the half-power frequencies such as to produce a reduction in both the lower and upper half-power frequencies.

The upper half-power frequency can be increased by decreasing the short-circuit inductance L_{sc_1}, which calls for an increase in the tightness of the coupling between the windings. This also increases the band width. However, an increase in the open-circuit inductance L_{oc_1} calls for an increase in the number of turns or in the size of the core. In any event, if L_{oc_1} is to be increased, without a corresponding increase in the resistance, the size of the transformer must be increased.

Perfect reproduction of the signal would require an output transformer of infinite size. However, economic considerations limit the size to relatively small values, and the resulting distortion in the wave form can be tolerated because of the insensitivity of the human ear to moderate changes in the relative amplitudes of the harmonic components of the signal. In addition, the human ear is unable to detect changes in the relative phase relations over a considerable range.

6-11 3-PHASE TRANSFORMER CONNECTIONS

Practically all of the electric power in commercial power systems is generated as three phase, although a large part of the total is consumed as single phase and in the form of direct current. However, there are a number of 3-phase voltage transformations between the generator terminals and buses in the distribution centers.

It is possible to convert 3-phase power from one voltage to another by a suitable connection of two or three single-phase transformers or by making use of one 3-phase transformer. When a 3-phase transformer or three single-phase transformers are available, several 3-phase arrangements are possible; among them are the delta-delta, the wye-wye, and the wye-delta or delta-wye connections.

Delta-delta connection

Figure 6-21 shows three single-phase transformers, assumed to be identical, with their primaries connected in delta and their secondaries connected in delta. A common physical arrangement of the three transformers is shown in Fig. 6-21(a); a schematic diagram typical for 3-phase delta circuits is shown in Fig. 6-21(b).

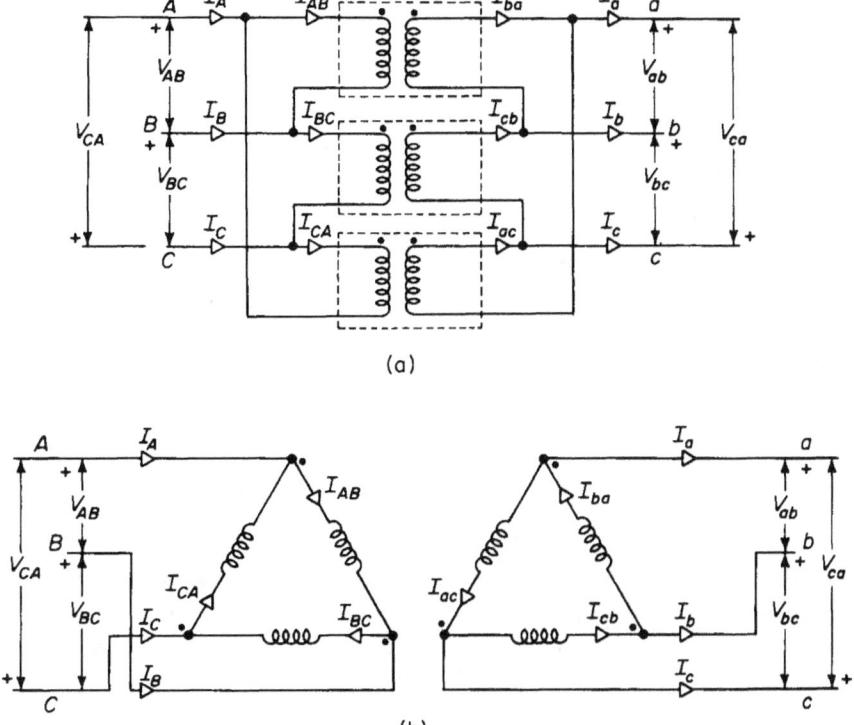

Figure 6-21. Delta-delta connection. (a) Common physical arrangement of three single-phase transformers; (b) schematic diagram.

The equivalent circuits of Figs. 6-11, 6-13, and 6-14 apply to each of the three transformers connected delta-delta with or without the ideal transformer. If the three transformers are identical and are operating under balanced 3-phase load and balanced 3-phase voltage conditions, each transformer carries one third of the 3-phase load.

It is evident from Fig. 6-21 that full line-to-line voltage exists across the windings of each transformer. Therefore, the secondary line-to-line voltages $\mathbf{V}_{ab}$, $\mathbf{V}_{bc}$, and $\mathbf{V}_{ca}$ are practically in phase with the corresponding primary line-to-line voltages V_{AB}, V_{BC} and V_{CA}. In addition, if the leakage impedance drops are neglected, the voltage ratios equal the turns ratio, i.e.

$$\frac{V_{AB}}{V_{ab}} = \frac{V_{BC}}{V_{bc}} = \frac{V_{CA}}{V_{ca}} = a \tag{6-84}$$

Figure 6-22 shows phasor diagrams for a bank of ideal transformers connected delta-delta and supplying a balanced unity power-factor load.

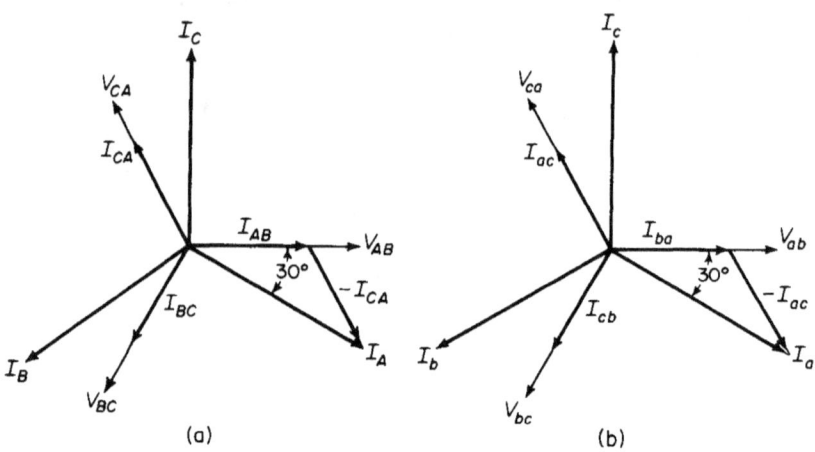

Figure 6-22. Phasor diagrams for delta-delta bank of ideal transformers supplying balanced noninductive load. (a) Primary; (b) secondary.

In the case of identical transformers, when the third harmonics in the exciting current are neglected, the line currents are $\sqrt{3}$ times the currents flowing in the windings under balanced conditions. This can be seen by referring to Fig. 6-21 and the phasor diagram of Fig. 6-22(a) as follows

$$\mathbf{I}_A = \mathbf{I}_{AB} - \mathbf{I}_{CA}$$

and

$$\mathbf{I}_{CA} = \mathbf{I}_{AB}\epsilon^{j120°}$$

from which

$$\mathbf{I}_A = \mathbf{I}_{AB}(1 - \epsilon^{j120°})$$
$$= \sqrt{3}\, \mathbf{I}_{AB}\epsilon^{-j30°} \tag{6-85}$$

Similarly

$$\mathbf{I}_B = \sqrt{3}\, \mathbf{I}_{BC}\epsilon^{-j30°} \tag{6-86}$$

and

$$\mathbf{I}_C = \sqrt{3}\, \mathbf{I}_{CA}\epsilon^{-j30°} \tag{6-87}$$

Sec. 6-11   3-PHASE TRANSFORMER CONNECTIONS   279

If the exciting current is neglected, then we have

$$\frac{\mathbf{I}_{ab}}{\mathbf{I}_{AB}} = \frac{\mathbf{I}_{bc}}{\mathbf{I}_{BC}} = \frac{\mathbf{I}_{ca}}{\mathbf{I}_{CA}} = \frac{\mathbf{I}_a}{\mathbf{I}_A} = \frac{\mathbf{I}_b}{\mathbf{I}_B} = \frac{\mathbf{I}_c}{\mathbf{I}_C} = a \qquad (6\text{-}88)$$

The delta-delta arrangement is restricted to applications in which neither the primary nor the secondary side requires a 3-phase neutral connection. It is generally used in moderate voltage systems because full line-to-line voltage exists across the windings, and in heavy current systems because the windings need to carry only $1/\sqrt{3}$ or 0.58 of the line current.

Wye-wye connection

Three single-phase transformers with their primaries and secondaries both connected in wye are shown in Figs. 6-23(a) and 6-23(b). The primary neutral is shown connected to the neutral of the source and the secondary neutral connected to that coming from the load. In many applications the neutral connection consists of ground. Connecting the primary neutral to the neutral of source assures balanced line-to-neutral voltage even if the load is unbalanced or if the transformers have unequal exciting admittances. The equivalent circuits of Fig. 6-11, 6-13, and 6-14 apply to each of the three transformers connected wye-wye with or without the ideal transformer just as they do in the delta-delta arrangement. Here also, if the transformers are identical and supply balanced 3-phase load, each transformer carries one-third of the 3-phase load.

It can be seen from Fig. 6-23 that the current in the transformer winding is the line current in the wye connection. The secondary currents $\mathbf{I}_a$, $\mathbf{I}_b$, and $\mathbf{I}_c$ are therefore practically in phase with the primary currents $\mathbf{I}_A$, $\mathbf{I}_B$, and $\mathbf{I}_C$ and if the exciting current is neglected, the current ratios are the reciprocals of the turns ratio, i.e.

$$\frac{\mathbf{I}_a}{\mathbf{I}_A} = \frac{\mathbf{I}_b}{\mathbf{I}_B} = \frac{\mathbf{I}_c}{\mathbf{I}_C} = a \qquad (6\text{-}89)$$

Also, if the leakage impedance is neglected, the voltage ratios equal the turns ratios, thus

$$\frac{\mathbf{V}_{AN}}{\mathbf{V}_{an}} = \frac{\mathbf{V}_{BN}}{\mathbf{V}_{bn}} = \frac{\mathbf{V}_{CN}}{\mathbf{V}_{cn}} = \frac{\mathbf{V}_{AB}}{\mathbf{V}_{ab}} = \frac{\mathbf{V}_{BC}}{\mathbf{V}_{bc}} = \frac{\mathbf{V}_{CA}}{\mathbf{V}_{ca}} = a \qquad (6\text{-}90)$$

Phasor diagrams are shown for the wye-wye arrangement in Fig. 6-24.

The wye connection is generally used in high-voltage applications because

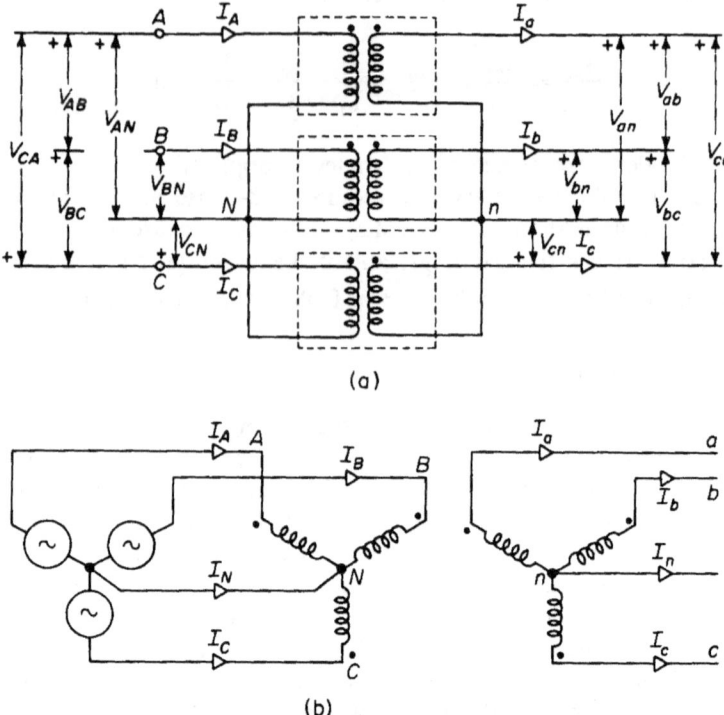

(a)

(b)

Figure 6-23. Wye-wye connection. (a) Common physical arrangement of three single-phase transformers; (b) schematic diagram showing primary neutral connected to the source and secondary going to the neutral of the load. The load is not shown.

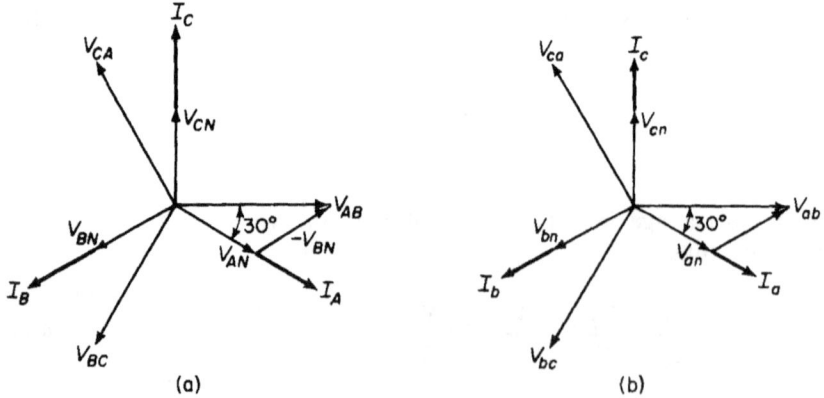

Figure 6-24. Phasor diagrams for wye-wye bank of ideal transformers supplying balanced noninductive load. (a) Primary; (b) secondary.

the voltage across the transformer winding is only $1/\sqrt{3}$ or 0.58 of the line-to-line voltage, which follows from Fig. 6-23 and the phasor diagrams in Fig. 6-24 as shown below

and
$$\mathbf{V}_{AB} = \mathbf{V}_{AN} - \mathbf{V}_{BN}$$

from which
$$\mathbf{V}_{BN} = \mathbf{V}_{AN}\epsilon^{-j120°}$$

$$\mathbf{V}_{AB} = \mathbf{V}_{AN}(1 - \epsilon^{-j120°})$$
$$= \sqrt{3}\,\mathbf{V}_{AN}\epsilon^{j30°} \quad (6\text{-}91)$$

Similarly
$$\mathbf{V}_{BC} = \sqrt{3}\,\mathbf{V}_{BN}\epsilon^{j30°} \quad (6\text{-}92)$$

and
$$\mathbf{V}_{CA} = \sqrt{3}\,\mathbf{V}_{CN}\epsilon^{j30°} \quad (6\text{-}93)$$

The wye-wye arrangement requires a neutral connection between the source and the primary of the transformers not only to assure balanced line-to-neutral voltage, but also to provide a path for the third-harmonic component in the exciting current of the transformers. Without the primary neutral connection, serious unbalances in the line-neutral voltage may result from (a) unequal exciting admittances among the three transformers and (b) unbalanced line-to-neutral loads in the secondary. Furthermore, if the third harmonics are suppressed in the exciting current, large third harmonic components may appear in the line-to-neutral voltages. The wye-wye arrangement is, therefore, a four-wire system if balanced voltages are to be assured. The delta-delta arrangement is, on the other hand, a 3-wire system.

Wye-delta connection

The wye-delta connection affords the advantage of the wye-wye connection without the resulting disadvantage of unbalanced voltages and third harmonics in the line-to-neutral voltages when operating without the neutral wire. The wye-delta arrangement is shown in Fig. 6-25. In high-voltage transmission systems, the high side of a transformer bank or of a 3-phase transformer is generally connected in wye, whereas the low side is connected in delta. The delta connection assures balanced line-to-neutral voltages on the wye side whether or not there is a neutral conductor on the wye side, and it provides a path for the third harmonic components in the exciting current independent of the neutral conductor.

The wye-delta or delta-wye transformation is not confined to applications in which the high-voltage side is connected in wye, but is also coming into general use in the 208/120-v system on the low side. In such systems, the low side is connected wye with the neutral point grounded. Single-phase loads

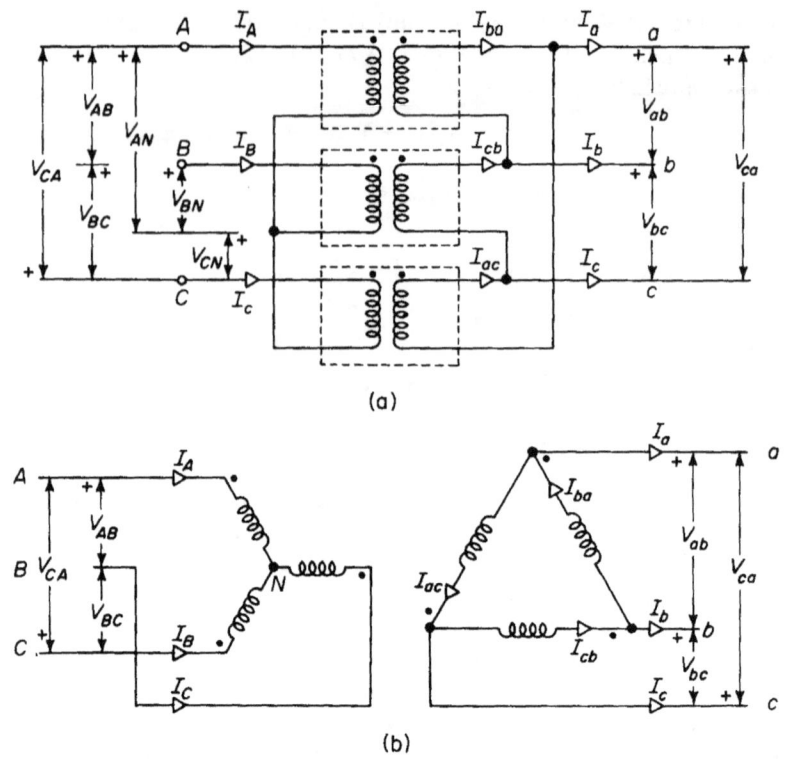

Figure 6-25. Wye-delta connection. (a) Common physical arrangement of three single-phase transformers; (b) schematic diagram.

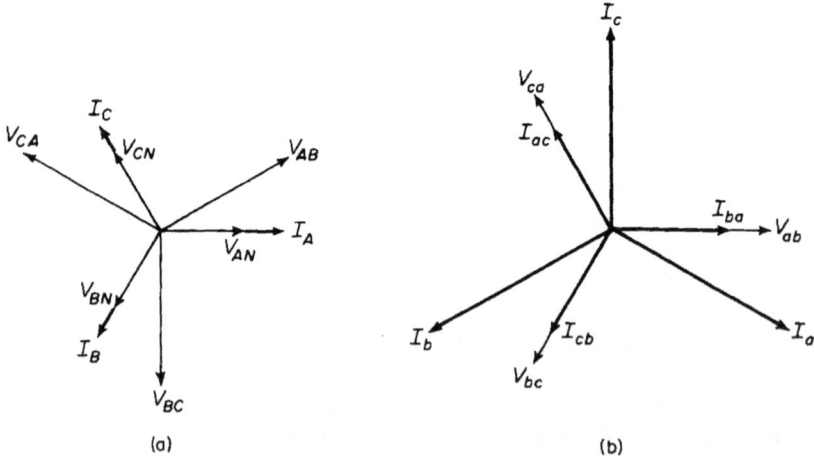

Figure 6-26. Phasor diagram for wye-delta arrangement of Figure 6-25 for ideal transformers supplying balanced noninductive load. (a) Primary wye-connected; (b) secondary delta-connected.

are connected line to neutral for 120-v operation, whereas 3-phase equipment, such as motors, are connected line to line for operation at 208 v.

Figure 6-26 shows the phasor diagram for a wye-delta transformation. From this diagram it is evident that there is a large phase angle between the line-to-line voltages on the wye side and the corresponding line-to-line voltages on the delta side. This angle is 30° with the phases as designated in Fig. 6-26. Angles of 90° and 150° are possible, depending on how the phases on the two sides are designated.

Open-delta or V-V connection

The open-delta, also known as the V-V connection, is a 3-phase arrangement that makes use of only two, instead of three, single-phase transformers,

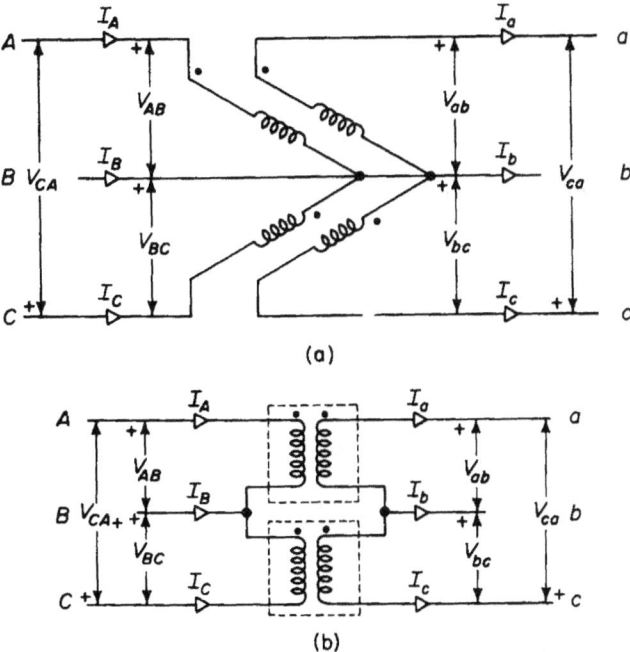

Figure 6-27. Open-delta or V-V connection. (a) Common physical arrangement; (b) schematic diagram.

as shown in Fig. 6-27. The open-delta connection is sometimes used in the case of instrument potential transformers for economy. It is also used initially in load centers, the full growth of which may require several years, at which time a third transformer is added for the conventional delta-delta operation.

The rating of two identical transformers operating delta-delta is $1/\sqrt{3}$ of the rating of three similar transformers, but connected delta-delta. It would seem that this rating should be $\frac{2}{3}$ instead of $1/\sqrt{3}$. However, the line current is also the current in the transformer winding and, if I is the rated current of each of the two transformers and V the rated voltage, then the rated 3-phase power is $\sqrt{3}VI$. In the case of three transformers, the line current is $\sqrt{3}$ times the current in the transformers windings when connected delta and when the load is balanced. The 3-phase rating of the three transformers is, therefore, $3VI$, where I is the rated current of the transformer winding.

If it were not for the leakage impedance of the transformers, the secondary voltages of an open-delta connection would be balanced when balanced 3-phase voltage is applied to the primary. This is evident when we let $\mathbf{V}_{AB}$, $\mathbf{V}_{BC}$, and $\mathbf{V}_{CA}$ be the balanced 3-phase primary applied voltages. Then

$$\mathbf{V}_{BC} = \mathbf{V}_{AB} \epsilon^{-j120°}$$

and

$$\mathbf{V}_{CA} = \mathbf{V}_{AB} \epsilon^{j120°}$$

If the transformer leakage impedances are neglected, and their ratio is a, then the secondary voltages are

$$\mathbf{V}_{ab} = \frac{\mathbf{V}_{AB}}{a}$$

$$\mathbf{V}_{bc} = \frac{\mathbf{V}_{BC}}{a} = \frac{\mathbf{V}_{AB}}{a} \epsilon^{-j120°} = \mathbf{V}_{ab} \epsilon^{-j120°}$$

but from Kirchhoff's law

$$\mathbf{V}_{ab} + \mathbf{V}_{bc} + \mathbf{V}_{ca} = 0$$

from which

$$\mathbf{V}_{ca} = -(\mathbf{V}_{ab} + \mathbf{V}_{bc})$$
$$= -\mathbf{V}_{ab}(1 + \epsilon^{-j120°}) = \mathbf{V}_{ab} \epsilon^{j120°}$$

showing that the secondary voltages V_{ab}, V_{bc}, and V_{ca} are equal in magnitude and displaced from each other by an angle of 120°. This is characteristic of balanced 3-phase systems, and is true only for ideal transformers. In the case of actual transformers, the secondary voltages will not be balanced exactly, even when balanced voltages are applied to the primary in an open-delta arrangement because there are only two transformers and, therefore, only two, instead of three, leakage impedance voltage drops.

Sec. 6-11   3-PHASE TRANSFORMER CONNECTIONS   285

3-phase transformers

The 3-phase transformer is one that incorporates the cores and windings for all three phases in one structure, an arrangement that is generally more economical (from the standpoint of both space and costs) than a combination of three single-phase transformers of the same total rating. A 3-phase transformer is shown in Fig. 6-28. Figure 6-29 shows the schematic arrangement commonly used for the core and windings in 3-phase transformers.

Figure 6-28. 3-phase core-type transformer rated 5,000 kva, 69,000/15,000 volts, 60 cycles. (Courtesy of Westinghouse Electric Corporation.)

The similarity between Fig. 6-29(a) and the construction of the core and windings in Fig. 6-28 is quite evident.

The core in Fig. 6-29(a) has only the three legs that carry the windings and does not require a fourth leg to serve as a return path for the flux

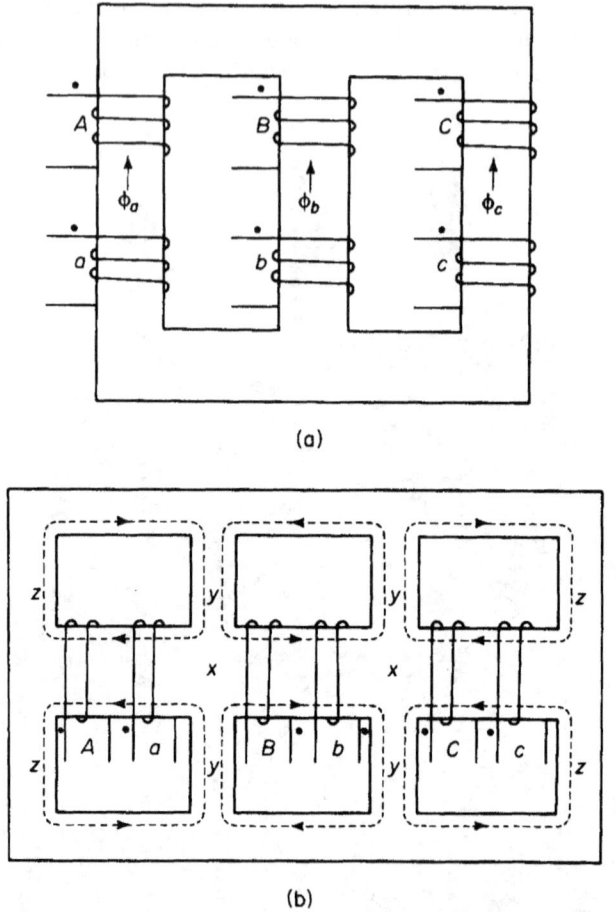

Figure 6-29. 3-phase transformers. (a) Core type; (b) shell type.

because the sum of the three fluxes ϕ_a, ϕ_b, and ϕ_c is zero for balanced 3-phase operation.

The middle windings in Fig. 6-29(b) have their polarities reversed with respect to the outer windings. Nevertheless, the polarities of the voltages across the primary and secondary windings must be consistent, relative to the polarity marks of the windings, with those shown in Figs. 6-21, 6-23, and 6-25, depending upon the kind of transformation involved—whether delta-delta, wye-wye, or wye-delta. By reversing the polarities of the middle windings, the magnetic fluxes from each adjacent phase *add* [in the regions x of Fig. 6-29(b)] to that of the middle phase rather than subtract from it. Subtracting from it would make for a larger core. This follows from the fact that the maximum instantaneous value of the sum of two equal magnetic fluxes 120° apart in time phase and undergoing a sinusoidal time variation equals the maximum instantaneous value of one of these fluxes, whereas the maximum instantaneous value of the difference between such fluxes equals the maximum instantaneous value of one of these fluxes multiplied by $\sqrt{3}$. Hence, if the polarities of the middle windings were the same as of the outer windings, the legs y would need to be $\sqrt{3}$ times as wide as the legs z in Fig. 6-29(b).

3-to-6-phase transformation

Large amounts of a-c power are transformed to d-c power by means of electronic rectifiers. A smoother voltage wave form is obtained on the d-c side as the number of phases is increased on the a-c input side. A greater number of phases also results in the reduction of objectionable harmonics in the alternating current. Therefore, 6-phase is preferable to 3-phase for rectification, and, consequently, there are many 6-phase rectifiers and some 12-phase rectifiers in larger installations. There are several methods of connecting transformers for obtaining 3-to-6-phase transformation. A simple arrangement known as the 6-phase star connection is shown in Fig. 6-30. The secondaries are provided with center taps, which are connected together to form the neutral of the 6-phase side. The primaries are shown connected in delta but they may also be connected in wye. The 6-phase forked-wye or double-zigzag connection is shown in Fig. 6-31. In this arrangement, each secondary winding is divided into three equal parts and the portions that are shown parallel to each other in Fig. 6-31 constitute one of the secondaries. A 6-phase transformation may be effected by means of three single-phase transformers or by one 3-phase transformer.

Third harmonics in 3-phase transformer operation

It was shown in Section 5-3 that the sinusoidal flux in iron cores requires a third-harmonic component in the exciting current, which, although small in relation to the rated current, may produce undesirable effects in 3-phase operation of transformers.

Consider three identical unloaded, single-phase transformers connected wye-wye to a 3-phase generator with their primary neutral connected to the generator neutral as shown in Fig. 6-32. The sum of the instantaneous

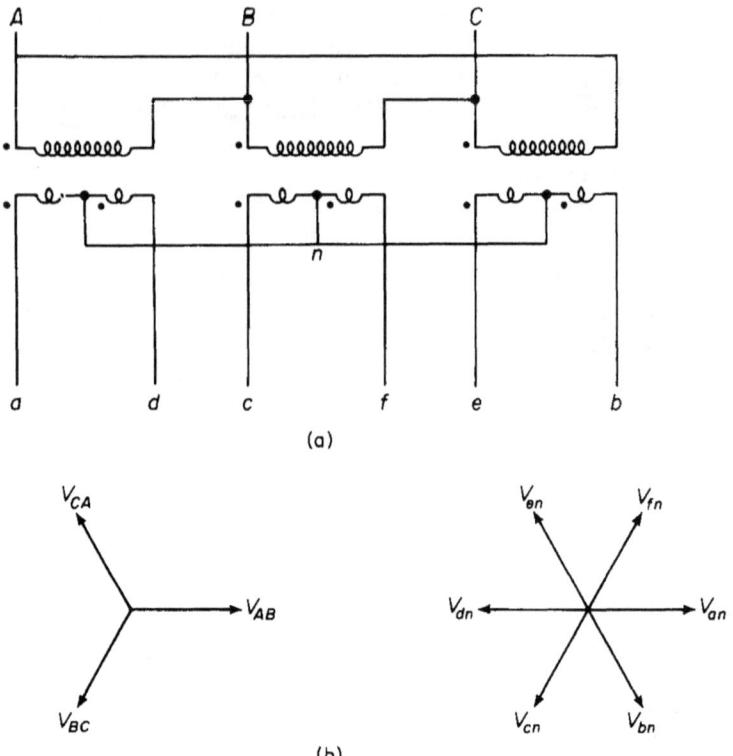

Figure 6-30. (a) 3-phase-delta to 6-phase-star connection, and (b) phasor diagram of primary and secondary voltages.

currents flowing in the primary must equal zero, i.e.

$$i_A + i_B + i_C + i_N = 0 \qquad (6\text{-}94)$$

The fundamental components, as well as harmonics—not including the third and multiples thereof, are 120° apart, and, being of equal amplitudes, their sum is

$$i_{A_h} + i_{B_h} + i_{C_h} = 0 \qquad (6\text{-}95)$$

where the subscript h stands for the order of the harmonics 1, 5, 7, 11, but not for 3, 9, 15, etc. It should be remembered that harmonics in the exciting current of an iron-core transformer are odd for sinusoidal flux when there is no d-c component of flux. It follows from Eqs. 6-94 and 6-95 that the neutral current in the unloaded transformers, or in such as deliver balanced sinusoidal 3-phase currents, is comprised of third-harmonic current, which

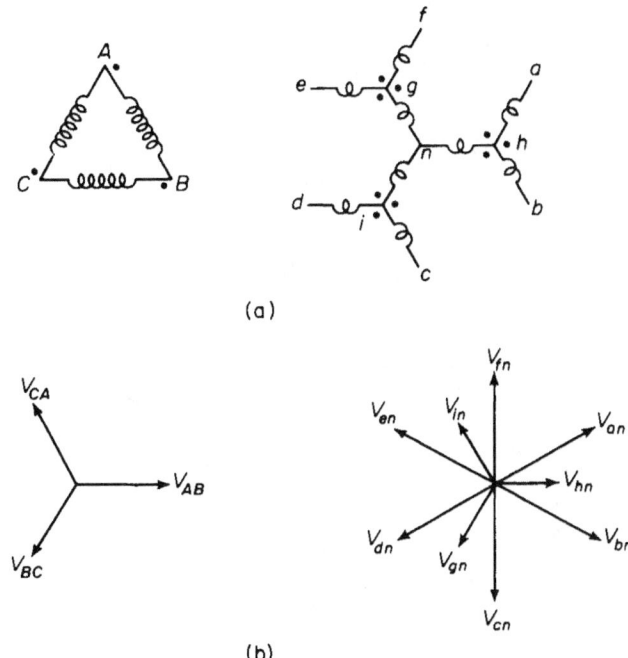

Figure 6-31. 6-phase forked-wye connection.

is the sum of the third harmonics in the three phases, thus

$$i_{A_3} + i_{B_3} + i_{C_3} + i_N = 0$$

or

$$i_N = -(i_{A_3} + i_{B_3} + i_{C_3}) = -3i_{A_3} \tag{6-96}$$

Figures 6-33(a), 6-33(b), and 6-33(c) show balanced 3-phase waves that are comprised of a fundamental and a third harmonic. The sum of the three

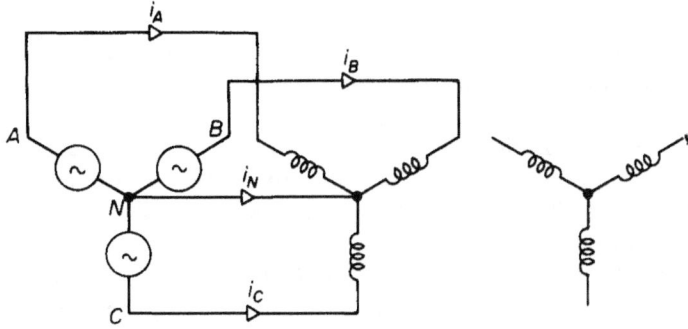

Figure 6-32. Exciting currents and neutral current in wye-wye transformer connection.

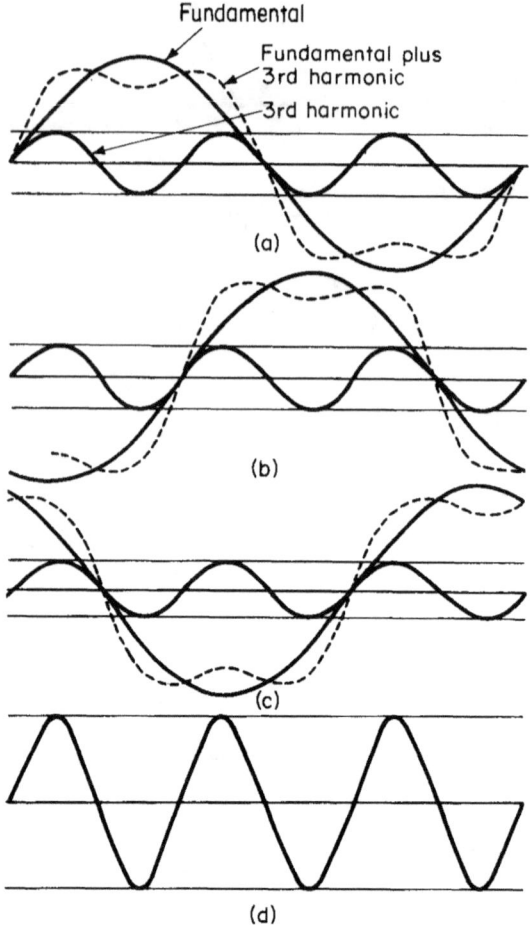

Figure 6-33. Balanced 3-phase waves containing a fundamental and a third harmonic component. (d) Sum of the three waves of (a), (b), and (c).

balanced waves is shown in Fig. 6-33(d) and is a pure third harmonic having an amplitude equal to three times that of the third harmonic in any one phase.

If the neutral connection between the transformer primaries and the generator is broken, then the path for the third-harmonic currents is interrupted and the third harmonics in the exciting current will be suppressed. As a result, the flux cannot be sinusoidal, as it will contain a third harmonic, which in turn produces a third harmonic in the transformer voltages. These third harmonics show up only in the line-to-neutral voltage if the transformers are identical, and will not appear in the line-to-line voltages because the line-to-line voltages are the phasor difference between the line-to-neutral voltages, i.e.

$$V_{AB} = V_{AN} - V_{BN}$$

The third harmonics in the line-to-neutral voltages of all three phases are equal and in phase with each other and, therefore, cancel in the line-to-line

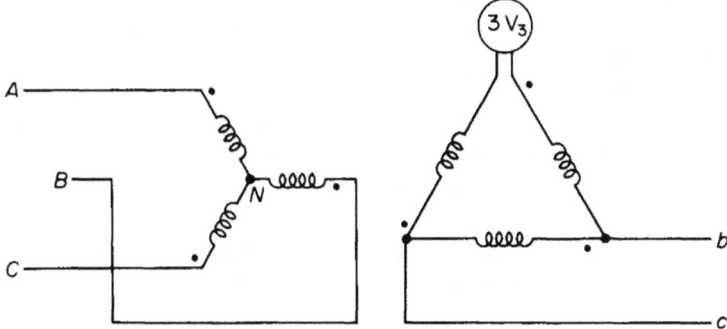

Figure 6-34. Third-harmonic voltage across an open corner in the delta on the secondary of a wye-delta connection with the primary neutral isolated.

voltages. This becomes evident when the difference is taken between any two of three waves a, b, and c of Fig. 6-33. Since

$$V_{AN_3} = V_{BN_3} = V_{CN_3}$$

and

$$V_{AB_3} = V_{AN_3} - V_{BN_3} = 0$$

When the primaries are connected in delta, the third-harmonic components in the current are free to flow, but will not show up in the line currents because the line currents are the differences between the currents flowing in the delta as shown in Fig. 6-21. The delta connection on the secondary side of a wye-delta arrangement also provides a path for the third-harmonic components in the exciting mmf. Figure 6-34 shows the primaries of a wye-delta arrangement connected in wye with the neutral isolated. One corner of the delta is shown open. Since the neutral is isolated, there is no return path on the primary side for the third harmonics in the exciting current, causing third harmonics to appear in the voltage across each primary winding. There will be corresponding third harmonics in the voltages across each secondary winding if one or more corners of the delta are open. The voltage appearing across the open corner of the delta in Fig. 6-34 is the sum of the voltages in the three secondary windings, and, if the exciting characteristics of the three phases are identical, the sum of the fundamentals, as well as that of all harmonics—except the third and its multiples—will be zero since these are all equal and 120° apart. The multiples of the third harmonics are usually negligible. The third harmonics are equal and in phase with each other. And the voltage across the open corner of the delta is three times the third-harmonic voltage in one phase of the secondary. Thus, if V_3 is the third harmonic voltage per phase in the delta, then $3V_3$ is the voltage across the open corner of the delta.

Closing the open corner of the delta in Fig. 6-34, for normal operation, short circuits the third-harmonic emf $3V_3$, causing a third-harmonic current to circulate in the delta, thus producing a substantially sinusoidal flux. If, in addition, the primary neutral is closed, the third-harmonic components of the mmf required by the sinusoidal flux divide between the primary and secondary, depending upon their relative third-harmonic leakage impedances.

Since the delta connection provides a path for the third-harmonic current, and because it assures balanced voltages, most 3-phase transformations include a delta winding, which makes the wye-delta or delta-wye arrangement very common. Where wye-wye transformation is required, it is quite common to incorporate a third winding, known as a tertiary, connected in delta. Generally, the rating of the delta-connected tertiary is considerably lower than that of the primary and secondary wye-connected windings.

6-12 PER UNIT QUANTITIES

The performance of a line of machines can be readily compared, and the analyses of complex power systems involving transformers of different ratios can be facilitated, by means of per unit quantities. Per unit quantities are decimal fractions that relate current, voltage, impedance, admittance, and power to certain base values of these respective quantities, i.e.

$$\text{per unit current} = \frac{\text{current in amperes}}{\text{base amperes}} \quad (6\text{-}97)$$

$$\text{per unit voltage} = \frac{\text{voltage in volts}}{\text{base volts}} \quad (6\text{-}98)$$

$$\text{per unit impedance} = \frac{\text{impedance in ohms}}{\text{base ohms}} \quad (6\text{-}99)$$

$$\text{per unit admittance} = \frac{\text{admittance in mhos}}{\text{base mhos}} \quad (6\text{-}100)$$

$$\text{per unit power} = \frac{\text{power in volt-amperes}}{\text{base volt-amperes}} \quad (6\text{-}101)$$

The values of the base quantities are dictated by convenience. In dealing with a specific piece of apparatus by itself, as in the case of a transformer, the rating of the transformer is generally taken as the base. In the transformer of Example 6-2, the following are the base quantities on the low-voltage side

base volt-amperes = 150,000
base volts = 240
base amperes = base volt-amperes ÷ base volts
= 625
base ohms = base volts ÷ base amperes
= base volt-amperes ÷ (base amperes)^2
= (base volts)^2 ÷ base volt-amperes
= 0.384
base mhos = base amperes ÷ base volts
= base volt-amperes ÷ (base volts)^2
= (base amperes)^2 ÷ base volt-amperes
= 2.604

Similarly, the base quantities on the high-voltage side are 150,000 v, 2,400 v, 62.5 amp, 38.4 ohms, and 0.02604 mho.

The leakage impedance of this transformer was found in Example 6-4 to be 1.008 ohm referred to the high-voltage side, and the per unit impedance from Eq. 6-99 is

$$Z_{eq} = \frac{1.008}{38.4} = 0.0263 \text{ per unit}$$

On the other hand, when the analysis is applied to a number of equipments in combination, a larger kva base is used.

The leakage impedance of the transformer referred to the low side is that referred to the high side divided by the impedance ratio, and, in this case, is 1.008 ÷ (10)^2 = 0.01008 ohm. The per unit impedance of the transformer referred to the low side is, therefore, 0.01008 ohm divided by the transformer leakage impedance referred to the low side, i.e., 0.384 ohm, which results in the same value as before, namely, 0.0263 per unit. This brings out an important advantage of the per unit system with regard to transformers, in that the per unit impedance has the same value whether referred to the high side or to the low side, regardless of the turns ratio of the transformer.

In the analysis of power systems encompassing several pieces of equipment of similar or of different ratings, it is generally more convenient to use a relatively large kva base, which may be several times the rating of any one of the generators or transformers in the system. The per unit values of generators and transformers are generally known in terms of their own rating and can be conveniently converted to any other base on the principle that the per unit value of an impedance, for a given base voltage, is proportional to the volt-ampere base. Thus, if an impedance has a per unit value of Z_{pu_1} for a base of VA_1 va, then its per unit value for a base of VA_2 va for the same voltage base, is expressed by

$$Z_{pu_2} = Z_{pu_1} \frac{VA_2}{VA_1} \qquad (6\text{-}102)$$

Thus, the per unit impedance of the transformer in Example 6-2 for a base of 10,000 kva would be

$$Z_{pu} = 0.0263 \times \frac{10,000,000}{150,000} = 17.55$$

Per unit impedance in 3-phase systems

In a 3-phase system, if

Z_{ohms} = line-to-neutral impedance in ohms per phase

and

VA = the 3-phase volt-ampere base

and

V = the base line-to-line voltage

then

$$Z_{pu} = Z_{ohms} \frac{VA}{(V)^2} \qquad (6\text{-}103)$$

There are cases in which transformers, generators, and transmission lines are operated at voltages somewhat different from rated value. Transformers are seldom operated at normal frequency with voltages that are substantially above rated values because of oversaturating the iron, which leads to excessive exciting current and to excessive core losses. However, transformers may operate at somewhat reduced voltages, as for example, 13.8-kv transformers operating at 13.2 kv and at normal frequency. In changing the per unit value from one base to another, the change in voltage base must be taken into account. Thus, if

$$Z_{pu_1} = Z_{ohms} \frac{VA_1}{(V_1)^2}$$

Sec. 6-13 MULTICIRCUIT TRANSFORMERS 295

is the per unit value of Z_{ohms} impedance for volt-ampere and voltage bases of VA_1 and V_1, then the same impedance will have a per unit value of

$$Z_{\text{pu}} = Z_{\text{ohms}} \frac{VA_2}{(V_2)^2}$$

for volt-ampere and voltage bases of VA_2 and V_2, from which it is evident that

$$Z_{\text{pu}_2} = Z_{\text{pu}_1} \frac{VA_2}{VA_1}\left(\frac{V_1}{V_2}\right)^2 \tag{6-103a}$$

Equation 6-103a is valid for single-phase systems as well as for 3-phase systems.

6-13 MULTICIRCUIT TRANSFORMERS

Three or more circuits that may have different voltages are frequently interconnected by means of one transformer, known as a multicircuit or multiwinding transformer. Such an arrangement is more economical than

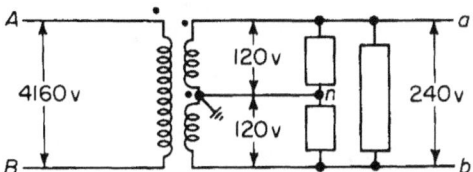

Figure 6-35. Distribution transformer.

one in which two or more 2-circuit or 2-winding transformers effect the same interconnection. For example, distribution transformers have a primary winding usually rated at 2,000 v or above, and generally have two 120-v secondaries connected in series with their common or neutral grounded as shown in Fig. 6-35. The secondaries thus supply a 240/120/120-v 3-wire circuit, in which the load comprised of the ordinary 120-v appliances is connected from line to neutral so that this load is divided about equally between the two 120-v windings. Appliances that operate at 240 v, such as electric stoves, are connected across the two windings in series. 3-winding transformers are also commonly used to supply electron tube circuits where one winding supplies the plate current at one voltage and another winding supplies current to the cathode heaters at a lower voltage than that applied to the plate circuit. Large power transformers used in 3-phase installations are frequently provided with three windings, the primary and secondary of which are connected for wye-wye operation, whereas the tertiary windings are connected in delta to maintain balanced voltage and to provide a path for the third harmonics in the exciting current.

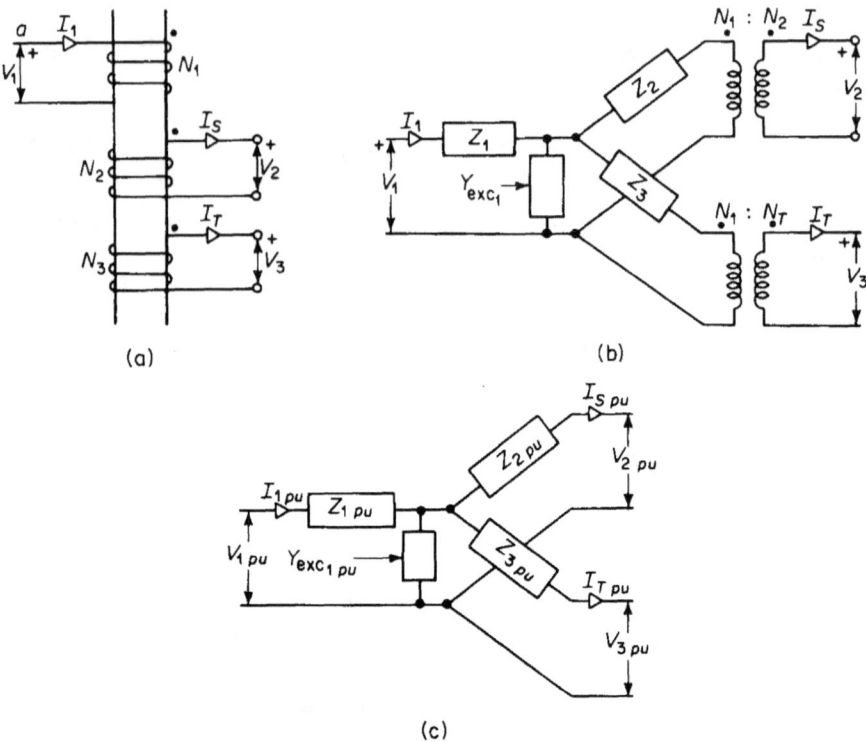

Figure 6-36. 3-winding transformer. (a) Circuit diagram, (b) equivalent circuit in ohms and in mhos with these quantities referred to the primary, (c) equivalent circuit in per unit.

The windings of a 3-circuit transformer are represented schematically in Fig. 6-36(a). If the transformer were ideal, i.e., had no leakage impedance and no exciting admittance, the following relationship would obtain

$$\frac{V_2}{V_1} = \frac{N_2}{N_1} \tag{6-104}$$

$$\frac{V_3}{V_1} = \frac{N_3}{N_1} \tag{6-105}$$

$$N_1 i_1 = N_2 i_S + N_3 i_T \tag{6-106}$$

where V_1, V_2, and V_3 are the primary, the secondary, and the tertiary terminal voltage respectively, and N_1, N_2, and N_3 are the turns in the respective windings. Further, i_1, i_S, and i_T are the instantaneous currents in the primary, the secondary, and the tertiary windings respectively. From Eq. 6-106 it follows that the effective ampere turns in the primary winding must be the phasor sum of the effective ampere turns in the secondary and tertiary windings. Hence

$$N_1 I_1 = N_2 I_S + N_3 I_T \qquad (6\text{-}107)$$

However, when the leakage impedance and the exciting admittance are taken into account the equivalent circuit of a three-circuit or three-winding transformer is as shown in Figs. 6-36(b) and 6-36(c).

Open-circuit and short-circuit tests

The open-circuit test, made in the same manner as the test for a 2-winding transformer, yields the data from which the exciting admittance can be determined. However, since there are three circuits, it is necessary to make three short-circuit tests to obtain the data for the equivalent circuits of Figs. 6-36(b) and 6-36(c). Assume the tertiary to have the lowest voltage rating, the primary to have the highest voltage of the three windings, and the secondary to have a voltage rating between those of the tertiary and the primary. The short-circuit tests are then usually made as follows

1. Voltage applied to the primary with the secondary winding short circuited and the tertiary open circuited. The exciting current is negligible and the measured impedance will be

$$Z_{12} = Z_1 + Z_2 \qquad (6\text{-}108)$$

2. Voltage applied to the primary with the tertiary short circuited and with the secondary open, from which

$$Z_{13} = Z_1 + Z_3 \qquad (6\text{-}109)$$

3. Voltage applied to the secondary with the tertiary short circuited and with the primary open, whence

$$Z_{23} = Z_2 + Z_3 \qquad (6\text{-}110)$$

and

$$Z_{23} = \left(\frac{N_1}{N_2}\right)^2 \frac{V_S}{I_{Ssc}} \tag{6-111}$$

where V_S is the voltage applied to the secondary and I_{Ssc} is the secondary current with the tertiary short circuited and the primary open.

The leakage impedances Z_1, Z_2, and Z_3, all of which are referred to the primary, are found as follows

$$Z_1 = \frac{Z_{12} + Z_{13} - Z_{23}}{2} \tag{6-112}$$

$$Z_2 = \frac{Z_{23} + Z_{12} - Z_{13}}{2} \tag{6-113}$$

$$Z_3 = \frac{Z_{13} + Z_{23} - Z_{12}}{2} \tag{6-114}$$

EXAMPLE 6-8: Three single-phase 3-winding transformers are connected for 3-phase operation at 66 kv, 11 kv, and 2.2 kv. All voltages are line to line. The 66-kv and 11-kv sides are connected in wye and the 2.2 kv side in delta. Assume the 66-kv side to be the primary, the 11-kv side to be the secondary, and the 2.2-kv side to be the tertiary. The following data were obtained on test.

(a) Short-circuit Test

Winding		Excited winding	
Excited	Short circuited	Volts	Amperes
Primary	Secondary	2,850	788
Primary	Tertiary	5,720	657
Secondary	Tertiary	529	1575

(b) Open-circuit Test

Winding	Volts	Amperes
Tertiary	2,200	146

Determine the values of reactance and admittance to be used in the equivalent circuits of Figs. 6-36(b) and 6-36(c). Neglect the resistance of the windings. Use a base of 30,000 kva per transformer or 90,000 kva as the 3-phase base for determining the per unit values.

Solution:
(a) The values of reactance are based on the short-circuit test and will be referred to the primary.

$$Z_{12} = j2{,}850 \div 788 = j3.62 \text{ ohms}$$

$$Z_{13} = j5{,}720 \div 657 = j8.72 \text{ ohms}$$

$$Z_{23} = \left(\frac{N_1}{N_2}\right)^2 j529 \div 1{,}575 \approx \left(\frac{66}{11}\right)^2 j529 \div 1{,}575 = j12.10 \text{ ohms}$$

$$Z_1 = \frac{Z_{12} + Z_{13} - Z_{23}}{2} = \frac{j3.62 + j8.72 - j12.10}{2} = j0.12 \text{ ohm}$$

$$Z_2 = \frac{Z_{23} + Z_{12} - Z_{13}}{2} = \frac{j12.10 + j3.62 - j8.72}{2} = j3.50 \text{ ohms}$$

$$Z_3 = \frac{Z_{13} + Z_{23} - Z_{12}}{2} = \frac{j8.72 + j12.10 - j3.62}{2} = j8.60 \text{ ohms}$$

The exciting admittance, if the core losses are neglected, when referred to the tertiary, is

$$Y_3 = -j\frac{146}{2{,}200} = -j0.0664 \text{ mho}$$

and

$$\left(\frac{N_3}{N_1}\right)^2 Y_3 \approx \left(\frac{2.2}{66/\sqrt{3}}\right)^2 (-j0.0664) = -j0.000221$$

referred to the primary.

(b) Since the impedances are all referred to the primary, the base impedance for 30,000 kva and for the primary voltage of $66 \text{ kv}/\sqrt{3}$ will suffice as a base for all three impedances.

$$Z_{\text{base}} = \frac{(\text{Volts})^2}{\text{Volt-amperes}} = \frac{(66{,}000/\sqrt{3})^2}{30{,}000{,}000} = \frac{(66)^2}{90} = 48.4 \text{ ohms}$$

Hence,
$$Z_{1\text{pu}} = j0.12 \div 48.4 = j0.00248$$
$$Z_{2\text{pu}} = j3.50 \div 48.4 = j0.0724$$
$$Z_{3\text{pu}} = j8.60 \div 48.4 = j0.178$$
$$Y_{\text{pu}} = -j0.000221 \times 48.4 = -j0.0107$$

Multicircuit transformers are not limited to three circuits but may have a greater number. When the number of circuits or windings exceeds three, the analysis and the equivalent circuits are considerably more complicated than for the 3-winding transformers and are, therefore, not included in this book.*

PROBLEMS

6-1 Figure 6-37 is a schematic diagram of an ideal transformer supplying a noninductive resistance $R = 10$ ohms. The primary winding has $N_1 = 100$ turns and the secondary winding has $N_2 = 50$ turns. The instantaneous voltage applied to the primary winding is $v_1 = 150$ v.

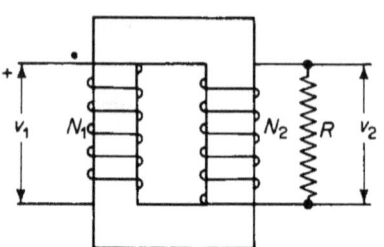

Figure 6-37. Ideal transformer for Problem 6-1.

(a) What is the value of the current and its direction through the resistance R?
(b) What is the value of the primary current?
(c) What is the direction of the flux in the core?
(d) What is the rate of change of
  (i) The flux in the primary winding?
(ii) The flux linkage with the primary winding?
(iii) The flux in the secondary winding?
(iv) The flux linkage with the secondary winding?
(e) The transformer and its connected load are replaced by a noninductive resistance which takes the same value of current at 150 v as that in the primary of the transformer. What is the value of the resistance in ohms?
(f) What is the ratio of the resistance in part (e) to that of the resistance R in the secondary? What is this ratio called?
(g) A polarity mark is shown near the upper terminal of the primary winding in Fig. 6-37. Where should the secondary polarity mark be placed?

6-2 The purpose of this problem is to show, firstly, that a sizeable reduction in frequency without a corresponding reduction in voltage leads to values of exciting current that can exceed the rated current of the transformer, and secondly, that a reduction of the voltage to a value that leads to a safe value of exciting current lowers the rating of the transformer.

A 1,500-w, 240/25-v, 400-cycle transformer operates at a maximum flux density of 36,000 lines per sq in. and the peak value of the exciting current is 0.20 amp.

(a) Use the magnetization curve of Fig. 3-11 and determine the peak value of the exciting current when the primary of this transformer is connected to 120-v 60-cycle source. Neglect resistance and leakage flux.

---

* For a discussion of transformers with more than three circuits, the reader is referred to Blume *et al.*, *Transformer Engineering* (New York: John Wiley & Sons, Inc., 1954), 117–133; and MIT Staff, *Magnetic Circuits and Transformers* (New York: John Wiley & Sons, Inc., 1943), Chap. V.

(b) Suppose this transformer were to operate at a frequency of 60 cps and at magnetic flux density $B_m = 90{,}000$ lines per sq in., and the current density in the windings is the same as for rated 400-cps operation. What would the rating of the transformer be in volt-amperes?

6-3 An audio output transformer operates from a 4,000-ohm resistance source delivering 5 w to a noninductive load of 10 ohms. Assume the transformer to be ideal and determine

(a) The turns ratio such that the load impedance referred to the primary has a value of 4,000 ohms.
(b) The secondary and primary currents.
(c) The primary applied voltage.

6-4 An ideal transformer is represented schematically in Fig. 6-38. Note the current directions and voltage polarities and draw a phasor diagram on a one-to-one ratio basis for these directions and polarities. Assume the load current to lag the load voltage by an angle of 30°.

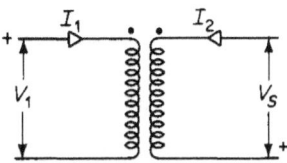

Figure 6-38. Ideal transformer for Problem 6-4.

6-5 The secondary winding of the transformer in Example 6-2 is short circuited and sufficient voltage, at rated frequency, is applied to the primary winding so that rated current flows in the secondary. On the basis of the exact equivalent circuit in Fig. 6-11(a) determine

(a) The secondary induced emf E_2.
(b) The primary induced emf E_1.
(c) The core-loss current on the basis that the exciting admittance is constant. (Actually, the exciting admittance is not constant; for one thing, it increases with increasing magnetization of the core.)
(d) The exciting current based on the admittance of part (c) above.
(e) The primary applied voltage V_1.
(f) The per unit impedance of the transformer on the basis that the per unit impedance is defined as the ratio of the voltage V_1, with rated current on short circuit in part (e) above, to the rated primary voltage.
(g) The per unit resistance of the transformer on the basis that the ratio of the I^2R losses with rated current to the volt-ampere rating of transformer is the per unit resistance.
(h) The mutual flux expresssed as a decimal fraction of the mutual flux at rated voltage and no load.
(i) The primary current and the secondary current when a short circuit is applied to the secondary with full rated voltage across the primary.

6-6 Repeat Problem 6-5,* using the equivalent circuit of Fig. 6-14(a).

* One purpose of Problems 6-5 and 6-6 is to demonstrate that, in the case of power transformers (low frequency), the sacrifice in accuracy is small when the approximate equivalent circuit is used instead of the exact equivalent circuit.

6-7 (a) Determine the current, real power, and reactive power when a 60-cycle emf of 240 v is applied to the secondary winding of the transformer in Example 6-2 when the primary winding is open.

(b) Determine the current, real power, and reactive power when a 60-cycle emf of 2,400 v is applied to the primary of the transformer above with the secondary open.

6-8 The results of an open-circuit test and short-circuit test at rated frequency on a 500-kva 42,000/2,400-v, 60-cycle transformer are as follows

Open-circuit Test—Voltage applied to Low-voltage Side

Volts	Amperes	Watts
2,400	18.1	3,785

Short-circuit Test—Low-voltage Side Short Circuited

Volts	Amperes	Watts
1,910	11.9	3,850

(a) Determine the following quantities referred to the high-voltage side of the transformer.
 (i) The exciting admittance, conductance, and susceptance.
 (ii) The equivalent leakage impedance.
 (iii) The equivalent resistance.
(b) Repeat part (a), referring all quantities to the low-voltage side.
(c) Determine the leakage reactance of each winding on the basis that the same amount of equivalent leakage flux links each winding.
(d) Determine the resistance of each winding on the basis of equal amounts of copper and equal current densities in the two windings.
(e) Repeat part (d), but on the basis that the length of the mean turn in the high-voltage winding is 40 percent greater than that in the low-voltage winding.

6-9 Determine the efficiency of the transformer in Problem 6-8 when delivering at 2,400 secondary v.

(a) Rated load at unity power factor.
(b) At $\frac{1}{4}$, $\frac{1}{2}$, $\frac{3}{4}$, 1, and $\frac{5}{4}$ rated load all at 0.80 power factor.

6-10 Determine the regulation of the transformer in Problem 6-8 when delivering rated load at 0.80 power factor

(a) Current lagging.
(b) Current leading.

6-11 The transformer in Problem 6-8 is a core type and has its high-voltage winding surrounding the low-voltage winding, which is next to the core. If the low-voltage winding were left unchanged, and the high-voltage winding were rewound for 69,000 v without a change in the amount of copper and without appreciable change in the space, as to both amount and configuration, what would be the

(a) Equivalent resistance of the transformer referred to
   (i) The low-voltage side?
   (ii) The high-voltage side?
(b) Equivalent leakage reactance of the transformer referred to
   (i) The low-voltage side?
   (ii) The high-voltage side?
(c) The exciting current at normal excitation (69,000/2,400-v operation) on
   (i) The low-voltage side?
   (ii) The high-voltage side?

6-12 A 240/120-v autotransformer delivers a current of 180 amp to a load connected to the low side. Neglect the exciting current and determine the current in each of the windings.

6-13 A 2,400/240-v transformer has an efficiency of 0.94 at rated load 0.80 power factor when operating as a 2-circuit or 2-winding transformer. This transformer is reconnected to operate as an autotransformer with the windings carrying their rated currents and feeding a 0.80 power factor load. Show a diagram of connections and determine the efficiency when the autotransformer is connected for operation

(a) 2,640/2,400 v.
(b) 2,400/2,640 v.
(c) 2,400/2,160 v.
(d) 2,640/240 v.

6-14 The data for an output transformer are as follows

Resistance of primary winding $R_1 = 225$ ohms
Resistance of secondary winding $R_2 = 0.52$ ohms
Primary short-circuit inductance $L_{sc_1} = 0.10$ h
Primary open-circuit inductance $L_{oc_1} = 5.85$ h
Turns ratio $a = 20$.

This transformer is connected between a source and the voice coil of a loud speaker. The source has an internal impedance of 2,000 ohms and the voice coil has an impedance of 8 ohms. Both of these impedances are practically noninductive. The power output of the transformer is 6 w. Assume the frequency to be of such a value that the exciting current and the leakage reactance of the transformer are negligible, and determine

(a) The secondary current in the transformer.
(b) The a-c component in the primary of the transformer.
(c) The internal a-c emf of the source.

6-15 Suppose that this transformer in Problem 6-14 were redesigned without any changes in the core and secondary winding, but that the turns in the primary windings were changed, keeping the amount of copper the same as before, so that the impedance of the voice coil exactly matches that of the source and the transformer. If the voltage and frequency were the same as in Problem 6-14

  (a) What would be the new power supplied to the loud speaker voice coil?
  (b) What percentage increase is this over the corresponding amount in Problem 6-14?

6-16 Determine for the transformer in Problem 6-14

  (a) The upper and lower half-power frequencies f_h and f_l.
  (b) The geometric mean frequency f_0.
  (c) The ratio f_h/f_l.

6-17 Repeat Problem 6-16 for a transformer having an identical core, the same amount of copper in the windings, but in which the number of turns in the primary and secondary is 20 percent greater than in the corresponding windings of the transformer in Problem 6-16.

6-18 Three single-phase transformers are to be connected for 4,000/440-v, 3-phase operation. The voltages are line to line. The 3-phase rating is to be 750 kva. There are four ways of connecting this transformer bank. Show a diagram of connections for each of the four ways and specify the voltage and current ratings of both windings in each transformer for each of the connections.

6-19 Figure 6-39 shows a bank of three single-phase, one-to-one ratio transformers with one side connected in wye to a 416-v, 3-phase source with the neutral

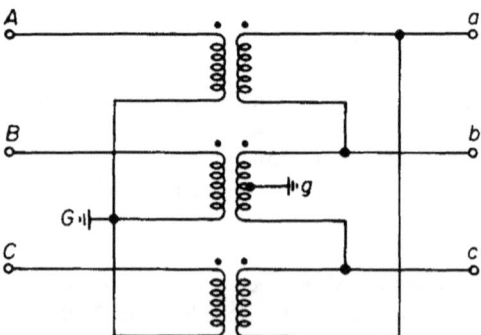

Figure 6-39. Wye-delta transformer bank for Problem 6-19.

grounded. The other side of the bank is connected in delta with the center tap of one phase grounded. The source is completely ungrounded. Determine

  (a) The line-to-ground voltages on the high and low sides of the transformer bank.

PROBLEMS   305

    (b) The voltages *A-a*, *A-b*, and *A-c*.
    (c) The voltages *A-a*, *A-b*, and *A-c* if the polarity of both windings in the middle transformer of Fig. 6-39 is reversed.
    (d) Repeat parts (a) and (b) if *C*-phase terminal of the 416-v side is grounded instead of the neutral.

6-20 A 12,000-v, 3-phase bus supplies (1) a bank of three single-phase transformers connected delta-delta and delivering a balanced 3-phase load of 5,000 kva, 0.90 power factor, current lagging, at 4,000 v and (2) a delta-wye connected, 3-phase transformer delivering a balanced 3-phase load of 2,500 kva, 0.71 power factor, current lagging, at 2,300 v. Determine the total load supplied to the two transformer arrangements by the 12,000-v bus. Neglect the exciting current and leakage impedance of the transformers.

6-21 Two identical 4,000/240-v, single-phase transformers are connected open delta and have a 3-phase rating of 11.53 kva. What is the 3-phase rating if another similar transformer is added for delta-delta operation?

6-22 Suppose that windings *A* and *C* in the 3-phase, core-type transformer in Fig. 6-29(a) were connected in parallel to a 100-v a-c source and windings *B*, *a*, *b*, and *c* were open circuited. How much voltage would be induced in the *B* winding if

    (a) The *A* and *C* windings have their marked (polarity) terminals connected together and their unmarked terminals connected together?
    (b) The marked terminal of *A* is connected to the unmarked terminal of *C* and the unmarked terminal of *A* is connected to the marked terminal of *C*? Assume the transformer to be ideal.

6-23 Describe the effect of exciting windings *A*, *B*, and *C* in parallel with

    (a) Like polarities of all three windings connected together, at rated voltage and frequency from a single-phase source.
    (b) The polarity of *B* winding reversed.

6-24 Repeat Problem 6-22 for the 3-phase shell-type transformer in Fig. 6-29(b), assuming the reluctance of the vertical components in the flux paths to equal that of the horizontal components. For example, in following around the flux path through the windings *B* and *b* in the center of the core, each vertical section marked *y* and each outer horizontal section has twice the reluctance of the horizontal center portion between points marked *x*.

6-25 Repeat Problem 6-23 for the 3-phase shell-type transformer in Fig. 6-29(b).

6-26 Is it possible to operate one or both of the transformers in Fig. 6-29 at rated voltage and frequency as a single-phase transformer without excessive exciting current by connecting all three primary windings in parallel with each other and all three secondary windings in parallel with each other?

6-27 Three single-phase transformers, each rated 2,400/240/120 v and 10 kva, have their primaries connected in delta and the center taps on their secondaries to form a neutral for 6-phase operation, as shown in Fig. 6-30. Determine the line-line voltages on the 6-phase side and the rated current in the 6-phase line wire.

6-28 A 6-phase forked-wye connection, with the primary connected 3-phase delta, converts 13,800 v 3-phase to 500 v line to line on the 6-phase side. The secondaries of each phase are divided into three equal sections. Determine the ratio of primary turns to all the turns in the secondary winding.

6-29 Three identical single-phase, 10,000-kva transformers are connected delta on the 22-kv side and wye on the 132-kv side with the neutral on the 132-kv side isolated. Balanced 3-phase voltages are applied to the 22-kv side such that

$$v_{AB} = \sqrt{2}(22,000 \cos 377t)$$

and the exciting current in the delta between A and B lines is

$$i_{exc_{AB}} = \sqrt{2}[3.2 \cos(377t - 81°) + 1.50 \cos(1,131t + 87°)$$
$$+ 0.4 \cos(1,885t - 94°)]$$

Express as functions of time

(a) The voltages v_{BC} and v_{CA} on the 22-kv side.
(b) The exciting currents $i_{exc_{BC}}$ and $i_{exc_{CA}}$ in the delta.
(c) The no-load current i_{exc_A}, i_{exc_B}, and i_{exc_C} in the 22-kv lines.

6-30 The three single-phase transformers of Problem 6-29 are operating without load and are excited from the 132-kv side with the neutral connected to the source. Assume a-phase, line-to-neutral voltage on the 132-kv side to be in phase with the line-to-line voltage between A and B phases on the 22-kv side. If the delta connection is opened

(a) How do the exciting currents i_{exc_a}, i_{exc_b}, and i_{exc_c} vary as functions of time?
(b) How does the neutral current i_n vary as a function of time?

6-31 The three single-phase transformers of Problems 6-29 and 6-30 are operating without load and are excited from the 132-kv side with the neutral isolated and the delta closed. Express the exciting currents i_{exc_a}, i_{exc_b}, and i_{exc_c} as functions of time.

6-32 The distribution transformer in Fig. 6-35 delivers 30 amp at 120 v, 1.00 power factor a to n; 40 amp at 120 v, 0.707 power factor, current lagging b to n; 25 amp at 240 v, 0.90 power factor, current lagging a to b. Neglect transformer impedance and admittance and calculate the primary current and the primary power factor.

6-33 The transformer bank in Example 6-8 delivers a load of 60,000 kva, 0.80 power factor, current lagging to a balanced 3-phase load on the 11-kv side. There is no load on the 2.2-kv side. Neglect resistance and determine the voltage on

(a) The 66-kv side.
(b) The 2.2-kv side.
if the load voltage is 11-kv line to line.

6-34 Determine for the transformer of Example 6-2 the following base quantities

(a) On the high side.
(b) On the low side using the rating of the transformer for the base power.
  (i) Current.
  (ii) Voltage.
  (iii) Impedance

6-35 Determine the per unit values of

(a) Exciting current.
(b) Resistance.
(c) Impedance.
(d) Leakage reactance.
of the transformer in Example 6-2, using the rating of the transformer as the base power.

6-36 Repeat Problem 6-35, but for a base kva of 450.

6-37 Determine the per unit resistance, reactance, and exciting current of the transformer in Problems 6-8 and 6-11, using the transformer rating as the base.

6-38 A 3-phase transformer rated at 150,000 kva, 138/13.8 kv has an impedance of 0.10 per unit and is operating at 132/13.2 kv. Calculate the per unit impedance for a 1,000,000-kva base and 132-kv base.

BIBLIOGRAPHY

Blume, L. F. et al., *Transformer Engineering* (2nd ed.) New York: John Wiley & Sons, Inc., 1954.

Gibbs, J. B., *Transformer Principles and Practice* (2nd ed.) New York: McGraw-Hill Book Company, 1950.

Kuhlman, J. H., *Design of Electrical Apparatus* (3rd ed.) New York: John Wiley & Sons, Inc., 1950.

Landee, R. W. et al., *Electronic Designers' Handbook*. New York: McGraw-Hill Book Company, 1957.

Lee, R., *Electronic Transformers and Circuits* (2nd ed.) New York: John Wiley & Sons, Inc., 1955.

MIT Staff, *Magnetic Circuits and Transformers*. New York: John Wiley & Sons, Inc., 1943.

Stigant, S. A. and A. L. Lacey, *J and P Transformer Book* (8th ed.) London: Johnson and Phillips, Ltd., 1941.

Westinghouse Electric Corporation, *Electrical Transmission and Distribution Reference Book* (4th ed.) (East Pittsburgh, Pa.: Westinghouse Electric Corporation, 1950), Chap. 5.

SATURABLE REACTORS

7

Although the nonlinear magnetic characteristics of ferromagnetic materials produce such undesirable effects as distortion of voltage and current waveforms, they are the basis for successful operation of certain types of devices. The saturable reactor is such a device, as it depends upon a high value of permeability in the unsaturated region and a low value of permeability in the saturated region of the magnetization curve. In general, the greater the departure of the magnetization from linearity, the more effective is the saturable reactor. An ideal magnetization curve for saturable reactor operation is shown in Fig. 7-1(b).

The nearly rectangular hysteresis loops of grain-oriented alloys of silicon-iron* and nickel-iron,† therefore, lead to operating characteristics that are superior to those of saturable reactors using core materials with the more rounded knees in the saturation curve. Figure 7-1(a) shows a hysteresis loop for a nickel-iron alloy. Although the saturable reactor has found some applications since about the turn of the century,‡ its use did not become widespread until the advent of the magnetic alloys mentioned above. The improvement in dry-type rectifiers by the use of such materials as silicon and selenium compounds gave further impetus to its development.

One of the widest applications of the saturable reactor is that

---

* W. P. Goss, "New Development in Electrical-Steels Characterized by Fine Grain Structure Approaching the Properties of a Single Crystal," *Trans. Am. Soc. Metals*, 33 (1935), 511–544.

† W. Morrill, "Oriented Crystals, Their Growth and Their Effects on Magnetic Properties," *Gen. Elec. Rev.*, 53 (1950), 16–21.

‡ See U.S. Patent No. 891,797 on "A Method of Automatic Regulation of Rectifiers and Rotary Converters," issued to F. B. Crocker in 1908 and applied for in 1904.

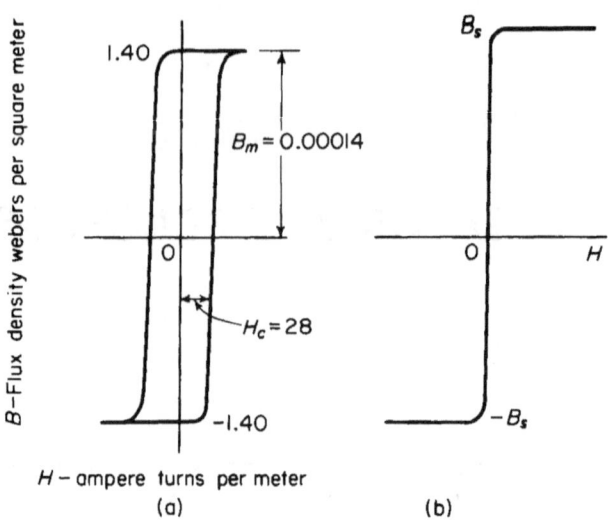

Figure 7-1. (a) 60-cycle hysteresis loop; (b) simplified magnetization curve of cold rolled 50%-nickel, 50%-iron alloy.

of a magnetic amplifier, by means of which large amounts of electric power can be controlled, with rapid response, by making use of much smaller amounts of control power. This makes for smaller control equipment and greater flexibility than if the output power were controlled directly. Some of these applications are motor speed control, thyratron-grid phase control, current in electric welders, amplification of small d-c voltages from thermocouples, and d-c transformers for measuring very large values of direct current. Saturable reactors also make it possible to multiply frequency with relatively simple circuitry and without rotation or electronic equipment. This text includes only a brief introduction to this subject, which has been treated thoroughly by other textbooks.*

7-1 MAGNETIC FREQUENCY MULTIPLIERS

Because of the simple relationships that govern the behavior of a saturable reactor in a frequency multiplying arrangement, the magnetic frequency

---

* For more comprehensive treatments see H. F. Storm, *Magnetic Amplifiers*. New York: John Wiley & Sons, Inc., 1955; A. G. Milnes, *Transductors and Magnetic Amplifiers*. London: Macmillan & Co., Ltd., 1957; G. E. Lynn, T. J. Pula, J. F. Ringelman, and F. G. Timmel, *Self-saturating Magnetic Amplifiers*. New York: McGraw-Hill Book Company, 1960; F. Kummel, *Regal Transduktoren*. Berlin: Springer-Verlag, 1961; and D. L. Lafuze, *Magnetic Amplifier Analysis* New York: John Wiley & Sons, Inc., 1962.

multiplier is treated here to serve as an introduction to the analysis of saturable reactor operation in wider applications. In the frequency multipliers under consideration, a polyphase circuit with an odd number of m phases, and having a frequency of f cps, supplies m saturable reactors connected in star to provide a single-phase source, operating at a frequency of mf cps. The load is connected between the neutral point of the source and that formed by the common connection of the saturable reactor as shown for the frequency tripler in Fig. 7-4(a).

The basic circuit of the frequency multiplier is shown for one phase in Fig. 7-2; the simplified magnetization curve in Fig. 7-1(b) is assumed for the core material. Although the core is represented as a toroid in Fig. 7-2, it may have other shapes as well. The design of the reactor must be such that it saturates at or above a certain angle α_f*, known as the firing angle, which is determined by

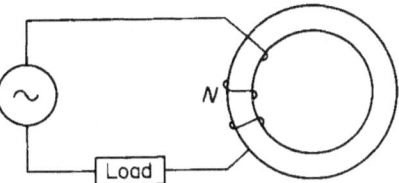

Figure 7-2. Basic circuit for frequency multiplier.

$$\alpha_f = \left(\frac{m-1}{m}\right)\pi \qquad (7\text{-}1)$$

On the basis of the simplified saturation curve in Fig. 7-1(b), the saturable reactor is an open circuit when unsaturated and a short circuit when saturated, thus operating as a synchronous switching device or a gate. This discussion is confined, for reasons of simplicity, to applications in which the load is noninductive, although inductive loads may be supplied as well.†

Wave forms of voltage, flux density, and current are shown in Fig. 7-3(a) with the idealized magnetization curve of Fig. 7-3(b) as a basis. The applied voltage is expressed by

$$v = \sqrt{2}\, V \sin \omega t \qquad (7\text{-}2)$$

The core saturates at $\omega t = \alpha_f$, the firing angle. The reactor, when the resistance and capacitance of its winding are neglected, becomes a short circuit after having been an open circuit, for $0 < \omega t < \alpha_f$ and full voltage is impressed on the load resistance. The current that results in the noninductive resistance is then defined by

$$i = 0 \quad \text{for} \quad 0 < \omega t \leqslant \alpha_f$$

---

* L. J. Johnson and S. E. Rauch, "Magnetic Frequency Multipliers," *Trans AIEE*, 73 Part 1 (1954), 448–451. See also, M. Camras, "A New Frequency Multiplier," *Electrical Engineering*, September 1962, 699–705.

† O. J. M. Smith and J. T. Salihi, "Analysis and Design of a Magnetic Frequency Multiplier," *Trans AIEE*, 74 Part 1 (1955), 99–106.

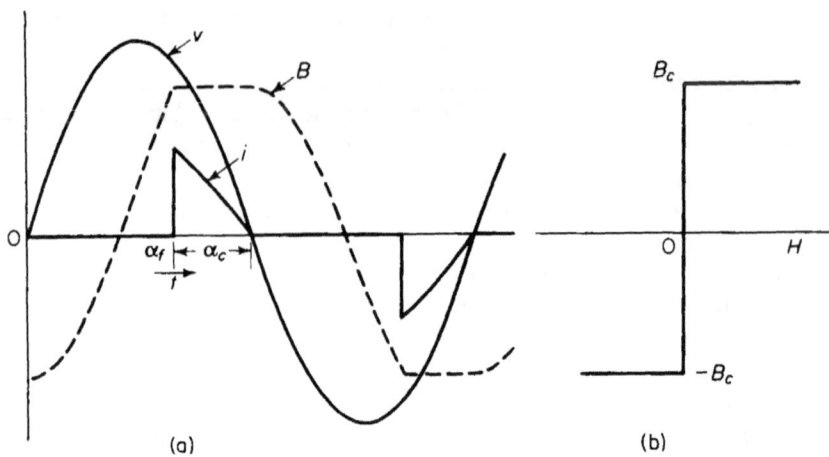

Figure 7-3. (a) Waveforms of voltage, flux density, and current for noninductive load; (b) idealized magnetization curve for multiplier.

and

$$i = \frac{\sqrt{2}V}{R} \sin \omega t \quad \text{for} \quad \alpha_f \leqslant \omega t \leqslant \pi \tag{7-3}$$

The angle α_c represents the interval during the half cycle in which conduction takes place and is called the *conduction angle*.

7-2 FREQUENCY TRIPLER

The frequency tripler affords a simple example of a frequency multiplier. It uses three saturable reactors supplied from a 3-phase source as shown in Fig. 7-4. The firing angle α_f must be adjusted so that only one reactor is saturated at a time, otherwise a line-to-line short circuit occurs. Then, according to Eq. 7-1, the firing angle must be at least $\alpha_f = 2\pi/3$ for a frequency tripler, since $m = 3$. Wave forms of currents in each phase and in the output are shown in Fig. 7-4(b). Because there is some inductance in practical devices, even when saturated, the corners in the current waves are not quite as sharp as those shown. The current in an inductive load having a Q of about 0.5 has fairly good wave form since the inductance tends to smooth out the wave.*

* *Ibid.*

Sec. 7-3 NONINDUCTIVE LOAD 313

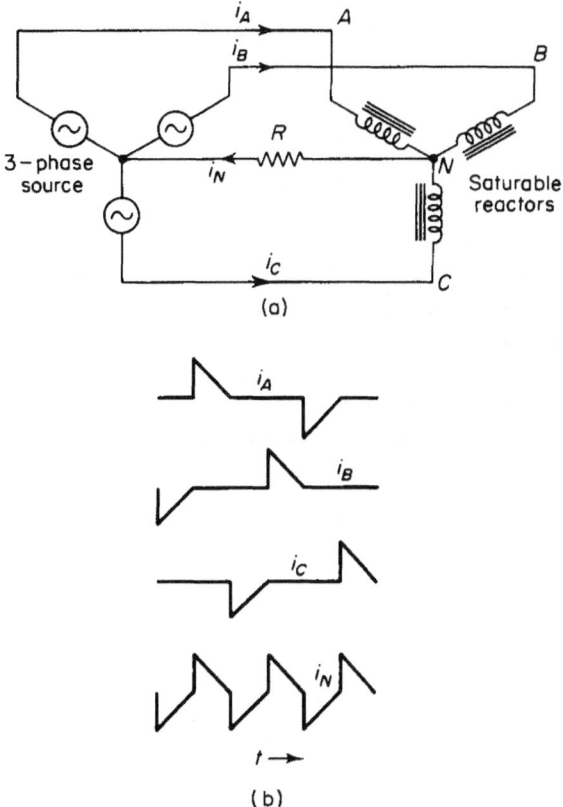

Figure 7-4. (a) Frequency tripler circuit; (b) wave forms of line currents and output current for noninductive load.

7-3 RELATIONSHIP BETWEEN APPLIED VOLTAGE AND CONDUCTION ANGLE FOR NONINDUCTIVE LOAD

The differential equation describing the operation of the basic circuit in Fig. 7-2 in the unsaturated or presaturated region is, on the basis of negligible winding resistance and the idealized magnetization curve

$$\frac{d\lambda}{dt} = \sqrt{2}\, V \sin \omega t \qquad (7\text{-}4)$$

If A_c is the effective core area, B_c the critical flux density, and N the number of turns in the winding, then, when these quantities are substituted in Eq. 7-4, the result is

$$\frac{dB}{dt} = \frac{\sqrt{2}\,V \sin \omega t}{A_c N}$$

Integrating

$$B = K_i + \frac{\sqrt{2}\,V}{A_c N \omega} \int_{\omega t_i}^{\omega t} \sin \omega t \, d(\omega t) \tag{7-5}$$

in which K_i is the initial flux density, i.e., at $t = t_i$. Wave forms of flux density, voltage, and current for resistance loading are shown in Fig. 7-3(a). At $\omega t = \pi$ the load current goes through zero and the voltage changes sign, which means that dB/dt must become negative also, and the reactor becomes unsaturated. The current, therefore, is extinguished at $\omega t = \pi$. The flux density starts to build up from a value of $-B_c$ at $\omega t = 0$, and t_i is the value of t at $\omega t = 0$, so that Eq. 7-5 can be put into the more definite form

$$B = -B_c + \frac{\sqrt{2}\,V}{A_c N \omega} \int_0^{\omega t} \sin \omega t \, d(\omega t) \tag{7-6}$$

Firing occurs when B reaches the saturated value $+B_c$, which occurs at $\omega t = \alpha_f$ in Fig. 7-3(a). When these quantities are substituted in Eq. 7-6, there results

$$B_c = -B_c + \frac{\sqrt{2}\,V}{A_c N \omega} \int_0^{\alpha_f} \sin \omega t \, d(\omega t)$$

$$= -B_c \frac{\sqrt{2}\,V}{A_c N \omega}(1 - \cos \alpha_f)$$

which yields

$$A_c N = \frac{V(1 - \cos \alpha_{ft})}{\sqrt{2}\,\omega\, B_c} \tag{7-7}$$

EXAMPLE 7-1: Three saturable reactors are supplied from a 3-phase, 4-wire, 208-v, 400-cycle source to provide single-phase power at a frequency of 1,200 cps to a noninductive load. The core of each reactor is comprised of toroidal laminations having an OD = 2 in. and ID = $1\frac{5}{8}$ in. stacked to a height of $\frac{1}{4}$ in. The stacking factor is 0.80. The magnetic characteristic of the core material is represented by the hysteresis loop in Fig. 7-1(a). Calculate the number of turns in the winding of each saturable reactor.

Solution: The number of turns is obtained from Eq. 7-7 by

$$N = \frac{V(1 - \cos \alpha_f)}{\sqrt{2}\,\omega B_c A_c}$$

where $V = 208/\sqrt{3} = 120$ v per phase
$\alpha_f = 2\pi/3$ radians
$\omega = 2\pi \times 400$ radians per sec
$B_f = 1.40$ webers per sq m
$A_c = \frac{1}{2}(2 - 1\frac{5}{8})(\frac{1}{4})(0.80)(2.54)^2(10^{-4})$
$= 2.42 \times 10^{-5}$ sq m

$$N = \frac{120(1 - \cos 2\pi/3)}{\sqrt{2}(2\pi)(400)(1.4)(2.42)(10^{-5})} = 1{,}490 \text{ turns}$$

7-4 SINGLE-CORE SATURABLE REACTOR WITH PREMAGNETIZATION

Saturable reactors that are used for controlling variable outputs are provided, in their simplest form, with a d-c winding that furnishes the premagnetization

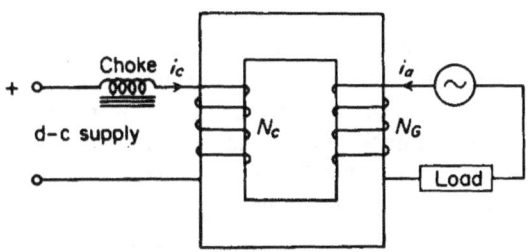

Figure 7-5. Single-core saturable reactor circuit.

and with an a-c winding between an a-c source and a load to which the current is controlled by varying the amount of premagnetization. Figure 7-5 is a schematic diagram of a single-core reactor with a d-c control winding of N_C turns and an a-c output winding, also known as the *gate winding*, of N_G turns. The core shown in Fig. 7-5 is rectangular; it may, however, be toroidal or any other convenient shape. To insure good efficiency and good regulation, the resistance of both windings must be low when referred to the load impedance. This means that unless the impedance of the control circuit or premagnetizing circuit is kept above a certain value by using, for example, an external choke coil, the control circuit will act as a short-circuited secondary with the a-c winding as the primary. Therefore, without a high impedance in the control circuit, the impedance of the reactor, viewed from the a-c winding, would be low, being practically equal to the leakage impedance. Consequently, the d-c component in the control winding would lose much of its effect on the output current, since the leakage flux paths are largely in air. A reactor or choke in series with the control circuit is a means for suppressing the a-c component of current in the control winding. Such a choke, if it is to be of reasonable size, must have a ferromagnetic core with an air gap to prevent saturation from the direct current. Even so, the size of the choke is comparable to that of the single-core saturable reactor itself. Practically all of the a-c voltage induced in the d-c winding appears across the choke coil.

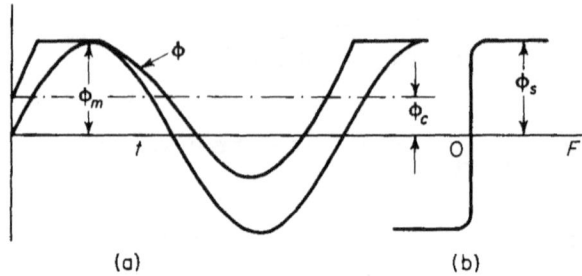

Figure 7-6. (a) a-c flux wave ϕ for $i_c = 0$ and for $i_c > 0$; (b) idealized magnetization curve.

When core losses are neglected, the idealized curve is again as shown in Fig. 7-6(a), having infinite permeability in the unsaturated region and zero differential permeability in the saturated region. The reactance is therefore infinite if the a-c flux is confined to the unsaturated region, i.e., when there is no premagnetization. On the other hand, the reactance is zero when the a-c flux is confined entirely to the saturated region by a sufficiently large d-c premagnetizing current, and the current in a noninductive load having a resistance R_L would then be

$$i = \frac{\sqrt{2}\,V}{R_L + R_G} \sin \omega t \tag{7-8}$$

Where R_G is the resistance of the a-c winding of the reactor and the voltage of the a-c source is

$$v = \sqrt{2}\,V \sin \omega t$$

It therefore appears that the saturable reactor affords a means of adjusting the output current from a very low value to a maximum of

$$I = \frac{V}{R_L + R_G} \tag{7-9}$$

Although a large range of output current is possible, the wave form of the current, while practically sinusoidal at maximum output, is quite distorted at the smaller outputs.

To demonstrate this, let the amplitude of the a-c flux be just below the saturated value ϕ_s when there is no premagnetization. Then the output current is zero and the entire voltage of the a-c source is impressed on the a-c winding of the reactor. And if the resistance of that winding is neglected, the amplitude of the a-c flux as shown in Fig. 7-6(b) is

$$\phi = \frac{V}{4.44fN_G} = \phi_s = A_c B_c \tag{7-10}$$

The output current for this condition is zero. When direct current is applied to the control winding, output current is furnished to the load. Because of the choke, the current in the control winding under steady-state conditions is

$$i_c = \frac{V_c}{R_c} \tag{7-11}$$

where V_c is the d-c source voltage and R_c the resistance of the control circuit including the choke.

In the unsaturated interval, the total mmf must be zero, and if N_G and N_C are the turns in the output and control windings with i_a the current in the output winding, then for the current directions shown in Fig. 7-1, the output current is

$$i_a = -\frac{N_C i_c}{N_G} \tag{7-12}$$

During the saturated interval the voltage induced in the a-c winding is zero, and the output current is

$$i_a = \frac{\sqrt{2} V}{R_L + R_G} \sin \omega t \tag{7-13}$$

where the source voltage is given by

$$v = \sqrt{2} V \sin \omega t$$

Since there is no d-c component in the voltage of the output circuit, there can be no d-c component in the output current under steady conditions, and the average value of the output current over a complete cycle must be zero, i.e.

$$\int_0^{2\pi} i_a d(\omega t) = 0 \tag{7-14}$$

On the basis of Eqs. 7-12, 7-13, and 7-14, the wave form of the output current is as shown in Fig. 7-7, from which it is found that

$$\frac{\sqrt{2}\,V}{R_L + R_G} \int_{\alpha_1}^{\pi} \sin \omega t \, d(\omega t)$$

$$= -\frac{\sqrt{2}\,V}{R_L + R_G} \int_{\pi}^{\alpha_2} \sin \omega t \, d(\omega t) + \frac{N_C i_c}{N_G}(2\pi + \alpha_1 - \alpha_2) \quad (7\text{-}15)$$

The left-hand side of Eq. 7-15 represents the area under the positive portion of the current wave and the right-hand side of the area above the negative

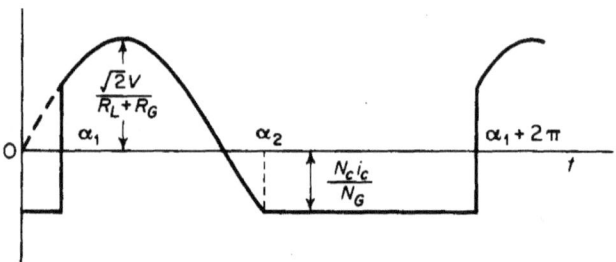

Figure 7-7. Wave form of output current from a single-core reactor.

portion of the current wave. The angle α_1 is the firing angle, i.e., the value of ωt at which the flux reaches the saturated value ϕ_s, and α_2 is the angle at which the iron becomes unsaturated.

When the integrations in Eq. 7-15 are performed, there results

$$\frac{\sqrt{2}\,V}{R_L + R_G}(\cos \alpha_1 - \cos \alpha_2) = \frac{N_C i_c}{N_G}(2\pi + \alpha_1 - \alpha_2) \quad (7\text{-}16)$$

From Fig. 7-7 it is apparent that

$$\frac{\sqrt{2}\,V}{R_L + R_G} \sin \alpha_2 = -\frac{N_C i_c}{N_G}$$

or

$$\sin \alpha_2 = -\frac{(R_L + R_G)N_C}{\sqrt{2}\,V N_G} i_c \quad (7\text{-}17)$$

Substitution of Eq. 7-17 in Eq. 7-16 yields

$$\cos \alpha_1 - \cos \arcsin\left[-\frac{(R_L + R_G)N_C i_c}{\sqrt{2}\, V N_G}\right]$$

$$= \frac{(R_L + R_G)N_C i_c}{\sqrt{2}\, V N_G}\left\{2\pi + \alpha_1 - \arcsin\left[-\frac{(R_L + R_G)N_C i_c}{\sqrt{2}\, V N_G}\right]\right\} \quad (7\text{-}18)$$

The angle α_1 can be evaluated from Eq. 7-18 by graphical methods or by means of suitable approximations.

It is important to remember that the results achieved in the equations above are based on the amplitude of the a-c flux ϕ_m being just below the saturated value ϕ_s.

The single-core premagnetized saturable reactor is seldom used because the wave form of the output current is unsymmetrical over a large part of the operating range. In addition, because of the choke in the control circuit, there is the disadvantage of a high time constant, where speed of response is important.

7-5 2-CORE SATURABLE REACTOR

The disadvantages of unsymmetrical current wave form and the need for a choke in the control circuit are overcome in the 2-core saturable reactor of Fig. 7-8(a), or its equivalent of Fig. 7-8(b). The 2-core reactor is comprised essentially of two identical transformers, I and II, each having a control winding of N_C turns and an output winding or gate winding of N_G turns. The control windings, energized from a d-c source, are connected in series with each other and with their polarities relative to those of the gate windings so that the fundamental component of the voltages, induced in them by the current in the gate windings, cancel. The gate windings may be connected in parallel with each other or in series with each other.

7-6 GATE WINDINGS IN PARALLEL

The gate windings of both saturable reactors in Figs. 7-8(a) and 7-8(b) are shown connected in parallel, which means that the voltage impressed on one must equal that impressed on the other.

Components of flux

If the resistances of the windings are neglected, the voltages induced in the two gate windings must equal each other, and if ϕ_{GI} and ϕ_{GII} are the

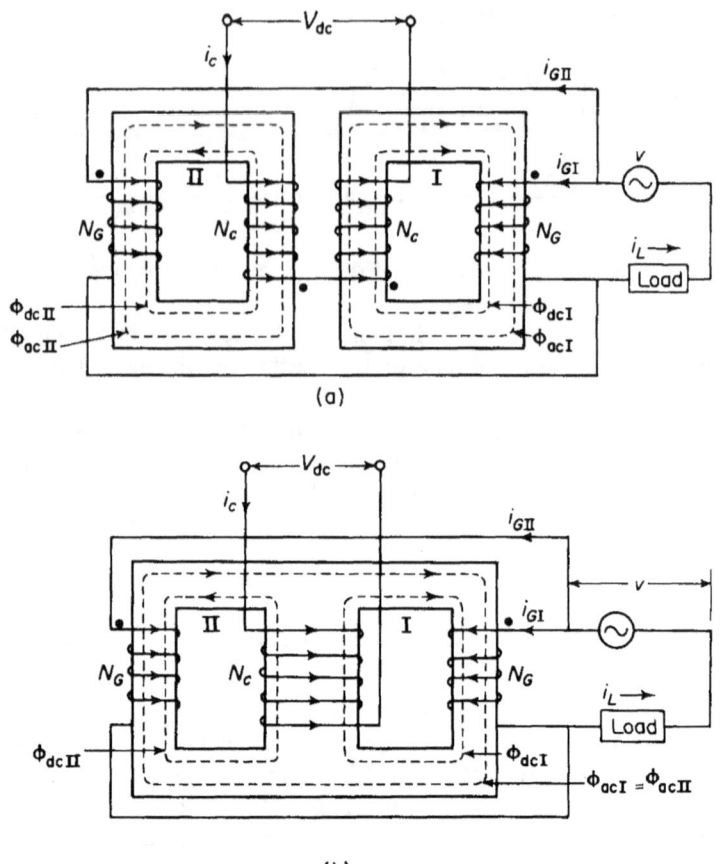

Figure 7-8. Saturable reactors with gate windings in parallel. (a) 2-core (b) 3-legged equivalent of 2-core.

fluxes linking gate windings I and II, we have

$$\frac{d\phi_{GI}}{dt} = \frac{d\phi_{GII}}{dt} = V_{GI} = V_{GII} \qquad (7\text{-}19)$$

The voltages induced in the control windings of Fig. 7-8(a) are equal and opposite, cancelling each other, thus eliminating the need for a choke in the control circuit, as required in the single-core reactor. In the case of the 3-legged reactor of Fig. 7-8(b), the voltages induced in the control winding by the fluxes linking the two gate windings also cancel each other. As a result, there is no a-c component in the current of the control circuit, so that i_c, the control current, is strictly d-c for steady-state operation when the gate

Sec. 7-6 GATE WINDINGS IN PARALLEL 321

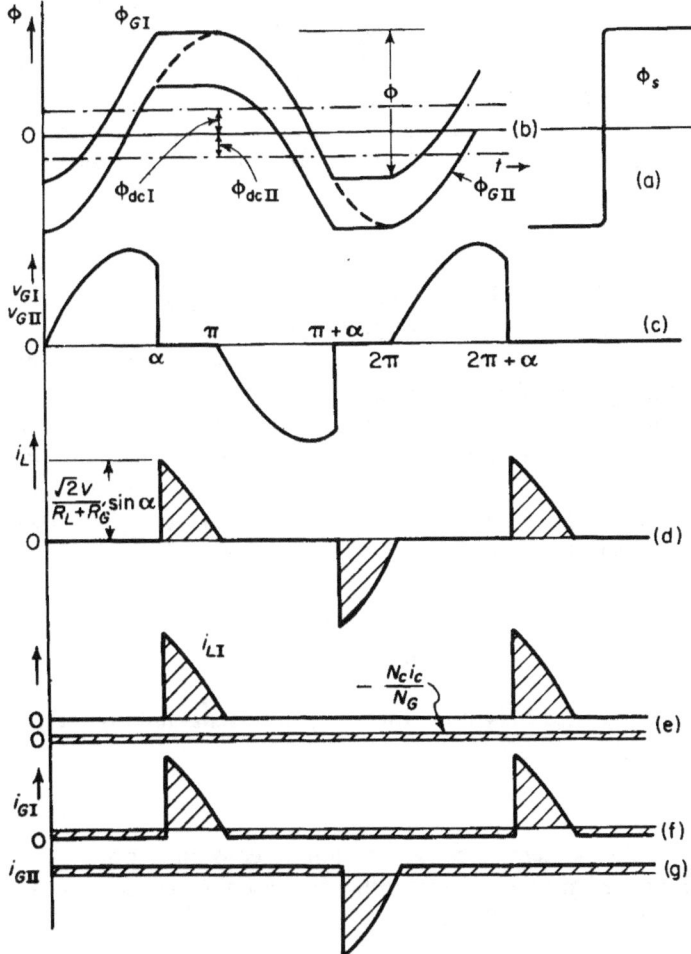

Figure 7-9. Wave forms for saturable reactor with gate windings in parallel and constant control current. (a) Idealized magnetization curve; (b) flux waves; (c) gate voltage $v_{GI} = v_{GII}$; (d) load current; (e) load component i_{LI} and steady component $N_c i_c/N_G$ in gate winding I; (f) resultant current in gate winding I; (g) resultant current in gate winding II.

windings are connected in parallel and when their resistance is appreciably lower than that of the control circuit.

Again, assume the magnetization curve for the cores to be as shown in Fig. 7-9(a) and the amplitude of the a-c flux, at zero premagnetization, to be just below the saturated value ϕ_s. At zero premagnetization, the control current i_c is zero and the zero axes of the a-c fluxes linking the two gate windings coincide. However, when there is d-c premagnetization, the zero axis for the a-c component ϕ_{acI} linking gate winding I is shifted upward, and the component ϕ_{acII} linking gate winding II is shifted downward, as shown in Fig. 7-9(b) for the polarities shown in Fig. 7-8. In the case of premagnetization, the a-c fluxes, although periodic, are not sinusoidal since the maximum value of the combined a-c and d-c fluxes cannot exceed the saturated value ϕ_s. If ϕ_{GI} and ϕ_{GII} are the total fluxes linking gate windings I and II, we have

$$\phi_{GI} = \phi_{acI} + \phi_{dcI} \qquad \phi_{GII} = \phi_{acII} - \phi_{dcII} \qquad (7\text{-}20)$$

where ϕ_{acI} and ϕ_{acII} are the a-c components and ϕ_{dcI} and ϕ_{dcII} are the d-c components of flux linking the respective gate windings.

The control current i_c is assumed to be a constant direct current, a condition that is satisfied if the equivalent resistance $(N_G/N_C)^2 R_c$ of the control circuit is much greater than the resistance $2R_G$ of the gate windings. The gate windings carry an unsymmetrical alternating current that contains no net d-c component as shown in Figs. 7-9(f) and (g). Then, for a given direct current in the control winding, the d-c flux components must be constant, and substitution of Eq. 7-20 in Eq. 7-19 yields

$$\frac{d\phi_{acI}}{dt} = \frac{d\phi_{acII}}{dt}$$

which can be satisfied only if the a-c flux components linking the gate windings are equal, i.e.

$$\phi_{acI} = \phi_{acII} \qquad (7\text{-}21)$$

since the a-c fluxes themselves can contain no constant term.

The d-c fluxes must equal each other because they are produced by equal mmfs in identical cores. Hence

$$\phi_{dcI} = \phi_{dcII} \qquad (7\text{-}22)$$

as shown in Fig. 7-9(b).

At the firing angle α, the flux ϕ_{GI} in core I reaches the saturated value ϕ_s and remains there until the instantaneous a-c voltage reverses its polarity, i.e., at $\omega t = \pi$. The voltage across the gate windings is therefore zero, since ϕ_{GII} also remains constant over the interval $\alpha < \omega t < \pi$. The a-c supply voltage during that interval is impressed across the load resistance. This action produces the wave form of the load voltage as shown in Fig. 7-9(d). Because the load is noninductive, the current through it has the same wave form as the load voltage.

The mechanism that holds the flux constant in the unsaturated core during the conducting angle is explained as follows: During the conducting

Sec. 7-6   GATE WINDINGS IN PARALLEL   323

period, in which core I is saturated, the emf induced in its gate winding and control winding is zero and both of these windings, then, are short circuits if their resistances are neglected. Therefore, control winding I acts as a short circuit on control winding II, which in turn acts as the secondary, while gate winding II acts as the primary of an ideal transformer that has its core unsaturated. Since core II is unsaturated while the constant control current i_c is flowing, there must be a corresponding current i'_G in gate winding II such that the resultant mmf acting on core II is zero, i.e.

$$N_G i'_G + N_C i_c = 0 \qquad (7\text{-}23)$$

Since the impedance of gate winding II is practically zero during this interval, this current i'_G circulates between the two gate windings. Next, consider the interval between π and $\pi + \alpha$ when both cores are unsaturated. The mmf exerted by the control current again demands an equal and opposing mmf in both gate windings and this constant component $i'_G = N_C i_c / N_G$ continues to circulate between gate windings during this interval. Now, at $\omega t = \pi + \alpha$, and the constant component $i'_G = N_C i_c / N_G$ again results, circulating in the same direction as before. As a result, during normal steady-state operation, there is a constant component of current circulating in one direction between the gate windings at all times in a direction such as to oppose the mmf of the control windings on both cores. The positive direction of this circulating d-c component is into the polarity-marked side of gate winding II and out of the polarity-marked side of gate winding I.

During the interval of saturation for core I, the load current I_L flows through gate winding I in a direction opposite to the constant circulating component. One-half cycle later core II saturates and the load current flows through gate winding II against the constant unidirectional circulating current. The wave forms of the load component and the constant component of i'_{GI} are shown in Fig. 7-9(e), that of the total current i_{GI} in gate winding I, which is the sum of these components, is shown in Fig. 7-9(f), and that of the total current i_{GII} in gate II is shown in Fig. 7-9(g).

Law of equal ampere turns

The average value of the current in either gate, when taken over an entire cycle, must be zero. This means that, in the case of gate winding I, the average values of the positive component i_L in Fig. 7-9(e) must equal the constant negative component $N_C i_c / N_G$.

If I_L is the *average* value of i_L, then we have

$$I_L N_G = N_C i_c \qquad (7\text{-}24)$$

The saturable reactor as a constant-current source

When operating within a range such that the amplitude of the a-c flux does not exceed the saturated value at zero premagnetization, the saturable reactor will furnish practically constant load current over a range of load resistance R_L determined by the value of the control current i_c and the a-c supply voltage V. The resistance of the output circuit is the sum of the load resistance and the gate circuit, i.e., $R_L + R'_G$. The resistance of the gate circuit R'_G equals the resistance of one gate winding R_G for the parallel connection because then only one winding carries load current at a time. When the gate windings are connected in series the resistance of the gate circuit R'_G is $2R_G$ or twice the resistance per gate winding. The *average* value of the rectified load current is

$$\begin{aligned} I_L &= \frac{1}{\pi} \int_\alpha^\pi i_L \, d(\omega t) \\ &= \frac{1}{\pi} \frac{\sqrt{2}V}{R_L + R'_G} \int_\alpha^\pi \sin \omega t \, d(\omega t) \\ &= \frac{\sqrt{2}V}{R_L + R'_G} \frac{1 + \cos \alpha}{\pi} \end{aligned} \quad (7\text{-}25)$$

A comparison of Eqs. 7-24 and 7-25 shows that

$$\frac{\sqrt{2}V}{R_L + R'_G} \frac{1 + \cos \alpha}{\pi} = \frac{N_C i_c}{N_G}$$

If V and i_c are held constant while the load resistance R_L is increased, the firing angle decreases, holding the load current constant with the limiting value of α equal to zero. For $\alpha = 0$, the saturable reactor is a short circuit in series with the load resistance throughout the voltage cycle, and the load current is simply a sinusoidal current expressed by

$$i = \frac{\sqrt{2}V}{R_L + R'_G} \sin \omega t$$

A reduction in the a-c voltage V has the same effect as an increase in the load resistance. The effect of inductive load on the operation of saturable reactors is not within the scope of this text, but has been treated in other sources.*

---

* H. F. Storm, *Magnetic Amplifiers*. New York: John Wiley & Sons, Inc., 1955; F. Kummel, *Regel Transduktoren*. Berlin: Springer-Verlag, 1961.

7-7 GATE WINDINGS IN SERIES

Figure 7-10 shows a saturable reactor with its gate windings connected in series with each other. The 2-core saturable reactor and its equivalent are capable of two modes of operation when the gate windings are connected in

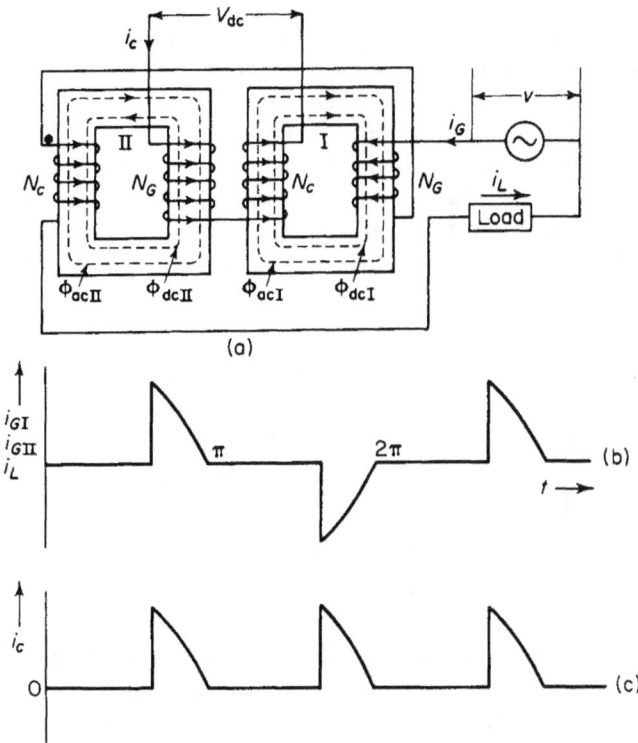

Figure 7-10. (a) Saturable reactor with gate windings in series; (b) load current; (c) control current for noninductive load.

series. In one of these, the impedance of the control circuit is low enough to permit free even-harmonics to circulate in that circuit. In the other mode of operation, the free even-harmonics are suppressed in the control circuit by connecting a high impedance, usually a choke, in series with the control windings or by providing leakage flux paths of small reluctance in the cores.

When the even-harmonics are allowed to circulate freely in the control circuit, the wave form of the output current is the same as that of the parallel-connected gate windings, whereas suppression of the even-harmonics in the control circuit produces an output current with a rectangular wave form if the magnetization curve is as assumed in Fig. 7-9(a) with the gate windings connected in series.

7-8 OPERATION WITH FREE EVEN-HARMONICS

The operating principle for the series connection is practically the same as that for the parallel connection of gate windings, as long as the impedance of the control circuit is low enough to permit the circulation of even-harmonics produced in the control circuit by the current in the output windings or gate windings. The sum of the voltages taken around the control circuit must be zero in accordance with Kirchhoff's Law. The impedance of the control circuit must be low enough to permit the circulation of free even-harmonics, and can therefore be neglected. Then, since the voltages across the two control windings must be equal at all times, and since the windings have equal numbers of turns, the flux linkages through them must change at the same rate, that is

$$\frac{d\phi_\text{I}}{dt} = \frac{d\phi_\text{II}}{dt}$$

which is the same as in the case of the parallel connection of gate windings. Thus, when one reactor saturates, the flux through both cores remains constant, and we have

$$\frac{d\phi_\text{I}}{dt} = \frac{d\phi_\text{II}}{dt} = 0$$

Because of symmetry, the saturation interval, or conduction interval, is the same, regardless of which reactor saturates under steady-state conditions. The load current, therefore, has the same wave form as for the parallel connection, i.e.

$$i_L = 0 \quad \text{for} \quad 0 < \omega t < \alpha$$

and

$$i_L = \frac{\sqrt{2}V}{R_L + R'_G} \sin \omega t \quad \text{for} \quad \alpha < \omega t < \pi$$

The a-c components of the fluxes in the two cores must therefore have identical wave forms, and, because of symmetry and by virtue of the polarities of the respective control windings, they are shifted by equal amounts, but in the opposite direction from the zero axis, just as shown in Fig. 7-9(b) for the parallel operation of gate windings.

Sec. 7-8 OPERATION WITH FREE EVEN-HARMONICS

With the series connection, the instantaneous currents through the gate windings must at all times be equal to each other and to the load current. Since the mmfs on each core must be zero when both are saturated (during which interval the current in the gate windings is zero), the current in the control winding must also be zero. However, when one core is saturated, its control winding acts as a short circuit on the control winding of the other core, which is unsaturated, and, through transformer action, causes the voltage across the gate winding of the unsaturated core to be zero. The net mmf acting on the unsaturated core must be zero so the ampere turns in the control winding (secondary) must equal the ampere turns in the gate winding (primary). Hence, with core I saturated

$$N_C i_c = N_G i_{GII}$$

and during the conducting interval the control current is

$$i_c = \frac{N_G}{N_C} \frac{\sqrt{2}V}{R_L + R'_G} \sin \omega t \quad \alpha < \omega t < \pi$$

But a half cycle later core II is saturated, and the current in the control circuit is expressed by

$$i_c = -\frac{N_G}{N_C} \frac{\sqrt{2}V}{R_L + R'_G} \sin \omega t \quad \pi + \alpha < \omega t < 2\pi$$

The wave forms of the load current, which is also the current in both gate windings and the current in the control circuit, are shown in Figs. 7-10(b) and 7-10(c).

The average value of the control current is the d-c component that is determined by the resistance of the control circuit and the d-c voltage applied to the control circuit. Hence

$$I_{cav} = \frac{V_{dc}}{R_C}$$

where R_C is the resistance control circuit. The rms value of the control current is actually greater when the gate windings carry current, being

$$I_{ceff} = \sqrt{I_{cav}^2 + I_{cac}^2} \qquad (7\text{-}27)$$

where I_{cac} is the rms value of the a-c component in the control current. It is important to note that the a-c component in the control current does not equal the rms value of the gate current or output current referred to the control winding.

7-9 POWER OUTPUT

It should be emphasized that I_L represents the average value of one-half of the load current wave. However, in expressing the power expended in a resistance, the square of the effective value of the current is multiplied by the value of the resistance. The ratio of effective to average value of a periodic current or voltage is called the *form factor* k_f, which is 1.11 for a sinusoid. The effective value of the load current is therefore

$$I_{Leff} = k_f I_L$$

and if the load resistance is R_L ohms, the output is

$$P_L = (k_f I_L)^2 R_L \quad \text{w} \tag{7-28}$$

For a given a-c source voltage V, the output current is a maximum when $\alpha = 0$, and, if the leakage reactance is neglected, the maximum output current is

$$k_f I_L = \frac{V}{R'_G + R_L} \tag{7-29}$$

where R'_G is the resistance of the gate circuit, which, in the case of the parallel connection, is the resistance of each gate winding, and for the series connection is twice the resistance of each gate winding. The a-c source voltage is assumed to be sinusoidal; therefore, the output current is sinusoidal when $\alpha = 0$ and $k_f = 1.11$. The maximum power output is, therefore, approximately

$$P_{L(\max)} = \left(\frac{V}{R'_G + R_L}\right)^2 R_L \tag{7-30}$$

Form factor

When the leakage impedance of the gate windings is neglected, the average value of the output current to a noninductive load resistance is expressed by Eq. 7-25, as follows

$$I_L = \frac{\sqrt{2}V}{R_L + R'_G}\left(\frac{1 + \cos \alpha}{\pi}\right) \tag{7-31}$$

The effective rms value of the load current is

$$I_{L\text{eff}} = \sqrt{\frac{1}{\pi} \int_0^\pi (i_L)^2 \, d(\omega t)}$$

$$= \sqrt{\frac{1}{\pi} \int_\alpha^\pi \left(\frac{\sqrt{2}V}{R_L + R_G'}\right)^2 \sin^2 \omega t \, d(\omega t)} \qquad (7\text{-}32)$$

$$= \frac{\sqrt{2}V}{R_L + R_G'} \sqrt{\frac{1}{4\pi}(2\pi - 2\alpha + \sin 2\alpha)}$$

and the form factor is the ratio of Eq. 7-32 to Eq. 7-31. This ratio can be reduced to

$$k_f = \frac{\sqrt{\frac{\pi}{4}(2\pi - 2\alpha + \sin 2\alpha)}}{1 + \cos \alpha} \qquad (7\text{-}33)$$

7-10 GAINS

Three general types of gain are associated with saturable reactors that are premagnetized with direct current. These are power gain, ampere-turn gain, and current gain.

Power gain

The power output to the load is

$$P_L = (k_f I_L)^2 R_L = I_L^2 R_L k_f^2$$

and the power input to the control circuit is

$$P_C = I_c^2 R_C$$

where I_c is the control current and R_C is the resistance of the control circuit.

An increase ΔI_c in the control current produces an increase ΔI_L in the load current with a corresponding increase of power ΔP_C applied to the control circuit, and ΔP_L in the output power to the load. Hence

$$\frac{\Delta P_L}{\Delta P_C} = \left(\frac{\Delta I_L}{\Delta I_c}\right)^2 \frac{R_L}{R_C} k_f^2$$

The power gain is defined by

$$K_p = \lim_{\Delta P_O \to 0} \frac{\Delta P_L}{\Delta P_C} = \left(\frac{dI_L}{dI_c}\right)^2 \frac{R_L}{R_C} k_f^2 \qquad (7\text{-}34)$$

Ampere-turn gain

By following a process similar to the one that led to the definition of power gain, the ampere-turn gain is defined by

$$K_{at} = \left(\frac{dI_L}{dI_c}\right)\frac{N_G}{N_C} \qquad (7\text{-}35)$$

which shows, on the basis of the law of equal ampere turns (Eq. 7-24), that for the series-connected gate windings

$$K_{at} = 1$$

and for parallel-connected gate windings

$$K_{at} = 2$$

Current gain

The current gain is defined, in terms of the ampere-turns gain on the basis of the law of equal ampere turns, by

$$K_I = \frac{dI_L}{dI_c} = K_{at} \frac{N_C}{N_G} \qquad (7\text{-}36)$$

7-11 STEADY-STATE OPERATION WITH SUPPRESSION OF EVEN-HARMONICS

Suppression of the free even-harmonics, in the control circuit of a saturable reactor in which the magnetization curve approaches the idealized shape of Fig. 7-9(a), leads to an almost rectangular wave form in the output current. Examples in which the even harmonics are suppressed are shown in Fig. 7-11. The saturable reactor in Fig. 7-11(a) receives its premagnetization from a permanent magnet instead of from a winding carrying direct current. One application is that of a constant d-c source that receives its supply from an a-c voltage source and in which the output is rectified. The rectangular wave form of the gate current makes for a low value of ripple in the d-c output. The permanent magnet in a saturable reactor, such as shown in Fig. 7-11(a), limits premagnetization to only one value, and therefore confines control to practically only one value of output current.

A single-phase, constant-current arrangement making use of a high series impedance Z_0 in series with the control circuit is shown in Fig. 7-11(b); a constant d-c output circuit* supplied from a 3-phase source with a high self-inductance L_0 in series with all three control circuits is shown in Fig. 7-11(c). The rectified output from a polyphase source has less ripple than from a single-phase source, which is one of the advantages of the 3-phase arrangement. Another advantage, in the case of large outputs, is that of placing a balanced load on the 3-phase supply.

The d-c current transformer shown in Fig. 7-11(d) is another example of saturable reactor operation with suppressed even-harmonics. This is an arrangement for measuring large values of direct current where the direct current to be measured is the premagnetized current. The alternating current supplied to the gate windings from the a-c source is rectified by means of the bridge-type rectifier, which supplies the d-c ammeter and which responds to the average value. The control winding, in the case of large values of direct current, consists of only one turn, i.e., one conductor linking the two cores as shown in Fig. 7-11(d). In this application the impedance of the d-c source is generally so high that the even-harmonics in the control circuit are suppressed. The d-c current transformer is particularly useful for measuring large values of direct current because it is smaller than the shunt that is ordinarily required in parallel with the d-c ammeter; the heat losses are also lower than in the case of a shunt. Moreover, there is the advantage that the measuring instruments are isolated from the d-c circuit.†

7-12 LOAD IMPEDANCE ZERO

Consider the saturable reactor in Fig. 7-11(b) with the impedance of its load at zero. For the assumed idealized magnetization curve, at the instant that the a-c flux goes through zero, both cores are saturated if there is any current in the control windings. However, for the current directions shown in Fig. 7-11(b), the total flux ϕ_I in core I cannot increase because it is already at the saturated value. The instantaneous a-c flux ϕ_{acII} for these current directions,

---

* T. Bonnema and G. R. Slemon, "A Controllable Low-Ripple Constant-Current Source," *Trans AIEE*, 80, Part I (1961), 3–8.

† W. A. Derr and E. J. Cham, "Magnetic-Amplifier Applications in d-c Conversion Stations," *Trans AIEE*, 72, Part III (1953), 229. See also, N. L. Kusters, W. J. M. Moore, P. N. Miljanic "A Current Comparator for the Precision Measurement of d-c Ratios," *Electrical Engineering*, March 1963, 204–210.

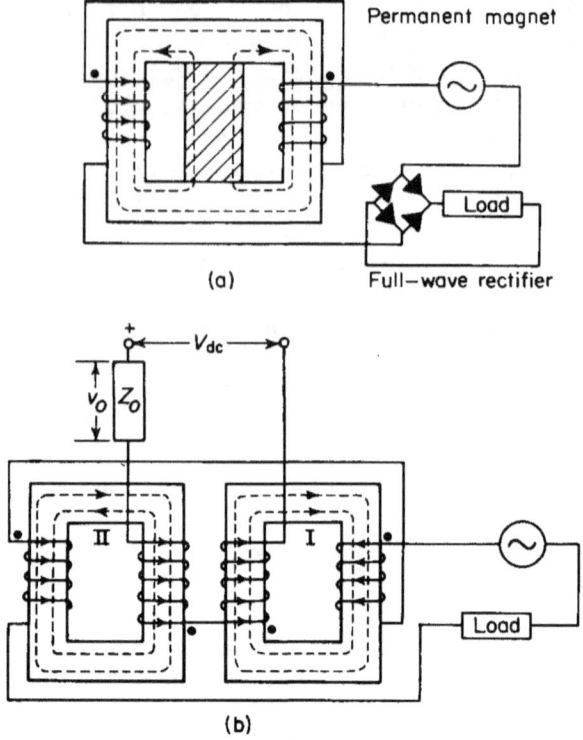

Figure 7-11. Saturable reactors with suppressed free-even harmonics. (a) Premagnetization supplied by permanent magnet; (b) high impedance in series with control circuit.

however, subtracts from the d-c flux in core II, causing this core to desaturate immediately, and its mmf must drop to zero and remain at that value until the a-c flux in core II attempts to reverse direction one half-cycle later. At that instant, the roles of cores I and II become interchanged; core I becoming unsaturated and core II remaining saturated during the second half-cycle as shown in Fig. 7-12. If the even-harmonics are suppressed completely in the control circuit, only one core at a time can be unsaturated, as each core remains saturated throughout alternate half-cycles. This is evident from the directions of currents and fluxes shown in Figs. 7-11(a) and 7-11(b). When the resistance of the gate windings is neglected and the load is short-circuited, the wave forms of the fluxes are as shown in Figs. 7-12(b) and 7-12(c).

During the first half-cycle of the condition shown in Fig. 7-12, core I is saturated and core II is unsaturated. The resultant mmf acting on core II must therefore be zero, and one half-cycle later core II is saturated with core I unsaturated, producing the wave form shown in Fig. 7-12(d), for the gate current.

Sec. 7-12   LOAD IMPEDANCE ZERO   333

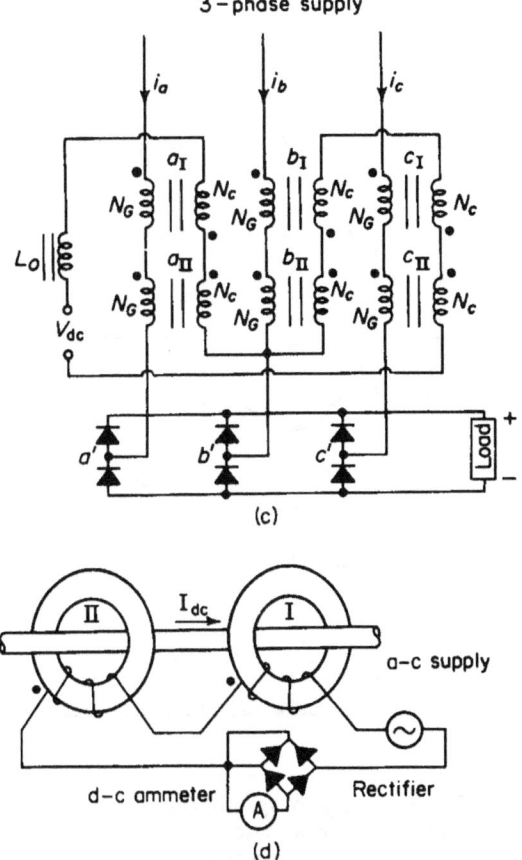

Figure 7-11 (cont.). (c) 3-phase arrangement with large inductance L_0 in series with control circuit; (d) d-c current transformer.

Law of equal ampere turns

The law of equal ampere turns is valid for operation in which the free even-harmonics are suppressed in the control circuit. This follows from the fact that the mmf in an unsaturated core, with the idealized magnetization characteristic, must be zero, i.e.

$$N_G i_G = N_C i_c$$

and the gate current is expressed by

$$i_G = \frac{N_C i_c}{N_G} \tag{7-37}$$

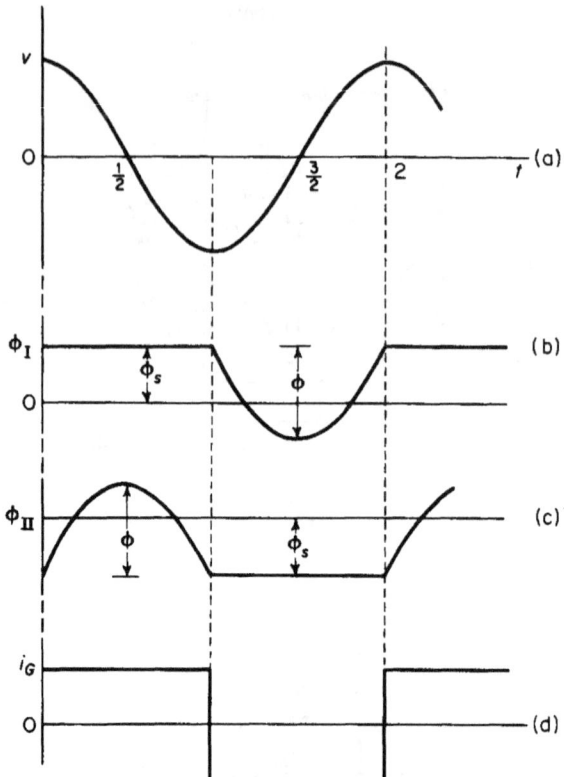

Figure 7-12. Wave forms of (a) a-c supply voltage; (b) and (c) fluxes linking gate windings; (d) output current of saturable reactor with zero output impedance when free even-harmonics are suppressed.

Since the control current i_c is a constant unidirectional current for a given premagnetization, the gate current must have a rectangular wave form for zero load impedance.

From Eq. 7-37 it may be concluded that, under practical conditions, minor variations in the magnitude and frequency will have little effect on the gate current for a given value of control current. The saturable reactor, when operating without the free even-harmonics in the control circuit, therefore provides a basically constant-current source.

Flux and voltage relations

The flux swing is related to the applied voltage in accordance with

$$\Delta \phi = \frac{V}{4.44 \, fN_G} \tag{7-38}$$

Sec. 7-13　　　FINITE LOAD RESISTANCE　　　335

The voltage induced in the control winding appears across the series impedance Z_0 in the control circuit, and for the assumed ideal conditions has an rms value of

$$V_c = \frac{N_C}{N_G} V \qquad (7\text{-}39)$$

7-13 FINITE LOAD RESISTANCE

The real power output of the a-c source with zero output impedance, when the resistance of the gate windings and the core losses are neglected, as under the assumed ideal conditions, must be zero. This is also evident from the current and voltage waves shown in Fig. 7-13(a), from which it is apparent that the average of the instantaneous power taken over a complete cycle is zero. When a noninductive load R_L is connected in series with the gate windings, the real power output of the source must equal $(k_I I_L)^2(R_L + 2R_G)$, where R_G is the resistance of each gate winding. The current wave is then advanced in phase from the wave shown in Fig. 7-13(a). If the value of $R_L + 2R_G$ is in the range for linear operation, i.e., low enough so that the law of equal ampere turns is practically satisfied, the current wave form will again be rectangular and of an amplitude determined by Eq. 7-37. The wave forms of applied voltage and output current including the fundamental component of output current for a noninductive load are shown in Fig. 7-14.

Since sinusoidal voltage is assumed, the real power output of the a-c source is

$$P_0 = V I_{L_1} \cos \alpha = (k_I I_L)^2 (R_L + 2R_G) \qquad (7\text{-}40)$$

where I_{L_1} is the rms value of the fundamental component in the gate current. It was shown in Section 5-4 that only the harmonics that are present in both the voltage and the current waves contribute to the real power. The applied voltage has no higher harmonics, and for that reason, only the fundamental in the current need be considered for the real power.

The values of V and I_{L_1} in Eq. 7-40 are rms. The fundamental component of a periodic rectangular wave has an amplitude that is $4/\pi$ times that of the rectangular wave. Therefore

$$\sqrt{2} I_{L_1} = \frac{4}{\pi} I_L$$

and

$$I_{L_1} = \frac{4}{\sqrt{2}\pi} I_L = \frac{2\sqrt{2}}{\pi} I_L \qquad (7\text{-}41)$$

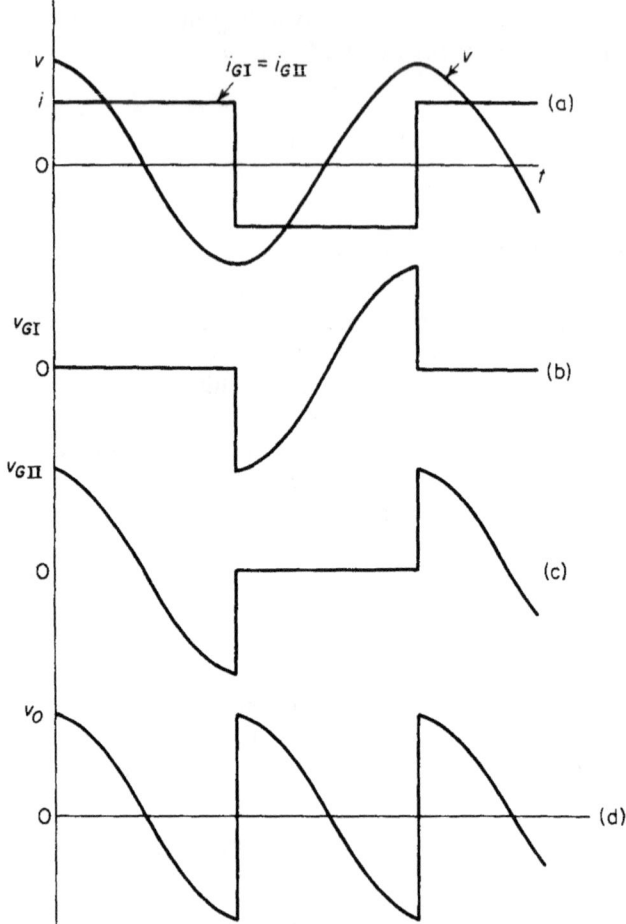

Figure 7-13. Wave forms of (a) a-c supply voltage and output current; (b) and (c) gate voltages, and (d) voltage across impedance in series with control circuit of saturable reactor supplying a noninductive load when free even-harmonics are suppressed.

The angle α, found by substituting Eq. 7-41 in Eq. 7-40, is

$$\alpha = \arccos \frac{\pi}{2\sqrt{2}} \frac{I_L}{V}(R_L + 2R_G) = \frac{1.11 I_L(R_L + 2R_G)}{V} \qquad (7\text{-}42)$$

where R_G is the resistance of each gate winding.

For a given value of a-c supply voltage and premagnetizing current, if the load resistance is increased beyond a certain value, the gate current is no longer entirely a function of the control current, but of the load resistance as well. This can be seen from the extreme case in which the load resistance R_L approaches infinity, causing the gate current to approach zero, regardless of the value of premagnetizing current i_c. With a given load resistance R_L, increasing the premagnetization causes the angle α to decrease and to shift the current wave to the left, which follows from Eq. 7-40. However, when the resistance reaches a value such that $(2V/R_L + 2R_G)\sin\alpha$ is less than

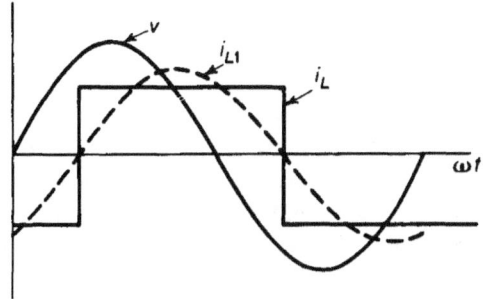

Figure 7-14. Source voltage and output current for resistance load when free even-harmonics are suppressed in control current.

$N_C i_c/N_G$, the wave form of the current is no longer rectangular, and the critical value of α is determined as follows

$$I_L(R_L + 2R_G) = \sqrt{2}V\sin\alpha_0 \tag{7-43}$$

from Eq. 7-49

$$I_L(R_L + 2R_G) = \frac{2\sqrt{2}}{\pi}V\cos\alpha_0 \tag{7-44}$$

Hence, from Eqs. 7-43 and 7-44, we get

$$\alpha_0 = \arctan\frac{2}{\pi} \tag{7-45}$$

and

$$\cos\alpha_0 = 0.844$$

Therefore, the range of load resistance over which the wave form of the output is rectangular is

$$0 < \frac{I_L(R_L + 2R_G)}{V} < \frac{2\sqrt{2}}{\pi} \times 0.844$$

$$0 < \frac{I_L(R_L + 2R_G)}{V} < 0.76 \tag{7-46}$$

where V is the rms value of the a-c supply voltage.

When the average value V_{av} instead of the rms value V of the sinusoidal supply voltage is used, Eq. 7-46 becomes

$$0 < \frac{I_L(R_L + 2R_G)}{V_{av}} \leq 0.844 \tag{7-47}$$

since $V = 1.11\, V_{av}$.

7-14 FREQUENCY DOUBLER

The wave form of the voltage induced in the control circuit, when the even-harmonics in the control current are suppressed, is comprised of even-harmonics of which the second harmonic (in terms of the a-c supply frequency) is the predominant one. The saturable reactor may be supplied

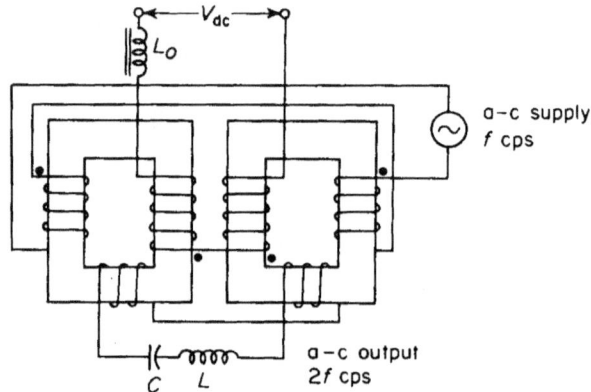

Figure 7-15. Frequency-doubling circuit.

with a third set of windings, one on each of the two cores, and connected with their polarities the same as the control windings as shown in Fig. 7-15. The even-harmonics in the voltage induced in the control circuit will also appear in that of the third set of windings, which supplies an L-C circuit tuned to resonance for the second harmonic and which serves as a supply of double-frequency power.

PROBLEMS

7-1 The resistance of the winding in each of the saturable reactors of Example 7-1 is 35 ohms and the resistance of the noninductive load is 140 ohms.

Calculate

(a) The max value of the current.
(b) The rms value of the current.
(c) The power output.
(d) The efficiency of the frequency tripler.

Neglect core losses.

7-2 Explain why a magnetic frequency multiplier operating on the principles discussed in Section 7-1 requires an odd number rather than an even number of saturable reactors.

7-3 A 120-v, 60-cycle, single-core saturable reactor has a d-c control winding of 900 turns and an a-c output winding of 225 turns. The load is a noninductive resistance of 59.5 ohms and the resistances are $R_c = 10.0$ ohms and $R_G = 0.5$ ohm for the control winding and output winding respectively. Assume the a-c flux density to be just below the saturated value and calculate the self-inductance of a choke in the control circuit such that the alternating current induced in the control winding is not to exceed 0.25 amp when the premagnetization is zero. Neglect the resistance of the control circuit.

7-4 Calculate the amplitude of the flux when the saturable reactor of Problem 7-3 is operated in series with the 59.5-ohm resistor from a 120-v, 60-cycle source with its control circuit open-circuited. Assume the shape of the magnetization curve in Fig. 7-3(b) for the core material.

7-5 Plot the wave form of the output current if the saturable reactor in Problems 7-3 and 7-4 is operated in series with the 59.5-ohm resistor from a 130-v, 60-cycle source with its control circuit open-circuited. What is the amplitude of the current?

7-6 Assume the choke in the control circuit of the saturable reactor in Problems 7-3 and 7-4 to have infinite inductive reactance.

(a) Calculate the output current if the premagnetization is of a value such that the core remains saturated throughout the entire voltage cycle, while the reactor is operated with its connected load from a 120-v, 60-cycle source.
(b) What is the minimum value of premagnetizing current for the condition of (a)?

7-7 (a) Calculate the angle α_2 at which the saturable reactor of Problems 7-3 and 7-4 becomes unsaturated when supplying the 59.5-ohm resistance load from a 120-v, 60-cycle source with a control current of 0.354 amp.
(b) Estimate the value of the firing angle α_1. Assume an infinite choke in the control circuit.

7-8 The core in the reactor of Problem 7-4 has a net cross-sectional area of 2.25 sq in. and a mean length of 11.75 in. Assume the hysteresis loop to be rectangular and the coercive force to be 72 amp turns per m for the material in the core, and calculate the current when the reactor operates with its connected load of 59.5 ohms from a 120-v, 60-cycle source with zero premagnetization. Plot the approximate wave form of the current.

7-9 The saturable reactor of Problems 7-3 and 7-4 is to operate from a 400-cycle system with the amplitude of the a-c flux just below the saturated value at zero premagnetization. Calculate

(a) The voltage of the source.
(b) The value of the noninductive load resistance if the current range is to be the same as for 60-cycle operation. What is the ratio of maximum power output to that for 60-cycle operation?

7-10 The saturable reactor of Problems 7-3 and 7-4 is to be rewound for operation at 120 v and 400 cps with the load specified in Problem 7-3.

(a) What changes, if any, should be made in one or both windings?
(b) How would the current rating be changed?
(c) How does the power output rating compare with that of Problem 7-9?

7-11 The cores in a 2-core saturable reactor are toroids of the following dimensions: ID = $2\frac{1}{2}$ in., OD = 5 in., and height = 2 in. The stacking factor is 0.80. The number of turns N_C in the control winding on each core is 900 turns. The core is assumed to saturate abruptly at B_s = 90,000 lines per sq in. and the magnetization curve to be horizontal in the saturated region and vertical in the unsaturated region. The control windings are connected in series. The gate windings are connected in parallel for operation from a 120-v, 60-cycle source. The rated value of the output current is 2 amp (average value, not rms). Calculate

(a) The number of turns in each gate winding if the flux density is just below saturated value at zero premagnetization.
(b) The rms value of rated current.
(c) The resistance of the load if the firing angle α is zero at rated output for a 60-cycle emf of 120 v if the resistance of the gate windings is neglected.
(d) The average value and rms value of the rated current of the control winding on the basis of (c).

7-12 The 2-core saturable reactor in Problem 7-11 is delivering rated current at 120 v, 60 cps to a noninductive resistance of 50 ohms. The resistance of the gate windings may be neglected. Calculate

(a) The firing angle α.
(b) The control current.
(c) The output current if the control current remains at the value in (b) while the load resistance is reduced to 10 ohms.
(d) The firing angle α for (c).
(e) The form factor for (a) and (c).

7-13 The reactor in Problem 7-12 is operating with a choke in its control circuit of such size that the free even-harmonics in the control circuit are negligible.

Neglect the resistance of the gate windings and calculate

(a) The maximum value of resistance for rectangular wave form of output current from a 120-v, 60-cycle supply.
(b) The voltage across the choke in the control circuit.

7-14 A 2-core saturable reactor is used in an arc welder. The number of turns on each core are $N_C = 300$ and $N_G = 20$. The gate windings are series-connected and supplied from a 90-v, 60-cycle source. The control windings are also series-connected and receive their current from a 30-v d-c source with a potentiometer voltage divider to adjust the control current. The voltage across the welding arc is nearly constant at 25 v rms. Calculate

(a) The d-c value of the control current for a welding current of 200 amp rms. (Assume sinusoidal current wave form.)
(b) The rms value of the control current.
(c) The amplitude of the voltage wave if the impedance of the control circuit is negligible.
(d) The power gain if the voltage applied to the control windings is 28 v and the arc voltage is 25 v. Neglect losses in the voltage divider and assume the arc voltage to be in phase with the arc current.
(e) The ampere turn gain.
(f) The current gain.

7-15 A d-c transformer, as shown in Fig. 7-11(d), for measuring direct currents up to 2,000 amp has a rated output of 1 amp d-c.

(a) Calculate the number of turns in each gate winding.
(b) The a-c supply is 120 v, 60 cps. Estimate the cross-sectional area of the core in square inches. Assume a stacking factor of 0.80.

BIBLIOGRAPHY

Extensive bibliographies are available in H. F. Storm, *Magnetic Amplifiers*. New York: John Wiley & Sons, Inc., 1955; and F. Kummel, *Regel Transduktoren*. Berlin; Springer-Verlag, 1961.

"Magnetic Amplifiers," prepared by Electronics and Development Division, Bureau of Ships, Department of the Navy, 1951. For sale by the Superintendent of Documents, U.S. Government Printing Office, Washington 25, D.C.

INDEX

Acoustic energy, 5
Air:
   dielectrics, 54
   as a magnetic insulator, 105
Air gaps:
   effect of, on core size, 219–229
      in iron-core reactors, 220–226
      to prevent d-c saturation, 227–229
      on stability, 220–226
      on wave form, 226–227
   in magnetic structures, 116–121
Alnico:
   demagnetization curve of, 137–138
   energy product curve of, 138
   permanent magnets from, 132
Alternating-current:
   harmonics and, 216
   magnets of, 111
   power in circuits of, 191
Aluminum:
   in capacitors, 67
   corona and, 77
   as a magnetic material, 109
   in transformers, 240
Ammeters, open-circuit test and, 259
Ammonium, in capacitors, 67
Ampere-turns:
   gain, 330
   law of equal, 323, 333–334
Ampere's circuital law, 101, 103
Arnold Engineering, 111
ASA Standards, 263
Atomic energy, 34–35
Atoms:
   of nonmagnetic materials, 107
   polarization and, 71

Audible corona, 77
Autotransformers, 268–269
AWG copper wire, 47, 118

Boric acid, in capacitors, 67

Cable oil, dielectric constants of, 75
Capacitance, 39–88
   concentric cylinders and, 54–56
   concentric spheres and, 50–52
   corona and, 75–77
   dielectrics and, 53–54, 59–62, 65–66, 68–75, 77–78
   electrostatic machines and, 81–83
   energy and, 18–20, 39–88
   graded insulation and, 56–57
   polarization and, 68–72
   problems, 83–87
Capacitor microphones, 81
Capacitors:
   electrolytic, 67
   energy stored in, 58–59, 62–63, 78–81
   force in, 78–81
   kva rating of, 63–64
   materials used in, 60, 66–68
   mechanical energy in, 78–81
   operating on direct-current, 62–63
   parallel-plate, 52–53
   power and, 63–64
   types of, 66–68
Carbon steel, permanent magnets from, 132
Castor oil, in capacitors, 60, 66
Ceramics, in capacitors, 68
Cgs electromagnetic system, 106

INDEX

Charge:
  electric, 89–93
  on an isolated sphere, 48–49
  static, 47–48
  voltage and, 49
Chemical energy, stored, 32–34
Chlorinated diphenyl, in capacitors, 60, 66
Chrome, permanent magnets from, 132
Circuits:
  coupled, 175–180
  electric, see Electric circuits
  equivalent, 217–219, 246–259
    approximate, 257–259
    conductance and, 247–248
    exact, 253–257
    exciting current and, 247–248
    leakage flux and, 249–253
    leakage reactance and, 249–253
    susceptance and, 247–248
  frequency-doubling, 338
  for a frequency multiplier, 311
  inductive, 163
    mutual, 172–178
    self-, 163–169
  magnetic, see Magnetic circuits
  open-, 259–262, 297–300
  in parallel, 152–154
  R-C, 18–20
  R-L, 15–16, 123
  R-L-C, 24–29
  in series, 152–154
  short-, 259–262, 297–300
  single-core saturable reactor, 315
Coal:
  conversion of, into energy, 2–3
  energy in, 34
Cobalt:
  as a magnetic material, 109
  permanent magnets from, 132
Coercive force, 131
Compliance, 31–32
Concentric cylinders, 54–56
Concentric spheres, 50–52
Condenser microphones, 81
Conductance, transformers and, 247
Conduction angle, voltage and, 313–315, 315
Conductors, static charge in, 47–48
Copper, in transformers, 240
Copper losses, 262 n.

2-Core saturable reactor, 319
Corona, 75–77
  audible, 77
  dielectrics and, 77
  electrodes and, 75
  ions and, 75–77
  radio interference and, 77
Coulomb's law, 40, 42–43, 95
Coulomb's law of force between point magnetic poles, 95
Coupling:
  coefficient of, 175–176
  energy and, 177–178
  force in, 178–180
  torque in, 178–180
Current:
  alternating, see Alternating-current
  constant load, 324
  core-loss, 215–217
  direct, see Direct-current
  eddy, 146–152
  exciting, transformers and, 247–248
  gain, in saturable reactors, 330
  magnetic fields and, 90–93
  magnetizing, 215–217
  nonoscillatory, 26–29
  oscillatory, 26
  per unit quantities and, 292
  ratio, in the ideal transformer, 241, 243–245
  reactors and, 214–217
  saturable reactors and, 311–312
  unidirectional, 227
  velocity and, 16

Delta-delta connection, 277–279
Demagnetization curve, 133–137
Dielectrics:
  air, 54
  characteristics of, 72–74
  classification of capacitors and, 66
  complex constants, 74–76
  configurations, 77–79, 81
  constant, 53–54
  corona and, 77
  energy stored in, 59–62
  liquid, 50–51, 54, 60
  polarization and, 68–72
  resistance in, 77–78
  solid, 54

INDEX

Dielectrics (cont.):
  strength values of, 65–66
  vacuums and, 83
  voltage and, 56–57
  volume of, 61–62
Dipoles, 71–72
Direct-current:
  capacitors operating on, 62–63
  harmonics and, 216
  magnets of, 111
  saturation, 227–229
Domains, of magnetic materials, 107
Double-zigzag connection, 287

Eddy currents, 146–152
Efficiency:
  defined, 11
  energy and, 11–12
Electrical energy, 9
Electrical systems, mechanical analogies of, 31–32
Electric charge:
  current and, 90
  magnetic field and, 89, 91–93
Electric circuits:
  inductive circuits and, 163
  magnetic circuits and, 105–106
  magnetic flux lines and, 121
Electric current, *described,* 90
Electric field, 39–40
  energy density of, 60
  flux, 40–42
  Gauss's theorem and, 45–47, 48, 50
  intensity, 42–43, 56, 59, 60
  potential and, 43–45
  static charge and, 47–48
  uniformly distributed charge and, 48–49
  voltage and, 43–45
Electrodes, corona and, 75
Electrolytic capacitors, 67
Electronic polarization, 71
Electronic rectifiers, 287
Electrons:
  flow of, 90
  motion and, 90
  polarization and, 71
  velocity of, 90
Electrostatic machines, 81–83
Electrostatic microphones, 81
Emf, induced, 238–239

Energy, 1–38
  acoustic, 5
  atomic, 34–35
  capacitance and, 18–20, 39–88
  in capacitors, stored, 58–59, 62–63
  chemically stored, 32–34
  in coal, 34
  coupling and, 177–178
  dielectrics and, 53–54, 59–62, 65–66, 68–75, 77–78
  efficiency and, 11–12
  electrical, 9
  electromagnetic, 163–201
  heat and, 3, 5, 8–9, 17–18, 20, 23, 39
  irreversible, 4–9, 39
  kinetic, 1, 3, 5–6, 12, 13
  law of conservation of, 3–4, 6, 11
  law of degradation of, 4–9
  in magnetic circuits, stored, 123–126
  magnetic fields and, 93–94
  magnetization curve and, 182–190
  mass and, 12–14, 21–24
  mechanical, in a capacitor, 78–81
  mechanical and electrical analogies, 31–32
  potential, 1, 3, 5–6, 12, 43–45
  power and, 10, 12, 30–31
  problems, 35–38
  R-C circuit and, 18–20
  resistance and, 15–16, 18–20
  reversible, 4, 5, 39
  R-L circuit and, 15–16
  R-L-C circuit and, 24–29
  rotating flywheel and, 29–30
  self-inductance and, 15–16
  solar, 1
  spring and, 17–18, 21–24
  storage batteries and, 34
  tangential force and, 10–11
  thermal, 9
  torque and, 10–11, 30–31
  viscous friction and, 12–14, 17–18, 21–24
  work and, 2–3
Energy product, in magnets, 137–140
Equivalent circuits, *see* Circuits, equivalent
Ethyleneglycol, in capacitors, 67
Even-harmonics, 326–328, 330–331

Faraday's Law, 238
Ferromagnetic materials, *see* Magnetic materials
First Law of Thermodynamics, 3–4, 6, 11
Flux:
  components of, in saturable reactors, 319–323
  linkages, 121
  magnetic force and, 126–130
  reactors and, 204–209, 311–312
  voltage and, in saturable reactors, 334–335
Flywheel, rotating, energy and, 29–30
Force:
  in a capacitor, 78–81
  in a circuit of variable self-inductance, 167–169
  coercive, 131
  friction and, 22–23
  in inductively coupled circuits, 178–180
  magnetic, 126–130
  magnetic field and, 93–95
  magnetization curve and, 182–190
  in nonlinear magnetic circuits, 180–190
  voltage and, 16
Forked-wye connection, 287
Fossil fuels, 1, 35
Fourier series, 209
Frequencies:
  radio, 271 n.
  variable, in transformers, 270–276
Frequency-doubling circuits, 338
Frequency triplers, 312–313
Friction:
  force and, 22–23
  heat and, 23
  loss of, 22–24
  motion and, 12
Fringing, 115

Gas, conversion of, into energy, 2–3
Gate windings:
  in parallel, 319–324
    components of flux in, 319–323
    constant-current source and, 324
    law of equal ampere turns and, 323
  in series, 325–326
Gauss's theorem, 45–47, 48, 50
Generators, electrostatic, 83

Harmonics, 209–212
  even-, 326–328, 330–331
  in 3-phase transformers, 287–292
HCl molecules, 71
Heat:
  energy and, 3, 5, 8–9, 17–18, 20, 23, 39
  friction and, 23
Hysteresis loop, 130–131
  magnetic, 109
  for a nickel-iron alloy, 309–310
  reactors and, 208
Hysteresis loss, 130, 142–146

Impedance:
  equivalent, 259–262
  per unit quantities and, 292
  ratio, in the ideal transformer, 241, 245–246
  saturable reactors and, 331–335
    flux and voltage relations, 334–335
    law of equal ampere turns, 333–334
Induced emf, 238–239
Inductance, 163–201
  circuits, 163
  force and, 178–180
  magnetic permeance and, 169–172
  magnetic reluctance and, 169
  mutual, 172–178
  nonlinear magnetic circuits and, 180–190
  problems, 193–201
  Q-factor in, 192–193
  reactance in, 190–191
  self-, 163–169
  torque and, 178–180
Instrument transformers, 270
Insulation, graded, 56–57
Ions, corona and, 75–77
Iron:
  air gap in reactors and, 219–229
  eddy currents and, 146–147
  inductance and, 180–182
  as a magnetic material, 107–109, 116–118
  permanent magnets and, 132
  in transformers, 239
Iron-core reactors, *see* Reactors, iron-core
Irreversible energy, 4–9, 39

Joule, the, 2, 10

INDEX

Kinetic energy, 1, 3, 5–6, 12, 13
Kirchhoff's Law, 326
Kva rating, of capacitors, 63–64

Laminations, eddy-current loss and, 146–147, 151
Law of Conservation of Energy, 3–4, 6, 11
Law of Degradation of Energy, 4–9
Leakage, magnetic, 114–115
Leakage flux, transformers and, 249–257
Leakage reactance, transformers and, 249–257
Lenz's Law, 111, 122–123, 238
Liquid, dielectric, 50–51, 54, 60
Load:
   losses, 262–263, 265
   noninductive, 313–315
   resistance, 335–338

Magnetic circuits, 89–161
   air gaps in, 116–121
   calculation of, without air gaps, 111–114
   cgs electromagnetic system and, 106
   core loss and, 142
      factors influencing, 151–152
   correction for fringing, 115
   demagnetization curve and, 133–137
   eddy-current loss and, 146–150
   the electric circuit and, 105–106
   energy product of, 137–140
   energy stored in, 123–126
   flux density and, 126–130
   flux linkages and, 121
   graphical solution for, with short air gap, 118–121
   hysteresis loop and, 130–131
   hysteresis loss and, 142–146
   iron and, 116–118
   Lenz's Law and, 122–123
   magnetic field and, 90–93
   magnetic flux and, 93–95
   magnetic leakage and, 114–115
   magnetic lines of force and, 93–95
   magnetic materials and, 107–111
   magnetism and, 89
   magnetomotive force and, 97–101
   Mixed English System of Units and, 106

Magnetic circuits (*cont.*):
   nonlinear, forces in, 180–190
   in parallel, 152–154
   permanent magnets and, 131–133
      operating characteristics of, 140–142
   problems, 155–160
   rotational hysteresis loss and, 146
   in series, 152–154
   the toroid and, 102–105
   unit magnet pole and, 95–97
Magnetic fields:
   about a straight wire, 90–93
   electric charges and, 89, 91–93
   energy and, 93–94
   force and, 93–95
   motion and, 89
   permanent magnets and, 96
   velocity and, 89
Magnetic flux, 93–95
Magnetic frequency multipliers, 310–312
Magnetic leakage, 114–115
Magnetic materials, 104, 105, 107–111, 116–117
   core losses of, 142
   demagnetization curve of, 133–137
   domains of, 107
   eddy-current loss of, 146–151
   energy product of, 137–140
   examples of, 107–109
   hysteresis loss of, 142–146
   permanent magnets from, 131–133
   permeability of, 109–111
Magnetic permeance, 169–172
Magnetic reluctance, 169–172
Magnetism, 89
Magnetization curve:
   force and energy relationships based on, 182–190
   for saturable reactor operation, 309–310
Magnetomotive force, 97–101
Magnets:
   of alternating-current, 111
   demagnetization curve and, 133–137
   of direct-current, 111
   energy product of, 137–140
   permanent, 96, 131–133
      operating characteristics of, 140–142
   poles of, 94, 95–97
   shapes of, 132

348 INDEX

Mass, energy and, 12–14, 21–24
Mechanical systems, electrical analogies of, 31–32
Microphones, capacitor, 81
Mineral oil, in capacitors, 60, 66
Mineral rutile TiO_2, in capacitors, 68
Mixed English System of Units, 106, 223–225
MKS system, 2, 10, 40, 53, 95, 104, 106
Mmf, 97–101
Motion:
   friction and, 12
   magnetic fields and, 89
Motors, electrostatic, 83
Multicircuit transformers, 295–300
Mutual inductance, 172–178
   coefficient of coupling and, 175–176

Nickel, as a magnetic material, 109
Nickel-iron, in saturable reactors, 309
Nitrobenzene, in capacitors, 66
Nitrogen, produced by corona, 77
Noninductive load, 313–315
Nonmagnetic materials, 107
Nonoscillatory current, 26–29

Ohm's Law, 169, 205
Oil, conversion of, into energy, 2–3
Open-circuit tests, 259–262, 297–300
Open-delta connection, 283–284
Oscillatory current, 26
Ozone O_3, produced by corona, 77

Parallel-plate capacitor, 52–53
Per unit quantities, 292–295
Permanent dipole polarization, 71
3-Phase transformers, 285–287
   connections for, 276–284
      delta-delta, 277–279
      open delta, 283–285
      V-V, 283–285
      wye-delta, 281–283
      wye-wye, 279–281
   per unit impedance in, 294–295
   third harmonics in, 287–292
Polarization:
   atoms and, 71
   dielectrics and, 68–72
   electrons and, 71
   mechanisms of, 71–72

Porcelain, dielectric constants of, 75
Potential energy, 1, 3, 5–6, 12, 43–45
Power:
   a-c transformed to d-c, 287
   capacitors and, 63–64
   energy and, 10, 12, 30–31
   gain, in saturable reactors, 329–330
   output, in saturable reactors, 328–329
   per unit quantities and, 292
   reactive, 191–192
   reactors and, 212–214, 328–330
Premagnetization, 315–319
Problems:
   capacitance, 83–87
   energy, 35–38
   inductance, 193–201
   magnetic circuits, 155–160
   reactors, 232–236, 338–341
   transformers, 232–236, 300–307
Pyranol, dielectric constants of, 75

Q-factor, 192–193

Radio-Electronics-Television Manufacturers Association, 68
Radio frequencies, 271 n.
Radio interference, corona and, 77
R-C circuits, 18–20
Reactive power, 191–192
Reactors:
   air-core, 204, 219
   air gaps and, effect of:
      on core size, 220–226
      to prevent d-c saturation, 227–229
      on stability, 220–226
      on wave form, 226–227
   iron core, excitation characteristics of, 203–236
      air gaps and, 219–229
      core-loss current and, 215–217
      effective current and, 214
      equivalent circuits and, 217–219
      flux and, 204–209
      harmonics and, 209–212
      magnetizing current and, 215–217
      power and, 212–214
      problems, 232–236
      rating of, 231–232
      time constant and, 229–231
      voltage current and, 204–209
   saturable, *see* Saturable reactors

INDEX

Resistance, energy and, 15–16, 18–20
Reversible energy, 4, 5, 39
R-L circuits, 15–16, 123
R-L-C circuits, 24–29
Rotating flywheel, energy and, 29–30
Rotational hysteresis loss, 146

Saturable reactors, 309–341
   ampere-turn gain and, 330
   applied voltage and, 313–315
   conduction angle and, 313–315
   2-core, 319
   current gain and, 330
   finite load resistance and, 335–338
   frequency doubler and, 338
   frequency tripler and, 312–313
   gate windings of:
      in parallel, 319–324
      in series, 325–326
   magnetic frequency multipliers and, 310–312
   noninductive load and, 313–315
   power gain and, 329–330
   power output and, 328–329
   problems, 338–341
   single-core, with premagnetization, 315–319
   steady-state operation, 330–331
   with load impedance at zero, 331–335
Second Law of Thermodynamics, 4–9
Selenium, in saturable reactors, 309
Self-inductance, 163–166
   the emf of, 164
   energy and, 15–16
   force in, 167–169
   torque in, 167–169
   variable, 166–169
Short-circuit tests, 259–262, 297–300
Silicon, in saturable reactors, 309
Silicone, in capacitors, 66
Silicon-iron, in saturable reactors, 309
Solid dielectrics, 54
Space-charge polarization, 72
Spring, energy and, 17–18, 21–24
Steel:
   core loss and, 151
   in transformers, 239–240
Steinmetz, Charles Proteus, 145
Storage batteries, 34
Sun, the, energy of, 1

Tangential force, 10–11, 30–31
Teflon, dielectric constants of, 75
Thermal energy, 9
Time constant, 14, 20, 229–231
Titanium oxide, in capacitors, 68
Toroid, the, 102–105
Torque, 10–11, 30–31
   in a circuit of variable self-inductance, 167–169
   defined, 10
   in inductively coupled circuits, 178–180
Transformer voltage formula, 241
Transformers, 237–308
   air gaps, effect of:
      on core size, 220–226
      to prevent d-c saturation, 227–229
      on stability, 220–226
      on wave form, 226–227
   auto-, 268–269
   current ratio in, 241, 243–245
   defined, 237–238
   efficiency of, 262–266
   equivalent circuits of, 246–259
      approximate, 257–259
      conductance and, 247–248
      exact, 253–257
      exciting current and, 247–248
      leakage flux and, 249–253
      leakage reactance and, 249–253
      susceptance and, 247–248
   induced emf and, 238–239
   instrument, 270
   iron core, excitation characteristics of, 203–236
      air gaps and, 219–229
      core-loss current and, 215–217
      effective current and, 214
      equivalent circuits and, 217–219
      flux and, 204–209
      harmonics and, 209–212
      magnetizing current and, 215–217
      power and, 212–214
      rating of, 231–232
      time constant and, 229–231
      voltage current and, 204–209
   losses in, 262–266
   multicircuit, 295–300
   open-circuit tests, 259–262, 297–300
   per unit quantities in, 292–295
   3-phase, 276–292
      connections for, 276–284

Transformers (*cont.*):
  3-phase (*cont.*):
    third harmonics in, 287–292
    6-phase, 287
    problems, 232–236, 300–307
    as responsible for the a-c power system, 237
    short-circuit tests, 259–262, 297–300
    two-winding, 239–241
    variable-frequency, 270–276
    voltage ratio in, 241–243
    voltage regulation and, 266–268
Tungsten:
  as a magnetic material, 109
  permanent magnets from, 132
Two-winding transformers, 239–241

Unit magnet pole, 95–97
Units:
  cgs electromagnetic system of, 106
  the joule, 2, 10
  Mixed English System of, 106, 223–225
  MKS system of, 2, 10, 40, 53, 95, 104, 106
  the watt, 10

Vacuums, dielectrics and, 83
Variable-frequency transformers, 270–276
Velocity:
  current and, 16

Velocity (*cont.*):
  of electrons, 90
  magnetic fields and, 89
Viscous friction, energy and, 12–14, 17–18, 21–24
Voltage:
  charge and, 49
  conduction angle and, 313–315
  dielectrics and, 56–57
  eddy-current loss and, 149–150
  energy and, 43–45
  flux and, in saturable reactors, 334–335
  force and, 16
  gradient, 56–57
  per unit quantities and, 292
  potential and, 43–45
  ratio, in the ideal transformer, 241–243
  reactors and, 204–209, 311–312, 320–321
  self-inductance circuits and, 164
  transformers and, 266–268, 270
Voltmeters, exciting current and, 259 n.
V-V connection, 283–284

Watt, the, 10
Wattmeters, exciting current and, 259 n.
Work, energy and, 2–3
Wye-delta connection, 281–283
Wye-wye conection, 279–281

www.ingramcontent.com/pod-product-compliance
Lightning Source LLC
Chambersburg PA
CBHW051851170526
45168CB00001B/57